ALGEBRA DEMYSTIFIED

Other Titles in the McGraw-Hill Demystified Series

Astronomy Demystified by Stan Gibilisco
Calculus Demystified by Steven G. Krantz
Physics Demystified by Stan Gibilisco

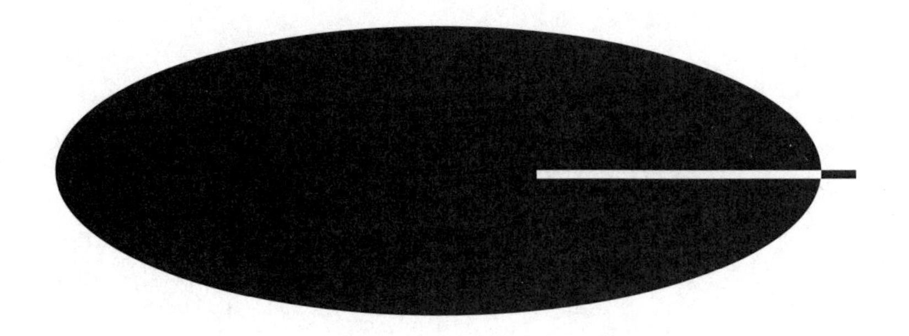

ALGEBRA DEMYSTIFIED

RHONDA HUETTENMUELLER

McGRAW-HILL
New York Chicago San Francisco Lisbon London Madrid
Mexico City Milan New Delhi San Juan Seoul
Singapore Sydney Toronto

The McGraw-Hill Companies

Cataloging-in-Publication Data is on file with the Library of Congress.

15 DOC/DOC 0 9 8 7 6 5

ISBN 0-07-138993-8

The sponsoring editor for this book was Scott Grillo and the production supervisor was Pamela Pelton. It was set in Times Roman by Keyword Publishing Services.

Printed and bound by RR Donnelley.

This book was printed on recycled, acid-free paper containing a minimum of 50% recycled, de-inked fiber.

To all those who struggle with math

CONTENTS

	Preface	ix
CHAPTER 1	Fractions	1
CHAPTER 2	Introduction to Variables	37
CHAPTER 3	Decimals	55
CHAPTER 4	Negative Numbers	65
CHAPTER 5	Exponents and Roots	79
CHAPTER 6	Factoring	113
CHAPTER 7	Linear Equations	163
CHAPTER 8	Linear Applications	197
CHAPTER 9	Linear Inequalities	285
CHAPTER 10	Quadratic Equations	319
CHAPTER 11	Quadratic Applications	353
	Appendix	417
	Final Review	423
	Index	437

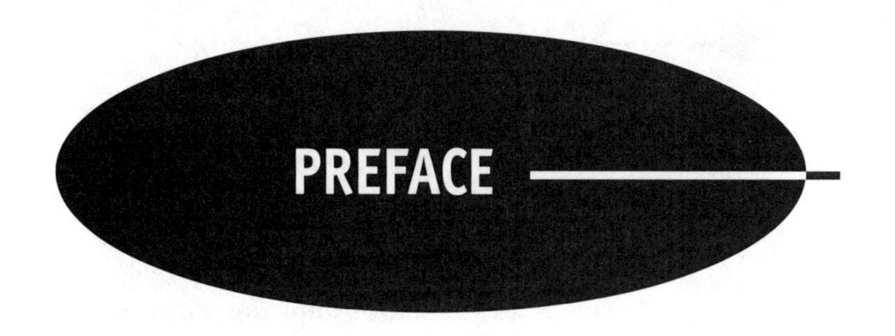

PREFACE

This book is designed to take the mystery out of algebra. Each section contains exactly one new idea—unlike most math books, which cover several ideas at once. Clear, brief explanations are followed by detailed examples. Each section ends with a few Practice problems, most similar to the examples. Solutions to the Practice problems are also given in great detail. The goal is to help you understand the algebra concepts while building your skills and confidence.

Each chapter ends with a Chapter Review, a multiple-choice test designed to measure your mastery of the material. The Chapter Review could also be used as a pretest. If you think you understand the material in a chapter, take the Chapter Review test. If you answer all of the questions correctly, then you can safely skip that chapter. When taking any multiple-choice test, work the problems before looking at the answers. Sometimes incorrect answers look reasonable and can throw you off. Once you have finished the book, take the Final Review, which is a multiple-choice test based on material from each chapter.

Spend as much time in each section as you need. Try not to rush, but do make a commitment to learning on a schedule. If you find a concept difficult, you might need to work the problems and examples several times. Try not to jump around from section to section as most sections extend topics from previous sections.

Not many shortcuts are used in this book. Does that mean you shouldn't use them? No. What you should do is try to find the shortcuts yourself. Once you have found a method that seems to be a shortcut, try to figure out *why* it works. If you understand how a shortcut works, you are less likely to use it incorrectly (a common problem with algebra students).

Because many find fraction arithmetic difficult, the first chapter is devoted almost exclusively to fractions. Make sure you understand the steps in this chapter because they are the same steps used in much of the rest of the book. For example, the steps used to compute $\frac{7}{36}+\frac{5}{16}$ are exactly those used to compute $\dfrac{2x}{x^2+x-2}+\dfrac{6}{x+2}$.

Even those who find algebra easy are stumped by word problems (also called "applications"). In this book, word problems are treated very carefully. Two important skills needed to solve word problems are discussed earlier than the word problems themselves. First, you will learn how to find quantitative relationships in word problems and how to represent them using variables. Second, you will learn how to represent multiple quantities using only one variable.

Most application problems come in "families"—distance problems, work problems, mixture problems, coin problems, and geometry problems, to name a few. As in the rest of the book, exactly one topic is covered in each section. If you take one section at a time and really make sure you understand why the steps work, you will find yourself able to solve a great many applied problems—even those not covered in this book.

Good luck.

RHONDA HUETTENMUELLER

ACKNOWLEDGMENTS

I want to thank my husband and family for their patience during the many months I worked on this project. I am also grateful to my students through the years for their thoughtful questions. Finally, I want to express my appreciation to Stan Gibilisco for his welcome advice.

Fractions

Fraction Multiplication

Multiplication of fractions is the easiest of all fraction operations. All you have to do is multiply straight across—multiply the numerators (the top numbers) and the denominators (the bottom numbers).

Example

$$\frac{2}{3} \cdot \frac{4}{5} = \frac{2 \cdot 4}{3 \cdot 5} = \frac{8}{15}.$$

Practice

1. $\dfrac{7}{6} \cdot \dfrac{1}{4} =$

2. $\dfrac{8}{15} \cdot \dfrac{6}{5} =$

3. $\frac{5}{3} \cdot \frac{9}{10} =$

4. $\frac{40}{9} \cdot \frac{2}{3} =$

5. $\frac{3}{7} \cdot \frac{30}{4} =$

Solutions

1. $\frac{7}{6} \cdot \frac{1}{4} = \frac{7 \cdot 1}{6 \cdot 4} = \frac{7}{24}$

2. $\frac{8}{15} \cdot \frac{6}{5} = \frac{8 \cdot 6}{15 \cdot 5} = \frac{48}{75}$

3. $\frac{5}{3} \cdot \frac{9}{10} = \frac{5 \cdot 9}{3 \cdot 10} = \frac{45}{30}$

4. $\frac{40}{9} \cdot \frac{2}{3} = \frac{40 \cdot 2}{9 \cdot 3} = \frac{80}{27}$

5. $\frac{3}{7} \cdot \frac{30}{4} = \frac{3 \cdot 30}{7 \cdot 4} = \frac{90}{28}$

Multiplying Fractions and Whole Numbers

You can multiply fractions by whole numbers in one of two ways:

1. The numerator of the product will be the whole number times the fraction's numerator, and the denominator will be the fraction's denominator.
2. Treat the whole number as a fraction—the whole number over one—then multiply as you would any two fractions.

Example

$$5 \cdot \frac{2}{3} = \frac{5 \cdot 2}{3} = \frac{10}{3}$$

or

$$5 \cdot \frac{2}{3} = \frac{5}{1} \cdot \frac{2}{3} = \frac{5 \cdot 2}{1 \cdot 3} = \frac{10}{3}$$

Practice

1. $\frac{6}{7} \cdot 9 =$
2. $8 \cdot \frac{1}{6} =$
3. $4 \cdot \frac{2}{5} =$
4. $\frac{3}{14} \cdot 2 =$
5. $12 \cdot \frac{2}{15} =$

Solutions

1. $\frac{6}{7} \cdot 9 = \frac{6 \cdot 9}{7} = \frac{54}{7}$ or $\frac{6}{7} \cdot \frac{9}{1} = \frac{6 \cdot 9}{7 \cdot 1} = \frac{54}{7}$
2. $8 \cdot \frac{1}{6} = \frac{8 \cdot 1}{6} = \frac{8}{6}$ or $\frac{8}{1} \cdot \frac{1}{6} = \frac{8 \cdot 1}{1 \cdot 6} = \frac{8}{6}$
3. $4 \cdot \frac{2}{5} = \frac{4 \cdot 2}{5} = \frac{8}{5}$ or $\frac{4}{1} \cdot \frac{2}{5} = \frac{4 \cdot 2}{1 \cdot 5} = \frac{8}{5}$
4. $\frac{3}{14} \cdot 2 = \frac{3 \cdot 2}{14} = \frac{6}{14}$ or $\frac{3}{14} \cdot \frac{2}{1} = \frac{3 \cdot 2}{14 \cdot 1} = \frac{6}{14}$

5. $12 \cdot \frac{2}{15} = \frac{12 \cdot 2}{15} = \frac{24}{15}$ or $\frac{12}{1} \cdot \frac{2}{15} = \frac{12 \cdot 2}{1 \cdot 15} = \frac{24}{15}$

Fraction Division

Fraction division is almost as easy as fraction multiplication. Invert (switch the numerator and denominator) the second fraction and the fraction division problem becomes a fraction multiplication problem.

Examples

$$\frac{2}{3} \div \frac{4}{5} = \frac{2}{3} \cdot \frac{5}{4} = \frac{10}{12}$$

$$\frac{3}{4} \div 5 = \frac{3}{4} \div \frac{5}{1} = \frac{3}{4} \cdot \frac{1}{5} = \frac{3}{20}$$

Practice

1. $\frac{7}{6} \div \frac{1}{4} =$
2. $\frac{8}{15} \div \frac{6}{5} =$
3. $\frac{5}{3} \div \frac{9}{10} =$
4. $\frac{40}{9} \div \frac{2}{3} =$
5. $\frac{3}{7} \div \frac{30}{4} =$
6. $4 \div \frac{2}{3} =$
7. $\frac{10}{21} \div 3 =$

Solutions

1. $\frac{7}{6} \div \frac{1}{4} = \frac{7}{6} \cdot \frac{4}{1} = \frac{28}{6}$

2. $\frac{8}{15} \div \frac{6}{5} = \frac{8}{15} \cdot \frac{5}{6} = \frac{40}{90}$

3. $\frac{5}{3} \div \frac{9}{10} = \frac{5}{3} \cdot \frac{10}{9} = \frac{50}{27}$

4. $\frac{40}{9} \div \frac{2}{3} = \frac{40}{9} \cdot \frac{3}{2} = \frac{120}{18}$

5. $\frac{3}{7} \div \frac{30}{4} = \frac{3}{7} \cdot \frac{4}{30} = \frac{12}{210}$

6. $4 \div \frac{2}{3} = \frac{4}{1} \div \frac{2}{3} = \frac{4}{1} \cdot \frac{3}{2} = \frac{12}{2}$

7. $\frac{10}{21} \div 3 = \frac{10}{21} \div \frac{3}{1} = \frac{10}{21} \cdot \frac{1}{3} = \frac{10}{63}$

Reducing Fractions

When working with fractions, you are usually asked to "reduce the fraction to lowest terms" or to "write the fraction in lowest terms" or to "reduce the fraction." These phrases mean that the numerator and denominator have no common factors. For example, $\frac{2}{3}$ is reduced to lowest terms but $\frac{4}{6}$ is not. Reducing fractions is like fraction multiplication in reverse. We will first use the most basic approach to reducing fractions. In the next section, we will learn a quicker method.

First write the numerator and denominator as a product of prime numbers. Refer to the Appendix if you need to review how to find the prime factorization of a number. Next collect the primes common to both the numerator and denominator (if any) at beginning of each fraction. Split each fraction into two fractions, the first with the common primes. Now the fraction is in the form of "1" times another fraction.

Examples

$$\frac{6}{18} = \frac{2 \cdot 3}{2 \cdot 3 \cdot 3} = \frac{(2 \cdot 3) \cdot 1}{(2 \cdot 3) \cdot 3} = \frac{2 \cdot 3}{2 \cdot 3} \cdot \frac{1}{3} = \frac{6}{6} \cdot \frac{1}{3} = 1 \cdot \frac{1}{3} = \frac{1}{3}$$

$$\frac{42}{49} = \frac{7 \cdot 2 \cdot 3}{7 \cdot 7} = \frac{7}{7} \cdot \frac{2 \cdot 3}{7} = 1 \cdot \frac{6}{7} = \frac{6}{7}$$

Practice

1. $\frac{14}{42} =$
2. $\frac{5}{35} =$
3. $\frac{48}{30} =$
4. $\frac{22}{121} =$
5. $\frac{39}{123} =$
6. $\frac{18}{4} =$
7. $\frac{7}{210} =$
8. $\frac{240}{165} =$
9. $\frac{55}{33} =$
10. $\frac{150}{30} =$

Solutions

1. $\frac{14}{42}=\frac{2\cdot 7}{2\cdot 3\cdot 7}=\frac{(2\cdot 7)\cdot 1}{(2\cdot 7)\cdot 3}=\frac{2\cdot 7}{2\cdot 7}\cdot\frac{1}{3}=\frac{14}{14}\cdot\frac{1}{3}=\frac{1}{3}$

2. $\frac{5}{35}=\frac{5}{5\cdot 7}=\frac{5\cdot 1}{5\cdot 7}=\frac{5}{5}\cdot\frac{1}{7}=\frac{1}{7}$

3. $\frac{48}{30}=\frac{2\cdot 2\cdot 2\cdot 2\cdot 3}{2\cdot 3\cdot 5}=\frac{(2\cdot 3)\cdot 2\cdot 2\cdot 2}{(2\cdot 3)\cdot 5}=\frac{2\cdot 3}{2\cdot 3}\cdot\frac{2\cdot 2\cdot 2}{5}=\frac{6}{6}\cdot\frac{8}{5}=\frac{8}{5}$

4. $\frac{22}{121}=\frac{2\cdot 11}{11\cdot 11}=\frac{11}{11}\cdot\frac{2}{11}=\frac{2}{11}$

5. $\frac{39}{123}=\frac{3\cdot 13}{3\cdot 41}=\frac{3}{3}\cdot\frac{13}{41}=\frac{13}{41}$

6. $\frac{18}{4}=\frac{2\cdot 3\cdot 3}{2\cdot 2}=\frac{2}{2}\cdot\frac{3\cdot 3}{2}=\frac{9}{2}$

7. $\frac{7}{210}=\frac{7}{2\cdot 3\cdot 5\cdot 7}=\frac{7\cdot 1}{7\cdot 2\cdot 3\cdot 5}=\frac{7}{7}\cdot\frac{1}{2\cdot 3\cdot 5}=\frac{1}{30}$

8. $\frac{240}{165}=\frac{2\cdot 2\cdot 2\cdot 2\cdot 3\cdot 5}{3\cdot 5\cdot 11}=\frac{(3\cdot 5)\cdot 2\cdot 2\cdot 2\cdot 2}{(3\cdot 5)\cdot 11}=\frac{3\cdot 5}{3\cdot 5}\cdot\frac{2\cdot 2\cdot 2\cdot 2}{11}$
$=\frac{15}{15}\cdot\frac{16}{11}=\frac{16}{11}$

9. $\frac{55}{33}=\frac{5\cdot 11}{3\cdot 11}=\frac{11\cdot 5}{11\cdot 3}=\frac{11}{11}\cdot\frac{5}{3}=\frac{5}{3}$

10. $\frac{150}{30}=\frac{2\cdot 3\cdot 5\cdot 5}{2\cdot 3\cdot 5}=\frac{(2\cdot 3\cdot 5)\cdot 5}{(2\cdot 3\cdot 5)\cdot 1}=\frac{2\cdot 3\cdot 5}{2\cdot 3\cdot 5}\cdot\frac{5}{1}=\frac{30}{30}\cdot 5=5$

Fortunately there is a less tedious method for reducing fractions to their lowest terms. Find the largest number that divides both the numerator and the denominator. This number is called the *greatest common divisor* (GCD) . Factor the GCD from the numerator and denominator and rewrite the fraction. In the previous examples and practice problems, the product of the common primes was the GCD.

Examples

$$\frac{32}{48} = \frac{16 \cdot 2}{16 \cdot 3} = \frac{16}{16} \cdot \frac{2}{3} = 1 \cdot \frac{2}{3} = \frac{2}{3}$$

$$\frac{45}{60} = \frac{15 \cdot 3}{15 \cdot 4} = \frac{15}{15} \cdot \frac{3}{4} = 1 \cdot \frac{3}{4} = \frac{3}{4}$$

Practice

1. $\frac{12}{38} =$
2. $\frac{12}{54} =$
3. $\frac{16}{52} =$
4. $\frac{56}{21} =$
5. $\frac{45}{100} =$
6. $\frac{48}{56} =$
7. $\frac{28}{18} =$
8. $\frac{24}{32} =$
9. $\frac{36}{60} =$
10. $\frac{12}{42} =$

Solutions

1. $\frac{12}{38} = \frac{2 \cdot 6}{2 \cdot 19} = \frac{2}{2} \cdot \frac{6}{19} = \frac{6}{19}$

2. $\frac{12}{54} = \frac{6 \cdot 2}{6 \cdot 9} = \frac{6}{6} \cdot \frac{2}{9} = \frac{2}{9}$

3. $\frac{16}{52} = \frac{4 \cdot 4}{4 \cdot 13} = \frac{4}{4} \cdot \frac{4}{13} = \frac{4}{13}$

4. $\frac{56}{21} = \frac{7 \cdot 8}{7 \cdot 3} = \frac{7}{7} \cdot \frac{8}{3} = \frac{8}{3}$

5. $\frac{45}{100} = \frac{5 \cdot 9}{5 \cdot 20} = \frac{5}{5} \cdot \frac{9}{20} = \frac{9}{20}$

6. $\frac{48}{56} = \frac{8 \cdot 6}{8 \cdot 7} = \frac{8}{8} \cdot \frac{6}{7} = \frac{6}{7}$

7. $\frac{28}{18} = \frac{2 \cdot 14}{2 \cdot 9} = \frac{2}{2} \cdot \frac{14}{9} = \frac{14}{9}$

8. $\frac{24}{32} = \frac{8 \cdot 3}{8 \cdot 4} = \frac{8}{8} \cdot \frac{3}{4} = \frac{3}{4}$

9. $\frac{36}{60} = \frac{12 \cdot 3}{12 \cdot 5} = \frac{12}{12} \cdot \frac{3}{5} = \frac{3}{5}$

10. $\frac{12}{42} = \frac{6 \cdot 2}{6 \cdot 7} = \frac{6}{6} \cdot \frac{2}{7} = \frac{2}{7}$

Sometimes the greatest common divisor is not obvious. In these cases you might find it easier to reduce the fraction in several steps.

Examples

$$\frac{3990}{6762} = \frac{6 \cdot 665}{6 \cdot 1127} = \frac{665}{1127} = \frac{7 \cdot 95}{7 \cdot 161} = \frac{95}{161}$$

$$\frac{644}{2842} = \frac{2 \cdot 322}{2 \cdot 1421} = \frac{322}{1421} = \frac{7 \cdot 46}{7 \cdot 203} = \frac{46}{203}$$

Practice

1. $\frac{600}{1280} =$

2. $\frac{68}{578} =$

3. $\frac{168}{216} =$

4. $\frac{72}{120} =$

5. $\frac{768}{288} =$

Solutions

1. $\frac{600}{1280} = \frac{10 \cdot 60}{10 \cdot 128} = \frac{60}{128} = \frac{4 \cdot 15}{4 \cdot 32} = \frac{15}{32}$

2. $\frac{68}{578} = \frac{2 \cdot 34}{2 \cdot 289} = \frac{34}{289} = \frac{17 \cdot 2}{17 \cdot 17} = \frac{2}{17}$

3. $\frac{168}{216} = \frac{6 \cdot 28}{6 \cdot 36} = \frac{28}{36} = \frac{4 \cdot 7}{4 \cdot 9} = \frac{7}{9}$

4. $\frac{72}{120} = \frac{12 \cdot 6}{12 \cdot 10} = \frac{6}{10} = \frac{2 \cdot 3}{2 \cdot 5} = \frac{3}{5}$

5. $\frac{768}{288} = \frac{4 \cdot 192}{4 \cdot 72} = \frac{192}{72} = \frac{2 \cdot 96}{2 \cdot 36} = \frac{96}{36} = \frac{4 \cdot 24}{4 \cdot 9} = \frac{24}{9} = \frac{3 \cdot 8}{3 \cdot 3} = \frac{8}{3}$

For the rest of the book, reduce fractions to their lowest terms.

Adding and Subtracting Fractions

When adding (or subtracting) fractions with the same denominators, add (or subtract) their numerators.

Examples

$$\frac{7}{9}-\frac{2}{9}=\frac{7-2}{9}=\frac{5}{9}$$

$$\frac{8}{15}+\frac{2}{15}=\frac{8+2}{15}=\frac{10}{15}=\frac{5\cdot 2}{5\cdot 3}=\frac{2}{3}$$

Practice

1. $\frac{4}{7}-\frac{1}{7}=$
2. $\frac{1}{5}+\frac{3}{5}=$
3. $\frac{1}{6}+\frac{1}{6}=$
4. $\frac{5}{12}-\frac{1}{12}=$
5. $\frac{2}{11}+\frac{9}{11}=$

Solutions

1. $\frac{4}{7}-\frac{1}{7}=\frac{4-1}{7}=\frac{3}{7}$
2. $\frac{1}{5}+\frac{3}{5}=\frac{1+3}{5}=\frac{4}{5}$
3. $\frac{1}{6}+\frac{1}{6}=\frac{1+1}{6}=\frac{2}{6}=\frac{1}{3}$
4. $\frac{5}{12}-\frac{1}{12}=\frac{5-1}{12}=\frac{4}{12}=\frac{1}{3}$
5. $\frac{2}{11}+\frac{9}{11}=\frac{2+9}{11}=\frac{11}{11}=1$

When the denominators are not the same, you have to rewrite the fractions so that they do have the same denominator. There are two common methods of doing this. The first is the easiest. The second takes more effort but can result in smaller quantities and less reducing. (When the denominators have no common divisors, these two methods are the same.)

The easiest way to get a common denominator is to multiply the first fraction by the second denominator over itself and the second fraction by the first denominator over itself.

Examples

In $\frac{1}{2}+\frac{3}{7}$ the first denominator is 2 and the second denominator is 7. Multiply $\frac{1}{2}$ by $\frac{7}{7}$ and multiply $\frac{3}{7}$ by $\frac{2}{2}$.

$$\frac{1}{2}+\frac{3}{7}=\left(\frac{1}{2}\cdot\frac{7}{7}\right)+\left(\frac{3}{7}\cdot\frac{2}{2}\right)=\frac{7}{14}+\frac{6}{14}=\frac{13}{14}$$

$$\frac{8}{15}-\frac{1}{2}=\left(\frac{8}{15}\cdot\frac{2}{2}\right)-\left(\frac{1}{2}\cdot\frac{15}{15}\right)=\frac{16}{30}-\frac{15}{30}=\frac{1}{30}$$

Practice

1. $\frac{5}{6}-\frac{1}{5}=$
2. $\frac{1}{3}+\frac{7}{8}=$
3. $\frac{5}{7}-\frac{1}{9}=$
4. $\frac{3}{14}+\frac{1}{2}=$
5. $\frac{3}{4}+\frac{11}{18}=$

Solutions

1. $\frac{5}{6}-\frac{1}{5}=\left(\frac{5}{6}\cdot\frac{5}{5}\right)-\left(\frac{1}{5}\cdot\frac{6}{6}\right)=\frac{25}{30}-\frac{6}{30}=\frac{19}{30}$

2. $\frac{1}{3}+\frac{7}{8}=\left(\frac{1}{3}\cdot\frac{8}{8}\right)+\left(\frac{7}{8}\cdot\frac{3}{3}\right)=\frac{8}{24}+\frac{21}{24}=\frac{29}{24}$

3. $\frac{5}{7}-\frac{1}{9}=\left(\frac{5}{7}\cdot\frac{9}{9}\right)-\left(\frac{1}{9}\cdot\frac{7}{7}\right)=\frac{45}{63}-\frac{7}{63}=\frac{38}{63}$

4. $\frac{3}{14}+\frac{1}{2}=\left(\frac{3}{14}\cdot\frac{2}{2}\right)+\left(\frac{1}{2}\cdot\frac{14}{14}\right)=\frac{6}{28}+\frac{14}{28}=\frac{20}{28}=\frac{5}{7}$

5. $\frac{3}{4}+\frac{11}{18}=\left(\frac{3}{4}\cdot\frac{18}{18}\right)+\left(\frac{11}{18}\cdot\frac{4}{4}\right)=\frac{54}{72}+\frac{44}{72}=\frac{98}{72}=\frac{49}{36}$

Our goal is to add/subtract two fractions with the same denominator. In the previous examples and practice problems, we found a common denominator. Now we will find the *least* common denominator (LCD). For example in $\frac{1}{3}+\frac{1}{6}$, we could compute

$$\frac{1}{3}+\frac{1}{6}=\left(\frac{1}{3}\cdot\frac{6}{6}\right)+\left(\frac{1}{6}\cdot\frac{3}{3}\right)=\frac{6}{18}+\frac{3}{18}=\frac{9}{18}=\frac{1}{2}.$$

But we really only need to rewrite $\frac{1}{3}$:

$$\frac{1}{3}+\frac{1}{6}=\left(\frac{1}{3}\cdot\frac{2}{2}\right)+\frac{1}{6}=\frac{2}{6}+\frac{1}{6}=\frac{3}{6}=\frac{1}{2}.$$

While 18 is *a* common denominator in the above example, 6 is the *smallest* common denominator. When denominators get more complicated, either by being large or having variables in them, you will find it easier to use the LCD to add or subtract fractions. The solution might require less reducing, too.

In the following practice problems one of the denominators will be the LCD; you only need to rewrite the other.

Practice

1. $\frac{1}{8}+\frac{1}{2}=$

2. $\frac{2}{3}-\frac{5}{12}=$

3. $\frac{4}{5}+\frac{1}{20}=$

4. $\frac{7}{30}-\frac{2}{15}=$

5. $\frac{5}{24}+\frac{5}{6}=$

Solutions

1. $\frac{1}{8}+\frac{1}{2}=\frac{1}{8}+\left(\frac{1}{2}\cdot\frac{4}{4}\right)=\frac{1}{8}+\frac{4}{8}=\frac{5}{8}$

2. $\frac{2}{3}-\frac{5}{12}=\left(\frac{2}{3}\cdot\frac{4}{4}\right)-\frac{5}{12}=\frac{8}{12}-\frac{5}{12}=\frac{3}{12}=\frac{1}{4}$

3. $\frac{4}{5}+\frac{1}{20}=\left(\frac{4}{5}\cdot\frac{4}{4}\right)+\frac{1}{20}=\frac{16}{20}+\frac{1}{20}=\frac{17}{20}$

4. $\frac{7}{30}-\frac{2}{15}=\frac{7}{30}-\left(\frac{2}{15}\cdot\frac{2}{2}\right)=\frac{7}{30}-\frac{4}{30}=\frac{3}{30}=\frac{1}{10}$

5. $\frac{5}{24}+\frac{5}{6}=\frac{5}{24}+\left(\frac{5}{6}\cdot\frac{4}{4}\right)=\frac{5}{24}+\frac{20}{24}=\frac{25}{24}$

There are a couple of ways of finding the LCD. Take for example $\frac{1}{12}+\frac{9}{14}$. We could list the multiples of 12 and 14—the first number that appears on each list will be the LCD:

12, 24, 36, 48, 60, 72, **84** and 14, 28, 42, 56, 70, **84**.

Because 84 is the first number on each list, 84 is the LCD for $\frac{1}{12}$ and $\frac{9}{14}$. This method works fine as long as your lists are not too long. But what if your denominators are 6 and 291? The LCD for these denominators (which is 582) occurs 97th on the list of multiples of 6.

We can use the prime factors of the denominators to find the LCD more efficiently. The LCD will consist of every prime factor in each denominator (at its most frequent occurrence). To find the LCD for $\frac{1}{12}$ and $\frac{9}{14}$ factor 12 and 14 into their prime factorizations: $12 = 2\cdot 2\cdot 3$ and $14 = 2\cdot 7$. There are two 2s and one 3 in the prime factorization of 12, so the LCD will have two 2s and one 3. There is one 2 in the prime factorization of 14, but this 2 is covered by the 2s from 12. There is one 7 in the prime factorization of 14, so the LCD will also have a 7 as a factor. Once you have computed the LCD, divide the LCD by each denominator. Multiply each fraction by this number over itself.

$\text{LCD} = 2\cdot 2\cdot 3\cdot 7 = 84$

$84 \div 12 = 7$: multiply $\frac{1}{12}$ by $\frac{7}{7}$ $\quad$ $84 \div 14 = 6$: multiply $\frac{9}{14}$ by $\frac{6}{6}$.

$$\frac{1}{12}+\frac{9}{14}=\left(\frac{1}{12}\cdot\frac{7}{7}\right)+\left(\frac{9}{14}\cdot\frac{6}{6}\right)=\frac{7}{84}+\frac{54}{84}=\frac{61}{84}$$

Examples

$$\frac{5}{6}+\frac{4}{15}$$

$6 = 2 \cdot 3$ and $15 = 3 \cdot 5$

The LCD is $2 \cdot 3 \cdot 5 = 30$; $30 \div 6 = 5$ and $30 \div 15 = 2$. Multiply $\frac{5}{6}$ by $\frac{5}{5}$ and $\frac{4}{15}$ by $\frac{2}{2}$.

$$\frac{5}{6}+\frac{4}{15}=\left(\frac{5}{6}\cdot\frac{5}{5}\right)+\left(\frac{4}{15}\cdot\frac{2}{2}\right)=\frac{25}{30}+\frac{8}{30}=\frac{33}{30}=\frac{11}{10}$$

$$\frac{17}{24}+\frac{5}{36}$$

$24 = 2 \cdot 2 \cdot 2 \cdot 3$ and $36 = 2 \cdot 2 \cdot 3 \cdot 3$

The LCD $= 2 \cdot 2 \cdot 2 \cdot 3 \cdot 3 = 72$; $72 \div 24 = 3$ and $72 \div 36 = 2$. Multiply $\frac{17}{24}$ by $\frac{3}{3}$ and $\frac{5}{36}$ by $\frac{2}{2}$.

$$\frac{17}{24}+\frac{5}{36}=\left(\frac{17}{24}\cdot\frac{3}{3}\right)+\left(\frac{5}{36}\cdot\frac{2}{2}\right)=\frac{51}{72}+\frac{10}{72}=\frac{61}{72}$$

Practice

1. $\frac{11}{12}-\frac{5}{18}=$

2. $\frac{7}{15}+\frac{9}{20}=$

3. $\frac{23}{24}+\frac{7}{16}=$

4. $\frac{3}{8}+\frac{7}{20}=$

5. $\frac{1}{6}+\frac{4}{15}=$

6. $\frac{8}{75}+\frac{3}{10}=$

7. $\frac{35}{54}-\frac{7}{48}=$

8. $\frac{15}{88}+\frac{3}{28}=$

9. $\frac{119}{180}+\frac{17}{210}=$

Solutions

1. $\frac{11}{12}-\frac{5}{18} = \left(\frac{11}{12}\cdot\frac{3}{3}\right)-\left(\frac{5}{18}\cdot\frac{2}{2}\right)=\frac{33}{36}-\frac{10}{36}=\frac{23}{36}$

2. $\frac{7}{15}+\frac{9}{20} = \left(\frac{7}{15}\cdot\frac{4}{4}\right)+\left(\frac{9}{20}\cdot\frac{3}{3}\right)=\frac{28}{60}+\frac{27}{60}=\frac{55}{60}=\frac{11}{12}$

3. $\frac{23}{24}+\frac{7}{16} = \left(\frac{23}{24}\cdot\frac{2}{2}\right)+\left(\frac{7}{16}\cdot\frac{3}{3}\right)=\frac{46}{48}+\frac{21}{48}=\frac{67}{48}$

4. $\frac{3}{8}+\frac{7}{20} = \left(\frac{3}{8}\cdot\frac{5}{5}\right)+\left(\frac{7}{20}\cdot\frac{2}{2}\right)=\frac{15}{40}+\frac{14}{40}=\frac{29}{40}$

5. $\frac{1}{6}+\frac{4}{15} = \left(\frac{1}{6}\cdot\frac{5}{5}\right)+\left(\frac{4}{15}\cdot\frac{2}{2}\right)=\frac{5}{30}+\frac{8}{30}=\frac{13}{30}$

6. $\frac{8}{75}+\frac{3}{10} = \left(\frac{8}{75}\cdot\frac{2}{2}\right)+\left(\frac{3}{10}\cdot\frac{15}{15}\right)=\frac{16}{150}+\frac{45}{150}=\frac{61}{150}$

7. $\frac{35}{54}-\frac{7}{48} = \left(\frac{35}{54}\cdot\frac{8}{8}\right)-\left(\frac{7}{48}\cdot\frac{9}{9}\right)=\frac{280}{432}-\frac{63}{432}=\frac{217}{432}$

8. $\frac{15}{88}+\frac{3}{28} = \left(\frac{15}{88}\cdot\frac{7}{7}\right)+\left(\frac{3}{28}\cdot\frac{22}{22}\right)=\frac{105}{616}+\frac{66}{616}=\frac{171}{616}$

9. $\frac{119}{180}+\frac{17}{210} = \left(\frac{119}{180}\cdot\frac{7}{7}\right)+\left(\frac{17}{210}\cdot\frac{6}{6}\right)=\frac{833}{1260}+\frac{102}{1260}=\frac{935}{1260}$
$=\frac{187}{252}$

Adding More than Two Fractions

Finding the LCD for three or more fractions is pretty much the same as finding the LCD for two fractions. Factor each denominator into its prime factorization and list the primes that appear in each. Divide the LCD by each denominator. Multiply each fraction by this number over itself.

Examples

$\frac{4}{5}+\frac{7}{15}+\frac{9}{20}$

Prime factorization of the denominators: $5=\mathbf{5}$
$15=\mathbf{3}\cdot 5$
$20=\mathbf{2}\cdot\mathbf{2}\cdot 5$

The LCD $=2\cdot 2\cdot 3\cdot 5=60$

$\frac{4}{5}+\frac{7}{15}+\frac{9}{20}=\left(\frac{4}{5}\cdot\frac{12}{12}\right)+\left(\frac{7}{15}\cdot\frac{4}{4}\right)+\left(\frac{9}{20}\cdot\frac{3}{3}\right)=\frac{48}{60}+\frac{28}{60}+\frac{27}{60}=\frac{103}{60}$

$\frac{3}{10}+\frac{5}{12}+\frac{1}{18}$

Prime factorization of the denominators: $10=\mathbf{2}\cdot\mathbf{5}$
$12=\mathbf{2}\cdot 2\cdot\mathbf{3}$
$18=2\cdot\mathbf{3}\cdot 3$

LCD $=2\cdot 2\cdot 3\cdot 3\cdot 5=180$

$\frac{3}{10}+\frac{5}{12}+\frac{1}{18}=\left(\frac{3}{10}\cdot\frac{18}{18}\right)+\left(\frac{5}{12}\cdot\frac{15}{15}\right)+\left(\frac{1}{18}\cdot\frac{10}{10}\right)=\frac{54}{180}+\frac{75}{180}$
$+\frac{10}{180}=\frac{139}{180}$

Practice

1. $\frac{5}{36}+\frac{4}{9}+\frac{7}{12}=$

2. $\frac{11}{24}+\frac{3}{10}+\frac{1}{8}=$

3. $\frac{1}{4}+\frac{5}{6}+\frac{9}{20}=$

4. $\frac{3}{35}+\frac{9}{14}+\frac{7}{10}=$

5. $\frac{5}{48}+\frac{3}{16}+\frac{1}{6}+\frac{7}{9}=$

Solutions

1. $\frac{5}{36}+\frac{4}{9}+\frac{7}{12}=\frac{5}{36}+\left(\frac{4}{9}\cdot\frac{4}{4}\right)+\left(\frac{7}{12}\cdot\frac{3}{3}\right)=\frac{5}{36}+\frac{16}{36}+\frac{21}{36}=\frac{42}{36}=\frac{7}{6}$

2. $\frac{11}{24}+\frac{3}{10}+\frac{1}{8}=\left(\frac{11}{24}\cdot\frac{5}{5}\right)+\left(\frac{3}{10}\cdot\frac{12}{12}\right)+\left(\frac{1}{8}\cdot\frac{15}{15}\right)=\frac{55}{120}+\frac{36}{120}+\frac{15}{120}$
$=\frac{106}{120}=\frac{53}{60}$

3. $\frac{1}{4}+\frac{5}{6}+\frac{9}{20}=\left(\frac{1}{4}\cdot\frac{15}{15}\right)+\left(\frac{5}{6}\cdot\frac{10}{10}\right)+\left(\frac{9}{20}\cdot\frac{3}{3}\right)=\frac{15}{60}+\frac{50}{60}+\frac{27}{60}=\frac{92}{60}$
$=\frac{23}{15}$

4. $\frac{3}{35}+\frac{9}{14}+\frac{7}{10}=\left(\frac{3}{35}\cdot\frac{2}{2}\right)+\left(\frac{9}{14}\cdot\frac{5}{5}\right)+\left(\frac{7}{10}\cdot\frac{7}{7}\right)=\frac{6}{70}+\frac{45}{70}+\frac{49}{70}$
$=\frac{100}{70}=\frac{10}{7}$

5. $\frac{5}{48}+\frac{3}{16}+\frac{1}{6}+\frac{7}{9}=\left(\frac{5}{48}\cdot\frac{3}{3}\right)+\left(\frac{3}{16}\cdot\frac{9}{9}\right)+\left(\frac{1}{6}\cdot\frac{24}{24}\right)+\left(\frac{7}{9}\cdot\frac{16}{16}\right)$
$=\frac{15}{144}+\frac{27}{144}+\frac{24}{144}+\frac{112}{144}=\frac{178}{144}=\frac{89}{72}$

Whole Number-Fraction Arithmetic

A whole number can be written as a fraction whose denominator is 1. With this in mind, we can see that addition and subtraction of whole numbers and fractions are nothing new. To add a whole number to a fraction, multiply the whole number by the fraction's denominator. Add this product to the fraction's numerator. The sum will be the new numerator.

Example

$$3+\frac{7}{8}=\frac{(3\cdot 8)+7}{8}=\frac{24+7}{8}=\frac{31}{8}$$

Practice

1. $4+\frac{1}{3}=$
2. $5+\frac{2}{11}=$
3. $1+\frac{8}{9}=$
4. $2+\frac{2}{5}=$
5. $3+\frac{6}{7}=$

Solutions

1. $4+\frac{1}{3}=\frac{(4\cdot 3)+1}{3}=\frac{12+1}{3}=\frac{13}{3}$
2. $5+\frac{2}{11}=\frac{(5\cdot 11)+2}{11}=\frac{55+2}{11}=\frac{57}{11}$
3. $1+\frac{8}{9}=\frac{(1\cdot 9)+8}{9}=\frac{17}{9}$

4. $2+\frac{2}{5}=\frac{(2\cdot 5)+2}{5}=\frac{10+2}{5}=\frac{12}{5}$

5. $3+\frac{6}{7}=\frac{(3\cdot 7)+6}{7}=\frac{21+6}{7}=\frac{27}{7}$

To subtract a fraction from a whole number multiply the whole number by the fraction's denominator. Subtract the fraction's numerator from this product. The difference will be the new numerator.

Example

$$2-\frac{5}{7}=\frac{(2\cdot 7)-5}{7}=\frac{14-5}{7}=\frac{9}{7}$$

Practice

1. $1-\frac{1}{4}=$

2. $2-\frac{3}{8}=$

3. $5-\frac{6}{11}=$

4. $2-\frac{4}{5}=$

Solutions

1. $1-\frac{1}{4}=\frac{(1\cdot 4)-1}{4}=\frac{3}{4}$

2. $2-\frac{3}{8}=\frac{(2\cdot 8)-3}{8}=\frac{16-3}{8}=\frac{13}{8}$

3. $5-\frac{6}{11}=\frac{(5\cdot 11)-6}{11}=\frac{55-6}{11}=\frac{49}{11}$

4. $2 - \frac{4}{5} = \frac{(2 \cdot 5) - 4}{5} = \frac{10 - 4}{5} = \frac{6}{5}$

To subtract a whole number from the fraction, again multiply the whole number by the fraction's denominator. Subtract this product from the fraction's numerator. This difference will be the new numerator.

Example

$$\frac{8}{3} - 2 = \frac{8 - (2 \cdot 3)}{3} = \frac{8 - 6}{3} = \frac{2}{3}$$

Practice

1. $\frac{12}{5} - 1 =$
2. $\frac{14}{3} - 2 =$
3. $\frac{19}{4} - 2 =$
4. $\frac{18}{7} - 1 =$

Solutions

1. $\frac{12}{5} - 1 = \frac{12 - (1 \cdot 5)}{5} = \frac{7}{5}$
2. $\frac{14}{3} - 2 = \frac{14 - (2 \cdot 3)}{3} = \frac{14 - 6}{3} = \frac{8}{3}$
3. $\frac{19}{4} - 2 = \frac{19 - (2 \cdot 4)}{4} = \frac{19 - 8}{4} = \frac{11}{4}$
4. $\frac{18}{7} - 1 = \frac{18 - (1 \cdot 7)}{7} = \frac{11}{7}$

Compound Fractions

Remember what a fraction is—the division of the numerator by the denominator. For example, $\frac{15}{3}$ is another way of saying "$15 \div 3$." A compound fraction, a fraction where the numerator or denominator or both are not whole numbers, is merely a fraction division problem. For this reason this section is almost the same as the section on fraction division.

Examples

$$\frac{\frac{2}{3}}{\frac{1}{6}} = \frac{2}{3} \div \frac{1}{6} = \frac{2}{3} \cdot \frac{6}{1} = \frac{12}{3} = 4$$

$$\frac{1}{\frac{2}{3}} = 1 \div \frac{2}{3} = 1 \cdot \frac{3}{2} = \frac{3}{2}$$

$$\frac{\frac{8}{9}}{5} = \frac{8}{9} \div 5 = \frac{8}{9} \cdot \frac{1}{5} = \frac{8}{45}$$

Practice

1. $\frac{3}{\frac{5}{9}} =$

2. $\frac{8}{\frac{4}{3}} =$

3. $\frac{\frac{5}{7}}{2} =$

4. $\frac{\frac{4}{11}}{2} =$

5. $\frac{\frac{10}{27}}{\frac{4}{7}} =$

Solutions

1. $\frac{3}{\frac{5}{9}} = 3 \div \frac{5}{9} = 3 \cdot \frac{9}{5} = \frac{27}{5}$

2. $\frac{8}{\frac{4}{3}} = 8 \div \frac{4}{3} = 8 \cdot \frac{3}{4} = \frac{24}{4} = 6$

3. $\frac{\frac{5}{7}}{2} = \frac{5}{7} \div 2 = \frac{5}{7} \cdot \frac{1}{2} = \frac{5}{14}$

4. $\frac{\frac{4}{11}}{2} = \frac{4}{11} \div 2 = \frac{4}{11} \cdot \frac{1}{2} = \frac{4}{22} = \frac{2}{11}$

5. $\frac{\frac{10}{27}}{\frac{4}{7}} = \frac{10}{27} \div \frac{4}{7} = \frac{10}{27} \cdot \frac{7}{4} = \frac{70}{108} = \frac{35}{54}$

Mixed Numbers and Improper Fractions

An improper fraction is a fraction whose numerator is larger than its denominator. For example, $\frac{6}{5}$ is an improper fraction. A mixed number consists of the sum of a whole number and a fraction. For example $1\frac{1}{5}$ (which is really $1 + \frac{1}{5}$) is a mixed number. We will practice going back and forth between the two forms.

To convert a mixed number into an improper fraction, first multiply the whole number by the fraction's denominator. Next add this to the numerator. The sum is the new numerator.

Examples

$$2\tfrac{6}{25} = \frac{(2 \cdot 25) + 6}{25} = \frac{50 + 6}{25} = \frac{56}{25}$$

$$1\tfrac{2}{9} = \frac{(1 \cdot 9) + 2}{9} = \frac{11}{9}$$

$$4\tfrac{1}{6} = \frac{(4 \cdot 6) + 1}{6} = \frac{24 + 1}{6} = \frac{25}{6}$$

Practice

1. $1\frac{7}{8} =$

2. $5\frac{1}{3} =$

3. $2\frac{4}{7} =$

4. $9\frac{6}{11} =$

5. $8\frac{5}{8} =$

Solutions

1. $1\frac{7}{8} = \frac{(1 \cdot 8) + 7}{8} = \frac{8 + 7}{8} = \frac{15}{8}$

2. $5\frac{1}{3} = \frac{(5 \cdot 3) + 1}{3} = \frac{15 + 1}{3} = \frac{16}{3}$

3. $2\frac{4}{7} = \frac{(2 \cdot 7) + 4}{7} = \frac{14 + 4}{7} = \frac{18}{7}$

4. $9\frac{6}{11} = \frac{(9 \cdot 11) + 6}{11} = \frac{99 + 6}{11} = \frac{105}{11}$

5. $8\frac{5}{8} = \frac{(8 \cdot 8) + 5}{8} = \frac{64 + 5}{8} = \frac{69}{8}$

There is a close relationship between improper fractions and division of whole numbers. First let us review the parts of a division problem.

$$\begin{array}{r} \textit{quotient} \\ \textit{divisor}\ \overline{)\textit{dividend}} \\ \cdots \\ \hline \textit{remainder} \end{array}$$

In an improper fraction, the numerator is the dividend and the divisor is the denominator. In a mixed number, the quotient is the whole number, the remainder is the new numerator, and the divisor is the denominator.

$$\begin{array}{rl} & \textit{quotient} \\ \textit{divisor} & \overline{)\textit{dividend}} \\ & \quad \cdots \\ & \overline{\textit{remainder}} \end{array} \qquad \begin{array}{r} 3 \\ 7\overline{)22} \\ -21 \\ \hline 1 \end{array} \qquad \frac{22}{7} = 3\frac{1}{7}$$

To convert an improper fraction to a mixed number, divide the numerator by the denominator. The remainder will be the new numerator and the quotient will be the whole number.

Examples

$$\frac{14}{5} \qquad \begin{array}{r} 2 \\ 5\overline{)14} \\ -10 \\ \hline 4 \end{array} \quad \text{new numerator}$$

$$\frac{14}{5} = 2\tfrac{4}{5}$$

$$\frac{21}{5} \qquad \begin{array}{r} 4 \\ 5\overline{)21} \\ -20 \\ \hline 1 \end{array} \quad \text{new numerator}$$

$$\frac{21}{5} = 4\tfrac{1}{5}$$

Practice

1. $\frac{13}{4}$
2. $\frac{19}{3}$
3. $\frac{39}{14}$
4. $\frac{24}{5}$
5. $\frac{26}{7}$

Solutions

1. $\frac{13}{4}$ $\quad 4\overline{)13}$ with quotient 3, -12, remainder 1

$$\frac{13}{4} = 3\tfrac{1}{4}$$

2. $\frac{19}{3}$ $\quad 3\overline{)19}$ with quotient 6, -18, remainder 1

$$\frac{19}{3} = 6\tfrac{1}{3}$$

3. $\frac{39}{14}$ $\quad 14\overline{)39}$ with quotient 2, -28, remainder 11

$$\frac{39}{14} = 2\tfrac{11}{14}$$

4. $\frac{24}{5}$ $\quad 5\overline{)24}$ with quotient 4, -20, remainder 4

$$\frac{24}{5} = 4\tfrac{4}{5}$$

5. $\frac{26}{7}$ $\quad 7\overline{)26}$ with quotient 3, -21, remainder 5

$$\frac{26}{7} = 3\tfrac{5}{7}$$

Mixed Number Arithmetic

You can add (or subtract) two mixed numbers in one of two ways. One way is to add the whole numbers then add the fractions.

$$4\tfrac{2}{3} + 3\tfrac{1}{2} = \left(4 + \tfrac{2}{3}\right) + \left(3 + \tfrac{1}{2}\right) = (4 + 3) + \left(\tfrac{2}{3} + \tfrac{1}{2}\right)$$
$$= 7 + \left(\tfrac{4}{6} + \tfrac{3}{6}\right) = 7 + \tfrac{7}{6} = 7 + 1 + \tfrac{1}{6} = 8\tfrac{1}{6}$$

The other way is to convert the mixed numbers into improper fractions then add.

$$4\tfrac{2}{3} + 3\tfrac{1}{2} = \frac{14}{3} + \frac{7}{2} = \frac{28}{6} + \frac{21}{6} = \frac{49}{6} = 8\tfrac{1}{6}$$

Practice

1. $2\tfrac{3}{7} + 1\tfrac{1}{2}$
2. $2\tfrac{5}{16} + 1\tfrac{11}{12}$
3. $4\tfrac{5}{6} + 1\tfrac{2}{3}$
4. $3\tfrac{4}{9} - 1\tfrac{1}{6}$
5. $2\tfrac{3}{4} + \tfrac{5}{6}$
6. $4\tfrac{2}{3} + 2\tfrac{1}{5}$
7. $5\tfrac{1}{12} - 3\tfrac{3}{8}$

Solutions

1. $2\tfrac{3}{7} + 1\tfrac{1}{2} = \frac{17}{7} + \frac{3}{2} = \frac{34}{14} + \frac{21}{14} = \frac{55}{14} = 3\tfrac{13}{14}$

2. $2\tfrac{5}{16} + 1\tfrac{11}{12} = \frac{37}{16} + \frac{23}{12} = \frac{111}{48} + \frac{92}{48} = \frac{203}{48} = 4\tfrac{11}{48}$

3. $4\tfrac{5}{6} + 1\tfrac{2}{3} = \frac{29}{6} + \frac{5}{3} = \frac{29}{6} + \frac{10}{6} = \frac{39}{6} = \frac{13}{2} = 6\tfrac{1}{2}$

4. $3\tfrac{4}{9} - 1\tfrac{1}{6} = \frac{31}{9} - \frac{7}{6} = \frac{62}{18} - \frac{21}{18} = \frac{41}{18} = 2\tfrac{5}{18}$

5. $2\tfrac{3}{4} + \tfrac{5}{6} = \frac{11}{4} + \frac{5}{6} = \frac{33}{12} + \frac{10}{12} = \frac{43}{12} = 3\tfrac{7}{12}$

6. $4\frac{2}{3} + 2\frac{1}{5} = \dfrac{14}{3} + \dfrac{11}{5} = \dfrac{70}{15} + \dfrac{33}{15} = \dfrac{103}{15} = 6\frac{13}{15}$

7. $5\frac{1}{12} - 3\frac{3}{8} = \dfrac{61}{12} - \dfrac{27}{8} = \dfrac{122}{24} - \dfrac{81}{24} = \dfrac{41}{24} = 1\frac{17}{24}$

When multiplying mixed numbers first convert them to improper fractions, and then multiply. Multiplying the whole numbers and the fractions is incorrect because there are really two operations involved—addition and multiplication:

$$1\tfrac{1}{2} \cdot 4\tfrac{1}{3} = \left(1 + \frac{1}{2}\right) \cdot \left(4 + \frac{1}{3}\right).$$

Convert the mixed numbers to improper fractions before multiplying.

$$1\tfrac{1}{2} \cdot 4\tfrac{1}{3} = \frac{3}{2} \cdot \frac{13}{3} = \frac{39}{6} = \frac{13}{2} = 6\tfrac{1}{2}$$

Practice

1. $1\frac{3}{4} \cdot 2\frac{1}{12} =$

2. $2\frac{2}{25} \cdot \frac{3}{7} =$

3. $2\frac{1}{8} \cdot 2\frac{1}{5} =$

4. $7\frac{1}{2} \cdot 1\frac{1}{3} =$

5. $\frac{3}{4} \cdot 2\frac{1}{4} =$

Solutions

1. $1\frac{3}{4} \cdot 2\frac{1}{12} = \dfrac{7}{4} \cdot \dfrac{25}{12} = \dfrac{175}{48} = 3\frac{31}{48}$

2. $2\frac{2}{25} \cdot \frac{3}{7} = \dfrac{52}{25} \cdot \dfrac{3}{7} = \dfrac{156}{175}$

3. $2\frac{1}{8} \cdot 2\frac{1}{5} = \dfrac{17}{8} \cdot \dfrac{11}{5} = \dfrac{187}{40} = 4\frac{27}{40}$

4. $7\frac{1}{2} \cdot 1\frac{1}{3} = \frac{15}{2} \cdot \frac{4}{3} = \frac{60}{6} = 10$

5. $\frac{3}{4} \cdot 2\frac{1}{4} = \frac{3}{4} \cdot \frac{9}{4} = \frac{27}{16} = 1\frac{11}{16}$

Division of mixed numbers is similar to multiplication in that you first convert the mixed numbers into improper fractions before dividing.

Example

$$\frac{3\frac{1}{3}}{1\frac{2}{5}} = \frac{\frac{10}{3}}{\frac{7}{5}} = \frac{10}{3} \div \frac{7}{5} = \frac{10}{3} \cdot \frac{5}{7} = \frac{50}{21} = 2\frac{8}{21}$$

Practice

1. $\frac{1\frac{3}{8}}{2\frac{1}{3}} =$

2. $\frac{\frac{7}{16}}{1\frac{2}{3}} =$

3. $\frac{1\frac{4}{15}}{1\frac{1}{4}} =$

4. $\frac{5\frac{1}{2}}{3} =$

5. $\frac{2\frac{1}{2}}{\frac{1}{2}} =$

Solutions

1. $\frac{1\frac{3}{8}}{2\frac{1}{3}} = \frac{\frac{11}{8}}{\frac{7}{3}} = \frac{11}{8} \div \frac{7}{3} = \frac{11}{8} \cdot \frac{3}{7} = \frac{33}{56}$

2. $\frac{\frac{7}{16}}{1\frac{2}{3}} = \frac{\frac{7}{16}}{\frac{5}{3}} = \frac{7}{16} \div \frac{5}{3} = \frac{7}{16} \cdot \frac{3}{5} = \frac{21}{80}$

3. $\dfrac{1\frac{4}{15}}{1\frac{1}{4}} = \dfrac{\frac{19}{15}}{\frac{5}{4}} = \dfrac{19}{15} \div \dfrac{5}{4} = \dfrac{19}{15} \cdot \dfrac{4}{5} = \dfrac{76}{75} = 1\frac{1}{75}$

4. $\dfrac{5\frac{1}{2}}{3} = \dfrac{\frac{11}{2}}{3} = \dfrac{11}{2} \div 3 = \dfrac{11}{2} \cdot \dfrac{1}{3} = \dfrac{11}{6} = 1\frac{5}{6}$

5. $\dfrac{2\frac{1}{2}}{\frac{1}{2}} = \dfrac{\frac{5}{2}}{\frac{1}{2}} = \dfrac{5}{2} \div \dfrac{1}{2} = \dfrac{5}{2} \cdot \dfrac{2}{1} = \dfrac{10}{2} = 5$

Recognizing Quantities and Relationships in Word Problems

Success in solving word problems depends on the mastery of three skills—"translating" English into mathematics, setting the variable equal to an appropriate unknown quantity, and using knowledge of mathematics to solve the equation or inequality. This book will help you develop the first two skills and some attention will be given to the third.

English	Mathematical Symbol
"Is," "are," "will be" (any form of the verb "to be") mean "equal"	$=$
"More than," "increased by," "sum of" mean "add"	$+$
"Less than," "decreased by," "difference of" mean "subtract"	$-$
"Of" means "multiply"	$\times$
"Per" means "divide"	$\div$
"More than" and "greater than" both mean the relation "greater than" although "more than" can mean "add"	$>$
"Less than," means the relation "less than" although it can also mean "subtract"	$<$

English	Mathematical Symbol
"At least," and "no less than," mean the relation "greater than or equal to"	$\geq$
"No more than," and "at most," mean the relation "less than or equal to"	$\leq$

We will ease into the topic of word problems by translating English sentences into mathematical sentences. We will not solve word problems until later in the book.

Examples

Five is two more than three.
$5 = 2 + 3$

Ten less six is four.
$10 - 6 = 4$

One half of twelve is six.
$\frac{1}{2} \times 12 = 6$

Eggs cost \$1.15 per dozen.
$1.15 / 12$ (this gives the price per egg)

The difference of sixteen and five is eleven.
$16 - 5 = 11$

Fourteen decreased by six is eight.
$14 - 6 = 8$

Seven increased by six is thirteen.
$7 + 6 = 13$

Eight is less than eleven.
$8 < 11$

Eight is at most eleven.
$8 \leq 11$

Eleven is more than eight.
$11 > 8$

Eleven is at least eight.
$11 \geq 8$

One hundred is twice fifty.
$100 = 2 \times 50$

Five more than eight is thirteen.
$5 + 8 = 13$

Practice

Translate the English sentence into a mathematical sentence.

1. Fifteen less four is eleven.
2. Seven decreased by two is five.
3. Six increased by one is seven.
4. The sum of two and three is five.
5. Nine more than four is thirteen.
6. One-third of twelve is four.
7. One-third of twelve is greater than two.
8. Half of sixteen is eight.
9. The car gets 350 miles per eleven gallons.
10. Ten is less than twelve.
11. Ten is no more than twelve.
12. Three-fourths of sixteen is twelve.
13. Twice fifteen is thirty.
14. The difference of fourteen and five is nine.
15. Nine is more than six.
16. Nine is at least six.

Solutions

1. $15 - 4 = 11$

2. $7 - 2 = 5$

3. $6 + 1 = 7$

4. $2 + 3 = 5$

5. $9 + 4 = 13$

6. $\frac{1}{3} \times 12 = 4$

7. $\frac{1}{3} \times 12 > 2$

8. $\frac{1}{2} \times 16 = 8$

9. $350 \div 11$ (miles per gallon)

10. $10 < 12$

11. $10 \leq 12$

12. $\frac{3}{4} \times 16 = 12$

13. $2 \times 15 = 30$

14. $14 - 5 = 9$

15. $9 > 6$

16. $9 \geq 6$

Chapter Review

1. Write $\frac{18}{5}$ as a mixed number.

(a) $1\frac{8}{5}$ (b) $3\frac{3}{5}$ (c) $8\frac{1}{5}$ (d) $2\frac{4}{5}$

2. $\frac{1}{6} + \frac{7}{15} =$

(a) $\frac{19}{30}$ (b) $\frac{8}{15}$ (c) $\frac{8}{23}$ (d) $\frac{1}{2}$

3. $\frac{1\frac{3}{4}}{2\frac{2}{5}} =$

(a) $\frac{21}{5}$ (b) $\frac{7}{5}$ (c) $\frac{21}{20}$ (d) $\frac{35}{48}$

4. $\frac{17}{18} - \frac{5}{12} =$

(a) $\frac{1}{18}$ (b) $\frac{19}{36}$ (c) 2 (d) $\frac{29}{36}$

5. Write $2\frac{4}{9}$ as an improper fraction.

(a) $\frac{24}{9}$ (b) $\frac{8}{9}$ (c) $\frac{22}{9}$ (d) $2\frac{4}{9}$ is an improper fraction

6. $\frac{3}{7} \cdot \frac{5}{6} =$

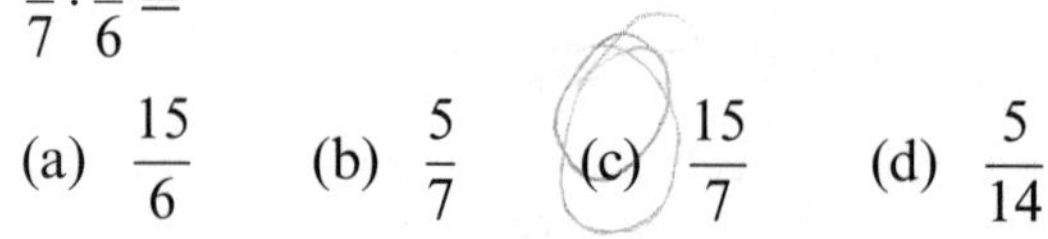

(a) $\frac{15}{6}$ (b) $\frac{5}{7}$ (c) $\frac{15}{7}$ (d) $\frac{5}{14}$

7. $\frac{1}{3} + \frac{2}{15} + \frac{1}{18} =$

(a) $\frac{1}{18}$ (b) $\frac{47}{90}$ (c) $\frac{4}{90}$ (d) $\frac{7}{15}$

8. $\frac{11}{12} + \frac{1}{2} =$

(a) $\frac{17}{12}$ (b) $\frac{3}{4}$ (c) $\frac{13}{14}$ (d) $\frac{13}{24}$

9. $\frac{2}{\frac{3}{4}} =$

(a) $\frac{3}{2}$ (b) $\frac{3}{8}$ (c) $\frac{4}{3}$ (d) $\frac{8}{3}$

10. $1\frac{1}{2} - \frac{1}{4} =$

(a) $\frac{17}{12}$ (b) $\frac{5}{12}$ (c) $\frac{5}{4}$ (d) $\frac{1}{3}$

11. $3\frac{1}{10} + 2\frac{4}{15} =$

(a) $\frac{13}{10}$ (b) $\frac{101}{30}$ (c) $\frac{161}{30}$ (d) $\frac{3}{10}$

12. $1\frac{3}{5} \cdot 2\frac{1}{8} =$

(a) $3\frac{2}{5}$ (b) $3\frac{3}{5}$ (c) $2\frac{3}{40}$ (d) $2\frac{3}{5}$

13. $3\frac{5}{6} \cdot \frac{1}{2} =$

(a) $2\frac{11}{12}$ (b) $1\frac{1}{4}$ (c) $3\frac{5}{12}$ (d) $1\frac{11}{12}$

14. $\frac{5}{24} + \frac{4}{21} + \frac{5}{16} =$

(a) $\frac{239}{336}$ (b) $\frac{14}{61}$ (c) $\frac{14}{24}$ (d) $\frac{41}{84}$

15. Write in mathematical symbols: Three more than two is five.

(a) $3 > 2 + 5$ (b) $3 + 2 = 5$

(c) $3 + 2 > 5$ (d) $3 + 5 > 2$

16. Write in mathematical symbols: Five is at least four.

(a) $4 > 5$ (b) $4 \geq 5$ (c) $5 > 4$ (d) $5 \geq 4$

Solutions

1. (b)	**2.** (a)	**3.** (d)	**4.** (b)
5. (c)	**6.** (d)	**7.** (b)	**8.** (a)
9. (d)	**10.** (c)	**11.** (c)	**12.** (a)
13. (d)	**14.** (a)	**15.** (b)	**16.** (d)

Introduction to Variables

A variable is a symbol for a number whose value is unknown. A variable might represent quantities at different times. For example if you are paid by the hour for your job and you earn \$10 per hour, letting x represent the number of hours worked would allow you to write your earnings as "$10x$." The value of your earnings *varies* depending on the number of hours worked. If an equation has one variable, we can use algebra to determine what value the variable is representing.

Variables are treated like numbers because they are numbers. For instance $2 + x$ means two plus the quantity x and $2x$ means two times the quantity x (when no operation sign is given, the operation is assumed to be multiplication). The expression $3x + 4$ means three times x plus four. This is not the same as $3x + 4x$ which is three x's plus four x's for a total of seven x's: $3x + 4x = 7x$.

If you are working with variables and want to check whether the expression you have computed is really equal to the expression with which you started, take some larger prime number, not a factor of anything else in the expression, and plug it into both the original expression and the last one. If the resulting numbers are the same, it is very likely that the first and last expressions are equal. For example you might ask "Is it true that $3x + 4 = 7x$?" Test $x = 23$:

$3(23)+4=73$ and $7(23)=161$, so we can conclude that in general $3x+4 \neq 7x$. (Actually for $x=1$, and only $x=1$, they are equal.)

This method for checking equality of algebraic expressions is not foolproof. Equal numbers do not *always* guarantee that the expressions are equal. Also be careful not to make an arithmetic error. The two expressions might be equal but making an arithmetic error might lead you to conclude that they are not equal.

Canceling with Variables

Variables can be canceled in fractions just as whole numbers can be.

Examples

$$\frac{2x}{x} = \frac{2}{1} \cdot \frac{x}{x} = 2$$

$$\frac{6x}{9x} = \frac{6}{9} \cdot \frac{x}{x} = \frac{6}{9} = \frac{2}{3}$$

$$\frac{7xy}{5x} = \frac{7y}{5} \cdot \frac{x}{x} = \frac{7y}{5}$$

When you see a plus or minus sign in a fraction, be *very careful* when you cancel. For example in the expression $\frac{2+x}{x}$, x cannot be canceled. The only quantities that can be canceled are factors. Many students mistakenly "cancel" the x and conclude that $\frac{2+x}{x} = \frac{2+1}{1} = 3$ or $\frac{2+x}{x} = 2$. These equations are false. If $\frac{2+x}{x}$ were equal to 2 or to 3, then we could substitute any value for x (except for 0) and we would get a true equation. Let's try $x = 19$: $\frac{2+19}{19} = \frac{21}{19}$.

We can see that $\frac{2+x}{x} \neq 2$ and $\frac{2+x}{x} \neq 3$. The reason that the x cannot be factored is that x is a term in this expression, not a factor. (A term is a quantity separated from others by a plus or minus sign.) If you must cancel the x out of $\frac{2+x}{x}$, you must rewrite the fraction:

$$\frac{2+x}{x} = \frac{2}{x} + \frac{x}{x} = \frac{2}{x} + 1.$$

Simply because a plus or minus sign appears in a fraction does not automatically mean that canceling is not appropriate. For instance $\frac{3+x}{3+x} = 1$ because any nonzero quantity divided by itself is one.

Examples

$$\frac{(2+3x)(x-1)}{2+3x} = \frac{2+3x}{2+3x} \cdot \frac{x-1}{1} = x - 1.$$

The reason $2 + 3x$ can be canceled is that $2 + 3x$ is a factor of $(2 + 3x)(x - 1)$.

$$\frac{2(x+7)(3x+1)}{2} = \frac{2}{2} \cdot \frac{(x+7)(3x+1)}{1} = (x+7)(3x+1)$$

$$\frac{15(x+6)(x-2)}{3(x-2)} = \frac{3 \cdot 5(x+6)(x-2)}{3(x-2)} = \frac{3(x-2)}{3(x-2)} \cdot \frac{5(x+6)}{1} = 5(x+6)$$

Practice

1. $\frac{3xy}{2x} =$

2. $\frac{8x}{4} =$

3. $\frac{30xy}{16y} =$

4. $\frac{72x}{18xy} =$

5. $\frac{x(x-6)}{2x} =$

6. $\frac{6xy(2x-1)}{3x} =$

7. $\dfrac{(5x+16)(2x+7)}{6(2x+7)} =$

8. $\dfrac{24x(y+8)(x+1)}{15x(y+8)} =$

9. $\dfrac{150xy(2x+17)(8x-3)}{48y} =$

Solutions

1. $\dfrac{3xy}{2x} = \dfrac{3y}{2} \cdot \dfrac{x}{x} = \dfrac{3y}{2}$

2. $\dfrac{8x}{4} = \dfrac{2x}{1} \cdot \dfrac{4}{4} = 2x$

3. $\dfrac{30xy}{16y} = \dfrac{15x}{8} \cdot \dfrac{2y}{2y} = \dfrac{15x}{8}$

4. $\dfrac{72x}{18xy} = \dfrac{4}{y} \cdot \dfrac{18x}{18x} = \dfrac{4}{y}$

5. $\dfrac{x(x-6)}{2x} = \dfrac{x-6}{2} \cdot \dfrac{x}{x} = \dfrac{x-6}{2}$

6. $\dfrac{6xy(2x-1)}{3x} = \dfrac{2y(2x-1)}{1} \cdot \dfrac{3x}{3x} = 2y(2x-1)$

7. $\dfrac{(5x+16)(2x+7)}{6(2x+7)} = \dfrac{5x+16}{6} \cdot \dfrac{2x+7}{2x+7} = \dfrac{5x+16}{6}$

8. $\dfrac{24x(y+8)(x+1)}{15x(y+8)} = \dfrac{8(x+1)}{5} \cdot \dfrac{3x(y+8)}{3x(y+8)} = \dfrac{8(x+1)}{5}$

9. $$\dfrac{150xy(2x+17)(8x-3)}{48y} = \dfrac{25x(2x+17)(8x-3)}{8} \cdot \dfrac{6y}{6y}$$
$$= \dfrac{25x(2x+17)(8x-3)}{8}$$

Operations on Fractions with Variables

Multiplication of fractions with variables is done in exactly the same way as multiplication of fractions without variables—multiply the numerators and multiply the denominators.

Examples

$$\frac{7}{10} \cdot \frac{3x}{4} = \frac{21x}{40}$$

$$\frac{24}{5y} \cdot \frac{3}{14} = \frac{24 \cdot 3}{5y \cdot 14} = \frac{12 \cdot 3}{5y \cdot 7} = \frac{36}{35y}$$

$$\frac{34x}{15} \cdot \frac{3}{16} = \frac{34x \cdot 3}{15 \cdot 16} = \frac{17x \cdot 1}{5 \cdot 8} = \frac{17x}{40}$$

Practice

1. $\frac{4x}{9} \cdot \frac{2}{7} =$
2. $\frac{2}{3x} \cdot \frac{6y}{5} =$
3. $\frac{18x}{19} \cdot \frac{2}{11x} =$
4. $\frac{5x}{9} \cdot \frac{4y}{3} =$

Solutions

1. $\frac{4x}{9} \cdot \frac{2}{7} = \frac{8x}{63}$
2. $\frac{2}{3x} \cdot \frac{6y}{5} = \frac{2 \cdot 6y}{3x \cdot 5} = \frac{2 \cdot 2y}{x \cdot 5} = \frac{4y}{5x}$

3. $\frac{18x}{19} \cdot \frac{2}{11x} = \frac{18x \cdot 2}{19 \cdot 11x} = \frac{18 \cdot 2}{19 \cdot 11} = \frac{36}{209}$

4. $\frac{5x}{9} \cdot \frac{4y}{3} = \frac{20xy}{27}$

At times, especially in calculus, you might need to write a fraction as a product of two fractions or of one fraction and a whole number or of one fraction and a variable. The steps you will follow are the same as in multiplying fractions—only in reverse.

Examples

$$\frac{3}{4} = \frac{3 \cdot 1}{1 \cdot 4} = \frac{3}{1} \cdot \frac{1}{4} = 3 \cdot \frac{1}{4}$$

(This expression is different from $3\frac{1}{4} = 3 + \frac{1}{4}$.)

$$\frac{x}{3} = \frac{1 \cdot x}{3 \cdot 1} = \frac{1}{3} \cdot \frac{x}{1} = \frac{1}{3}x$$

$$\frac{3}{x} = \frac{3 \cdot 1}{1 \cdot x} = \frac{3}{1} \cdot \frac{1}{x} = 3 \cdot \frac{1}{x}$$

$$\frac{7}{8x} = \frac{7 \cdot 1}{8 \cdot x} = \frac{7}{8} \cdot \frac{1}{x}$$

$$\frac{7x}{8} = \frac{7 \cdot x}{8 \cdot 1} = \frac{7}{8}x$$

$$\frac{x+1}{2} = \frac{1 \cdot (x+1)}{2 \cdot 1} = \frac{1}{2}(x+1)$$

$$\frac{2}{x+1} = \frac{2 \cdot 1}{1 \cdot (x+1)} = \frac{2}{1} \cdot \frac{1}{x+1} = 2\frac{1}{x+1}$$

Practice

Separate the factor having a variable from the rest of the fraction.

1. $\frac{4x}{5} =$

2. $\frac{7y}{15} =$

3. $\frac{3}{4t} =$

4. $\frac{11}{12x} =$

5. $\frac{x-3}{4} =$

6. $\frac{5}{7(4x-1)} =$

7. $\frac{10}{2x+1} =$

Solutions

1. $\frac{4x}{5} = \frac{4 \cdot x}{5 \cdot 1} = \frac{4}{5} \cdot \frac{x}{1} = \frac{4}{5}x$

2. $\frac{7y}{15} = \frac{7 \cdot y}{15 \cdot 1} = \frac{7}{15} \cdot \frac{y}{1} = \frac{7}{15}y$

3. $\frac{3}{4t} = \frac{3 \cdot 1}{4 \cdot t} = \frac{3}{4} \cdot \frac{1}{t}$

4. $\frac{11}{12x} = \frac{11 \cdot 1}{12 \cdot x} = \frac{11}{12} \cdot \frac{1}{x}$

5. $\frac{x-3}{4} = \frac{1 \cdot (x-3)}{4 \cdot 1} = \frac{1}{4}(x-3)$

6. $\frac{5}{7(4x-1)} = \frac{5 \cdot 1}{7 \cdot (4x-1)} = \frac{5}{7} \cdot \frac{1}{4x-1}$

7. $\frac{10}{2x+1} = \frac{10 \cdot 1}{1 \cdot (2x+1)} = 10 \cdot \frac{1}{2x+1}$

Fraction Division and Compound Fractions

Division of fractions with variables can become multiplication of fractions by inverting the second fraction. Because compound fractions are really only fraction division problems, rewrite the compound fraction as fraction division then as fraction multiplication.

Example

$$\frac{\frac{4}{5}}{\frac{x}{3}} = \frac{4}{5} \div \frac{x}{3} = \frac{4}{5} \cdot \frac{3}{x} = \frac{12}{5x}$$

Practice

1. $\frac{\frac{15}{32}}{\frac{x}{6}} =$

2. $\frac{\frac{9x}{14}}{\frac{2x}{17}} =$

3. $\frac{\frac{4}{3}}{\frac{21x}{8}} =$

4. $\frac{\frac{3}{8y}}{\frac{1}{y}} =$

5. $\frac{\frac{2}{x}}{\frac{7}{11y}} =$

6. $\frac{\frac{12x}{7}}{\frac{23}{6y}} =$

7. $\frac{3}{\frac{2}{9y}} =$

8. $\frac{\frac{x}{2}}{5} =$

Solutions

1. $\frac{\frac{15}{32}}{\frac{x}{6}} = \frac{15}{32} \div \frac{x}{6} = \frac{15}{32} \cdot \frac{6}{x} = \frac{15 \cdot 6}{32 \cdot x} = \frac{15 \cdot 3}{16 \cdot x} = \frac{45}{16x}$

2. $\frac{\frac{9x}{14}}{\frac{2x}{17}} = \frac{9x}{14} \div \frac{2x}{17} = \frac{9x}{14} \cdot \frac{17}{2x} = \frac{9x \cdot 17}{14 \cdot 2x} = \frac{9 \cdot 17}{14 \cdot 2} = \frac{153}{28}$

3. $\frac{\frac{4}{3}}{\frac{21x}{8}} = \frac{4}{3} \div \frac{21x}{8} = \frac{4}{3} \cdot \frac{8}{21x} = \frac{32}{63x}$

4. $\frac{\frac{3}{8y}}{\frac{1}{y}} = \frac{3}{8y} \div \frac{1}{y} = \frac{3}{8y} \cdot \frac{y}{1} = \frac{3 \cdot y}{8y \cdot 1} = \frac{3}{8}$

5. $\frac{\frac{2}{x}}{\frac{7}{11y}} = \frac{2}{x} \div \frac{7}{11y} = \frac{2}{x} \cdot \frac{11y}{7} = \frac{22y}{7x}$

6. $\frac{\frac{12x}{7}}{\frac{23}{6y}} = \frac{12x}{7} \div \frac{23}{6y} = \frac{12x}{7} \cdot \frac{6y}{23} = \frac{72xy}{161}$

7. $\frac{3}{\frac{2}{9y}} = \frac{3}{1} \div \frac{2}{9y} = \frac{3}{1} \cdot \frac{9y}{2} = \frac{27y}{2}$

8. $\frac{\frac{x}{2}}{5} = \frac{x}{2} \div \frac{5}{1} = \frac{x}{2} \cdot \frac{1}{5} = \frac{x}{10}$

Adding and Subtracting Fractions with Variables

When adding or subtracting fractions with variables, treat the variables as if they were prime numbers.

Examples

$$\frac{6}{25} + \frac{y}{4} = \frac{6}{25} \cdot \frac{4}{4} + \frac{y}{4} \cdot \frac{25}{25} = \frac{24}{100} + \frac{25y}{100} = \frac{24 + 25y}{100}$$

$$\frac{2t}{15}+\frac{14}{33}=\frac{2t}{15}\cdot\frac{11}{11}+\frac{14}{33}\cdot\frac{5}{5}=\frac{22t}{165}+\frac{70}{165}=\frac{22t+70}{165}$$

$$\frac{18}{x}-\frac{3}{4}=\frac{18}{x}\cdot\frac{4}{4}-\frac{3}{4}\cdot\frac{x}{x}=\frac{72}{4x}-\frac{3x}{4x}=\frac{72-3x}{4x}$$

$$\frac{9}{16t}+\frac{5}{12}$$

$16t=2\cdot 2\cdot 2\cdot 2\cdot t$ and $12=2\cdot 2\cdot 3$ so the LCD $=2\cdot 2\cdot 2\cdot 2\cdot 3\cdot t$
$=48t$

$48t\div 16t=3$ and $48t\div 12=4t$

$$\frac{9}{16t}+\frac{5}{12}=\frac{9}{16t}\cdot\frac{3}{3}+\frac{5}{12}\cdot\frac{4t}{4t}=\frac{27}{48t}+\frac{20t}{48t}=\frac{27+20t}{48t}$$

$$\frac{71}{84}-\frac{13}{30x}$$

$84=2\cdot 2\cdot 3\cdot 7$ and $30x=2\cdot 3\cdot 5\cdot x$

LCD $=2\cdot 2\cdot 3\cdot 5\cdot 7\cdot x=420x$ and
$420x\div 84=5x$ and $420x\div 30x=14$

$$\frac{71}{84}-\frac{13}{30x}=\frac{71}{84}\cdot\frac{5x}{5x}-\frac{13}{30x}\cdot\frac{14}{14}=\frac{355x}{420x}-\frac{182}{420x}=\frac{355x-182}{420x}$$

Practice

Do not try to reduce your solutions. We will learn how to reduce fractions like these in a later chapter.

1. $\frac{4}{15}+\frac{2x}{33}=$
2. $\frac{x}{48}-\frac{7}{30}=$
3. $\frac{3}{x}+\frac{4}{25}=$
4. $\frac{2}{45x}+\frac{6}{35}=$

5. $\dfrac{11}{150x}+\dfrac{7}{36}=$

6. $\dfrac{2}{21x}+\dfrac{3}{98x}=$

7. $\dfrac{1}{12x}+\dfrac{5}{9y}=$

8. $\dfrac{19}{51y}-\dfrac{1}{6x}=$

9. $\dfrac{7x}{24y}+\dfrac{2}{15y}=$

10. $\dfrac{3}{14y}+\dfrac{2}{35x}=$

Solutions

1. $\dfrac{4}{15}+\dfrac{2x}{33}=\dfrac{4}{15}\cdot\dfrac{11}{11}+\dfrac{2x}{33}\cdot\dfrac{5}{5}=\dfrac{44}{165}+\dfrac{10x}{165}=\dfrac{44+10x}{165}$

2. $\dfrac{x}{48}-\dfrac{7}{30}=\dfrac{x}{48}\cdot\dfrac{5}{5}-\dfrac{7}{30}\cdot\dfrac{8}{8}=\dfrac{5x}{240}-\dfrac{56}{240}=\dfrac{5x-56}{240}$

3. $\dfrac{3}{x}+\dfrac{4}{25}=\dfrac{3}{x}\cdot\dfrac{25}{25}+\dfrac{4}{25}\cdot\dfrac{x}{x}=\dfrac{75}{25x}+\dfrac{4x}{25x}=\dfrac{75+4x}{25x}$

4. $\dfrac{2}{45x}+\dfrac{6}{35}=\dfrac{2}{45x}\cdot\dfrac{7}{7}+\dfrac{6}{35}\cdot\dfrac{9x}{9x}=\dfrac{14}{315x}+\dfrac{54x}{315x}=\dfrac{14+54x}{315x}$

5. $\dfrac{11}{150x}+\dfrac{7}{36}=\dfrac{11}{150x}\cdot\dfrac{6}{6}+\dfrac{7}{36}\cdot\dfrac{25x}{25x}=\dfrac{66}{900x}+\dfrac{175x}{900x}=\dfrac{66+175x}{900x}$

6. $\dfrac{2}{21x}+\dfrac{3}{98x}=\dfrac{2}{21x}\cdot\dfrac{14}{14}+\dfrac{3}{98x}\cdot\dfrac{3}{3}=\dfrac{28}{294x}+\dfrac{9}{294x}=\dfrac{37}{294x}$

7. $\dfrac{1}{12x}+\dfrac{5}{9y}=\dfrac{1}{12x}\cdot\dfrac{3y}{3y}+\dfrac{5}{9y}\cdot\dfrac{4x}{4x}=\dfrac{3y}{36xy}+\dfrac{20x}{36xy}=\dfrac{3y+20x}{36xy}$

8. $\dfrac{19}{51y}-\dfrac{1}{6x}=\dfrac{19}{51y}\cdot\dfrac{2x}{2x}-\dfrac{1}{6x}\cdot\dfrac{17y}{17y}=\dfrac{38x}{102xy}-\dfrac{17y}{102xy}=\dfrac{38x-17y}{102xy}$

9. $\frac{7x}{24y} + \frac{2}{15y} = \frac{7x}{24y} \cdot \frac{5}{5} + \frac{2}{15y} \cdot \frac{8}{8} = \frac{35x}{120y} + \frac{16}{120y} = \frac{35x + 16}{120y}$

10. $\frac{3}{14y} + \frac{2}{35x} = \frac{3}{14y} \cdot \frac{5x}{5x} + \frac{2}{35x} \cdot \frac{2y}{2y} = \frac{15x}{70xy} + \frac{4y}{70xy} = \frac{15x + 4y}{70xy}$

Word Problems

Often the equations used to solve word problems should have only one variable, and other unknowns must be written in terms of one variable. The goal of this section is to get you acquainted with setting your variable equal to an appropriate unknown quantity, and writing other unknown quantities in terms of the variable.

Examples

Andrea is twice as old as Sarah.

Because Andrea's age is being compared to Sarah's, the easiest thing to do is to let x represent Sarah's age:

Let $x =$ Sarah's age.

Andrea is twice as old as Sarah, so Andrea's age $= 2x$. We could have let x represent Andrea's age, but we would have to re-think the statement as "Sarah is *half* as old as Andrea." This would mean Sarah's age would be represented by $\frac{1}{2}x$.

John has eight more nickels than Larry has.

The number of John's nickels is being compared to the number of Larry's nickels, so it is easier to let x represent the number of nickels Larry has.

Let $x =$ the number of nickels Larry has.

$x + 8 =$ the number of nickels John has.

A used car costs \$5000 less than a new car.

Let $x =$ the price of the new car.

$x - 5000 =$ the price of the used car

A box's length is three times its width.

Let $x =$ width (in the given units).

$3x =$ length (in the given units)

Jack is two-thirds as tall as Jill.

Let $x =$ Jill's height (in the given units).

$\frac{2}{3}x =$ Jack's height (in the given units)

From 6 pm to 6 am the temperature dropped 30 degrees.

Let $x =$ temperature (in degrees) at 6 pm.

$x - 30 =$ temperature (in degrees) at 6 am

One-eighth of an employee's time is spent cleaning his work station.

Let $x =$ the number of hours he is on the job.

$\frac{1}{8}x =$ the number of hours he spends cleaning his work station

$10,000 was deposited between two savings accounts, Account A and Account B.

Let $x =$ amount deposited in Account A.

How much is left to represent the amount invested in Account B? If x dollars is taken from $10,000, then it must be that $10{,}000 - x$ dollars is left to be deposited in Account B.

Or if x represents the amount deposited in Account B, then $10{,}000 - x$ dollars is left to be deposited in Account A.

A wire is cut into three pieces of unequal length. The shortest piece is $\frac{1}{4}$ the length of the longest piece, and the middle piece is $\frac{1}{3}$ the length of the longest piece.

Let $x =$ length of the longest piece.

$\frac{1}{3}x =$ length of the middle piece
$\frac{1}{4}x =$ length of the shortest piece

A store is having a one-third off sale on a certain model of air conditioner.

Let $x =$ regular price of the air conditioner. Then $\frac{2}{3}x =$ sale price of the air conditioner.

We cannot say that the sale price is $x - \frac{1}{3}$ because $\frac{1}{3}$ is not "one-third off the price of the air conditioner;" it is simply "one-third." "One-third the price of the air conditioner" is represented by $\frac{1}{3}x$. "One-third off the price of the air conditioner" is represented by

$$x - \frac{1}{3}x = \frac{x}{1} - \frac{x}{3} = \frac{3x}{3} - \frac{x}{3} = \frac{2x}{3} = \frac{2}{3}x.$$

Practice

1. Tony is three years older than Marie.
   Marie's age = ________
   Tony's age = ________
2. Sandie is three-fourths as tall as Mona.
   Mona's height (in the given unit of measure) = ________
   Sandie's height (in the given unit of measure) = ________
3. Michael takes two hours longer than Gina to compute his taxes.
   Number of hours Gina takes to compute her taxes = ________
   Number of hours Michael takes to compute his taxes = ________
4. Three-fifths of a couple's net income is spent on rent
   Net income = ________
   Amount spent on rent = ________
5. A rectangle's length is four times its width
   Width (in the given unit of measure) = ________
   Length (in the given unit of measure) = ________
6. Candice paid $5000 last year in federal and state income taxes.
   Amount paid in federal income taxes = ________
   Amount paid in state income taxes = ________

7. Nikki has \$8000 in her bank, some in a checking account, some in a certificate of deposit (CD).
   Amount in checking account = _______
   Amount in CD = _______
8. A total of 450 tickets were sold, some adult tickets, some children's tickets.
   Number of adult tickets sold = _______
   Number of children's tickets sold =_______
9. A boutique is selling a sweater for three-fourths off retail.
   Retail selling price = _______
   Sale price = _______
10. A string is cut into three pieces of unequal length. The shortest piece is $\frac{1}{5}$ as long as the longest piece. The mid-length piece is $\frac{1}{2}$ the length of the longest piece.
    Length of the longest piece (in the given units) = _______
    Length of the shortest piece (in the given units) = _______
    Length of the mid-length piece (in the given units) = _______

Solutions

1. Marie's age = x
   Tony's age = $x+3$
2. Mona's height (in the given unit of measure) = x
   Sandie's height (in the given unit of measure) = $\frac{3}{4}x$
3. Number of hours Gina takes to compute her taxes = x
   Number of hours Michael takes to compute his taxes = $x+2$
4. Net income = x
   Amount spent on rent = $\frac{3}{5}x$
5. Width (in the given unit of measure) = x
   Length (in the given unit of measure) = $4x$
6. Amount paid in federal income taxes = x
   Amount paid in state income taxes = $5000 - x$
   (Or x = amount paid in state income taxes and $5000 - x$ = amount paid in federal taxes)
7. Amount in checking account = x
   Amount in CD = $8000 - x$
   (Or x = amount in CD and $8000 - x$ = amount in checking account)
8. Number of adult tickets sold = x
   Number of children's tickets sold = $450 - x$

(Or x = number of children's tickets and $450 - x$ = number of adult tickets)

9. Retail selling price = $\underline{\quad x \quad}$

   Sale price $= \frac{1}{4}x$ or $\frac{x}{4}$ $\left(x - \frac{3}{4}x = \frac{4x}{4} - \frac{3x}{4} = \frac{4x - 3x}{4} = \frac{x}{4}\right)$

10. Length of the longest piece (in the given units) = $\underline{\quad x \quad}$

    Length of the shortest piece (in the given units) = $\underline{\quad \frac{1}{5}x \quad}$

    Length of the mid-length piece (in the given units) = $\underline{\quad \frac{1}{2}x \quad}$

Chapter Review

1. Reduce to lowest terms: $\frac{3}{6x}$

(a) $\frac{1}{2}$ (b) $\frac{x}{2}$ (c) $\frac{1}{x}$ (d) $\frac{1}{2x}$

2. Rewrite as a product of a number and a variable: $\frac{7x}{15}$

(a) $\frac{7}{15}x$ (b) $\frac{7}{15} \cdot \frac{1}{x}$ (c) $\frac{7}{15} + x$ (d) $\frac{7}{15} - x$

3. $\frac{x}{4} - \frac{1}{3} =$

(a) $\frac{x - 1}{12}$ (b) $x - 1$ (c) $\frac{3x - 4}{12}$ (d) $\frac{4x - 3}{12}$

4. The length of a square is twice the length of a smaller square. If x represents the length of the smaller square, then the length of the larger square is

(a) $\frac{x}{2}$ (b) $\frac{1}{2x}$ (c) $\frac{2}{x}$ (d) $2x$

5. $\frac{10}{3} \cdot \frac{2x}{15} =$

(a) $\frac{4x}{3}$ (b) $\frac{4x}{9}$ (c) $\frac{10x}{3}$ (d) $\frac{2x}{45}$

6. $\frac{\frac{2}{3}}{\frac{4x}{9}} =$

(a) $\frac{8x}{27}$ (b) $\frac{3x}{2}$ (c) $\frac{3}{2x}$ (d) $\frac{1}{2x}$

7. A movie on DVD is on sale for $\frac{1}{3}$ off its regular price. If x represents the movie's regular price, then the sale price is

(a) $\frac{1}{3}x$ (b) $x - \frac{1}{3}$ (c) $\frac{1}{3} - x$ (d) $\frac{2}{3}x$

8. Rewrite as a product of a number and a variable: $\frac{3}{4x}$

(a) $\frac{3}{4}x$ (b) $\frac{4}{3}x$ (c) $\frac{3}{4} \cdot \frac{1}{x}$ (d) $\frac{4}{3} \cdot \frac{1}{x}$

9. $\frac{x}{4} \cdot \frac{3}{2x} =$

(a) $\frac{3}{8}$ (b) $\frac{3}{2}$ (c) $\frac{3x}{2}$ (d) $\frac{3x}{8}$

10. $\frac{\frac{7}{4}}{2x} =$

(a) $\frac{7}{8x}$ (b) $7x$ (c) $\frac{7}{2x}$ (d) $\frac{2}{7x}$

11. $\frac{4y}{12y} =$

(a) $\frac{1}{2}$ (b) $\frac{1}{3}$ (c) $\frac{y}{2}$ (d) $\frac{y}{3}$

12. $\frac{2}{15x} + \frac{7}{18} =$

(a) $\frac{4}{11x}$ (b) $\frac{47}{90}$ (c) $\frac{12 + 35x}{90x}$ (d) $\frac{37}{15}$

13. Suppose \$6000 is invested in two stocks—Stock A and Stock B. If x represents the amount invested in Stock A, then the amount invested in Stock B is

(a) $x - 6000$ (b) $x + 6000$ (c) $\frac{x}{6000}$ (d) $6000 - x$

14. Reduce to lowest terms: $\frac{10(x+3)(x-2)}{15(x-2)}$

(a) $\frac{2(x+3)}{3}$ (b) $2x$ (c) $\frac{10(x+3)}{15}$ (d) $2(x+1)$

15. Rewrite as a product of a number and algebraic factor (one with a variable in it): $\frac{x-4}{2}$

(a) $1 \cdot (x-1)$ (b) $\frac{1}{2}(x-4)$ (c) $\frac{1}{2}(x-2)$

(d) $1 \cdot (x-2)$

Solutions

1. (d)	**2.** (a)	**3.** (c)	**4.** (d)
5. (b)	**6.** (c)	**7.** (d)	**8.** (c)
9. (a)	**10.** (a)	**11.** (b)	**12.** (c)
13. (d)	**14.** (a)	**15.** (b)	

Decimals

A decimal number is a fraction in disguise: $0.48 = \frac{48}{100}$ and $1.291 = 1\frac{291}{1000}$ or $\frac{1291}{1000}$. The number in front of the decimal point is the whole number (if there is one) and the number behind the decimal point is the numerator of a fraction whose denominator is a power of ten. The denominator will consist of 1 followed by one or more zeros. The number of zeros is the same as the number of digits behind the decimal point.

$2.8 = 2\frac{8}{10}$ (one decimal place—one zero)

$0.7695 = \frac{7695}{10{,}000}$ (four decimal places—four zeros)

Practice

Rewrite as a fraction. If the decimal number is more than 1, rewrite the number both as a mixed number and as an improper fraction.

1. $1.71 =$
2. $34.598 =$

3. $0.6 =$

4. $0.289421 =$

Solutions

1. $1.71 = 1\frac{71}{100} = \dfrac{171}{100}$

2. $34.598 = 34\frac{598}{1000} = \dfrac{34{,}598}{1000}$

3. $0.6 = \dfrac{6}{10}$

4. $0.289421 = \dfrac{289{,}421}{1{,}000{,}000}$

There are two types of decimal numbers, *terminating* and *nonterminating*. The above examples and practice problems are terminating decimal numbers. A nonterminating decimal number has infinitely many nonzero digits following the decimal point. For example, 0.333333333 ... is a nonterminating decimal number. Some nonterminating decimal numbers represent fractions—$0.333333333\ldots = \frac{1}{3}$. But some nonterminating decimals, like $\pi = 3.1415926654\ldots$ and $\sqrt{2} = 1.414213562\ldots$, do not represent fractions. We will be concerned mostly with terminating decimal numbers in this book.

You can add as many zeros at the end of a terminating decimal number as you want because the extra zeros cancel away.

$$0.7 = \frac{7}{10}$$
$$0.70 = \frac{70}{100} = \frac{7 \cdot 10}{10 \cdot 10} = \frac{7}{10}$$
$$0.700 = \frac{700}{1000} = \frac{7 \cdot 100}{10 \cdot 100} = \frac{7}{10}$$

Adding and Subtracting Decimal Numbers

In order to add or subtract decimal numbers, each number needs to have the same number of digits behind the decimal point. If you write the problem vertically, you can avoid the common problem of adding the numbers incorrectly. For instance, 1.2 + 3.41 is *not* 4.43. The "2" needs to be added to the "4," not to the "1."

$$\begin{array}{r} 1.20 \\ +4.43 \\ \hline 5.63 \end{array}$$ (Add as many zeros at the end as you need.)

510.3 − 422.887 becomes

$$\begin{array}{r} 510.300 \\ -422.887 \\ \hline 87.413 \end{array}$$

Practice

Rewrite as a vertical problem and solve.

1. 7.26 + 18.1
2. 5 − 2.76
3. 15.01 − 6.328
4. 968.323 − 13.08
5. 28.56 − 16.7342
6. 0.446 + 1.2
7. 2.99 + 3

Solutions

1. 7.26 + 18.1

$$\begin{array}{r} 7.26 \\ +18.10 \\ \hline 25.36 \end{array}$$

2. 5 − 2.76

$$\begin{array}{r} 5.00 \\ -2.76 \\ \hline 2.24 \end{array}$$

3. 15.01 − 6.328

$$\begin{array}{r} 15.010 \\ -\ \underline{\ 6.328} \\ 8.682 \end{array}$$

4. 968.323 − 13.08

$$\begin{array}{r} 968.323 \\ -\ \underline{\ 13.080} \\ 955.243 \end{array}$$

5. 28.56 − 16.7342

$$\begin{array}{r} 28.5600 \\ -\underline{16.7342} \\ 11.8258 \end{array}$$

6. 0.446 + 1.2

$$\begin{array}{r} 0.446 \\ +\underline{1.200} \\ 1.646 \end{array}$$

7. 2.99 + 3

$$\begin{array}{r} 2.99 \\ +\underline{3.00} \\ 5.99 \end{array}$$

Multiplying Decimal Numbers

To multiply decimal numbers, perform multiplication as you would with whole numbers. Then count the number of digits that follow the decimal point or points in the factors. This total will be the number of digits that follow the decimal point in the product.

Examples

$12.\underline{83} \times 7.\underline{91}$ The product will have four digits following the decimal point.

$3499.\underline{782} \times 19.\underline{41}$ The product will have five digits following the decimal point.

$14 \times 3.\underline{55}$ The product will have two digits following the decimal point. (The second digit behind the decimal point is 0: $49.70 = 49.7$.)

Practice

1. $3.2 \times 1.6 =$
2. $4.11 \times 2.84 =$
3. $8 \times 2.5 =$
4. $0.153 \times 6.8 =$
5. $0.0351 \times 5.6 =$

Solutions

1. $3.2 \times 1.6 = 5.12$
2. $4.11 \times 2.84 = 11.6724$
3. $8 \times 2.5 = 20.0 = 20$
4. $0.153 \times 6.8 = 1.0404$
5. $0.0351 \times 5.6 = 0.19656$

Decimal Fractions

Fractions having a decimal number in their numerator and/or denominator can be rewritten as fractions without decimal points. Multiply the numerator and denominator by a power of 10—the same power of 10—large enough so that the decimal point becomes unnecessary.

$$\frac{1.3}{3.9} = \frac{1.3 \cdot 10}{3.9 \cdot 10} = \frac{13}{39} = \frac{1}{3}$$

To determine what power of 10 you will need, count the number of digits behind each decimal point.

$$\frac{1.28}{4.6} \begin{array}{l} \leftarrow \text{Two digits behind the decimal point} \\ \leftarrow \text{One digit behind the decimal point} \end{array}$$

You will need to use $10^2 = 100$. $\frac{1.28 \cdot 100}{4.6 \cdot 100} = \frac{128}{460} = \frac{32}{115}$

Examples

$$\frac{7.1}{2.285} = \frac{7.1 \cdot 1000}{2.285 \cdot 1000} = \frac{7100}{2285} = \frac{1420}{457}$$

$$\frac{6}{3.14} = \frac{6 \cdot 100}{3.14 \cdot 100} = \frac{600}{314} = \frac{300}{157}$$

Practice

1. $\frac{4.58}{2.15} =$

2. $\frac{3.6}{18.11} =$

3. $\frac{2.123}{5.6} =$

4. $\frac{8}{2.4} =$

5. $\frac{6.25}{5} =$

6. $\frac{0.31}{1.2} =$

7. $\frac{0.423}{0.6} =$

Solutions

1. $\frac{4.58}{2.15} = \frac{4.58 \cdot 100}{2.15 \cdot 100} = \frac{458}{215}$

2. $\frac{3.6}{18.11} = \frac{3.6 \cdot 100}{18.11 \cdot 100} = \frac{360}{1811}$

3. $\frac{2.123}{5.6} = \frac{2.123 \cdot 1000}{5.6 \cdot 1000} = \frac{2123}{5600}$

4. $\frac{8}{2.4} = \frac{8 \cdot 10}{2.4 \cdot 10} = \frac{80}{24} = \frac{10}{3}$

5. $\frac{6.25}{5} = \frac{6.25 \cdot 100}{5 \cdot 100} = \frac{625}{500} = \frac{5}{4}$

6. $\frac{0.31}{1.2} = \frac{0.31 \cdot 100}{1.2 \cdot 100} = \frac{31}{120}$

7. $\frac{0.423}{0.6} = \frac{0.423 \cdot 1000}{0.6 \cdot 1000} = \frac{423}{600} = \frac{141}{200}$

Division with Decimals

We can use the method in the previous section to rewrite decimal division problems as whole number division problems. Rewrite the division problem as a fraction, clear the decimal, and then rewrite the fraction as a division problem without decimal points.

Examples

$1.2\,\overline{)6.03}$ is another way of writing $\frac{6.03}{1.2}$.

$\frac{6.03}{1.2} = \frac{6.03 \cdot 100}{1.2 \cdot 100} = \frac{603}{120}$ and $\frac{603}{120}$ becomes $120\,\overline{)603}$.

$0.51\,\overline{)3.7}$ becomes $\frac{3.7}{0.51} = \frac{3.7 \cdot 100}{0.51 \cdot 100} = \frac{370}{51}$ which becomes $51\,\overline{)370}$.

$8\overline{)12.8}$ becomes $\frac{12.8}{8}=\frac{12.8\cdot 10}{8\cdot 10}=\frac{128}{80}$ which becomes $80\overline{)128}$.

You could reduce your fraction and get an even simpler division problem.

$\frac{128}{80}=\frac{8}{5}$ which becomes $5\overline{)8}$.

Practice

Rewrite as a division problem without decimal points.

1. $6.85\overline{)15.11}$
2. $0.9\overline{)8.413}$
3. $4\overline{)8.8}$
4. $19.76\overline{)60.4}$
5. $3.413\overline{)7}$

Solutions

1. $6.85\overline{)15.11}$ $\frac{15.11}{6.85}=\frac{15.11\cdot 100}{6.85\cdot 100}=\frac{1511}{685}$ becomes $685\overline{)1511}$
2. $0.9\overline{)8.413}$ $\frac{8.413}{0.9}=\frac{8.413\cdot 1000}{0.9\cdot 1000}=\frac{8413}{900}$ becomes $900\overline{)8413}$
3. $4\overline{)8.8}$ $\frac{8.8}{4}=\frac{8.8\cdot 10}{4\cdot 10}=\frac{88}{40}$ becomes $40\overline{)88}$

   Reduce to get a simpler division problem.

   $\frac{88}{40}=\frac{11}{5}$ becomes $5\overline{)11}$
4. $19.76\overline{)60.4}$ $\frac{60.4}{19.76}=\frac{60.4\cdot 100}{19.76\cdot 100}=\frac{6040}{1976}$

   Reduce to get a simpler division problem: $\frac{6040}{1976}=\frac{755}{247}$

   which becomes $247\overline{)755}$

5. $3.413\,\overline{)7}$ $\quad \dfrac{7}{3.413} = \dfrac{7 \cdot 1000}{3.413 \cdot 1000} = \dfrac{7000}{3413}$ becomes $3413\,\overline{)7000}$

Chapter Review

1. $\dfrac{4.9}{2.71} =$

(a) $\dfrac{49}{271}$ (b) $\dfrac{4900}{271}$ (c) $\dfrac{49}{2710}$ (d) $\dfrac{490}{271}$

2. $4.78 =$

(a) $\dfrac{478}{100}$ (b) $\dfrac{478}{10}$ (c) $\dfrac{478}{1000}$ (d) $\dfrac{478}{10{,}000}$

3. $3.2 \times 2.11 =$

(a) 0.6752 (b) 67.52 (c) 6.752 (d) 67.52

4. $4.2 - 1.96 =$

(a) 2.06 (b) 2.24 (c) 2.60 (d) 3.34

5. $5 \times 4.4 =$

(a) 22.2 (b) 22.0 (c) 20.2 (d) 2.2

6. Rewrite $1.2\overline{)5.79}$ without decimal points.

(a) $120\overline{)579}$ (b) $12\overline{)579}$ (c) $120\overline{)5790}$ (d) $12\overline{)5790}$

7. $1.1 + 3.08 =$

(a) 4.09 (b) 3.19 (c) 4.18 (d) 3.18

8. $\dfrac{0.424}{1.5} =$

(a) $\dfrac{424}{15}$ (b) $\dfrac{4240}{1500}$ (c) $\dfrac{424}{150}$ (d) $\dfrac{424}{1500}$

9. 0.016 =

(a) $\frac{16}{1000}$ (b) $\frac{16}{10}$ (c) $\frac{16}{100}$ (d) $\frac{160}{1000}$

Solutions

1. (d) **2.** (a) **3.** (c) **4.** (b)
5. (b) **6.** (a) **7.** (c) **8.** (d)
9. (a)

CHAPTER 4

Negative Numbers

A negative number is a number smaller than zero. Think about the readings on a thermometer. A reading of $-10°$ means the temperature is $10°$ *below* $0°$ and that the temperature would need to warm up $10°$ to reach $0°$. A reading of $10°$ means the temperature would need to cool down $10°$ to reach $0°$.

Let us use a test example to discover some facts about arithmetic with negative numbers. Suppose you are taking a test where one point is awarded for each correct answer and one point is deducted for each incorrect answer. If you miss the first three problems, how many would you need to answer correctly to bring your score to 10? You would need to answer three correctly to bring your answer up to zero, then you would need to answer 10 more correctly to bring your score to 10; you would need to answer 13 correctly to bring a score of -3 to 10: $10 = -3 + 3 + 10 = -3 + 13$. Now suppose you miss the first eight problems and get the next two answers correct. You now only need to answer six more correctly to reach zero: $-8 + 2 = -6$.

When adding a negative number to a positive number (or a positive number to a negative number), take the difference of the numbers. The sign on the sum will be the same as the sign of the "larger" number. If no sign appears in front of a number, the number is positive.

Examples

$-82 + 30 =$ ____. The difference of 82 and 30 is 52. Because 82 is larger than 30, the sign on 82 will be used on the sum: $-82 + 30 = -52$.

$-125 + 75 =$ ____. The difference of 125 and 75 is 50. Because 125 is larger than 75, the sign on 125 will be used on the sum: $-125 + 75 = -50$.

$-10 + 48 =$ ____. The difference of 48 and 10 is 38. Because 48 is larger than 10, the sign on 48 will be used on the sum: $-10 + 48 = 38$.

Practice

1. $-10 + 8 =$
2. $-65 + 40 =$
3. $13 + -20 =$
4. $-5 + 9 =$
5. $-24 + 54 =$
6. $-6 + 19 =$
7. $-71 + 11 =$
8. $40 + -10 =$
9. $12 + -18 =$

Solutions

1. $-10 + 8 = -2$
2. $-65 + 40 = -25$
3. $13 + -20 = -7$

4. $-5+9=4$

5. $-24+54=30$

6. $-6+19=13$

7. $-71+11=-60$

8. $40+-10=30$

9. $12+-18=-6$

Returning to the test example, suppose you have gotten the first five correct but missed the next two. Of course, your score would be $5-2=3$. Suppose now that you missed more than five, your score would then become a negative number. If you missed the next seven problems, you will have lost credit for all five you got correct plus another two: $5-7=-2$. When subtracting a larger positive number from a smaller positive number, take the difference of the two numbers. The difference will be negative.

Examples

$410-500=-90$ $\quad$ $10-72=-62$

Be careful what you call these signs; a negative sign in front of a number indicates that the number is smaller than zero. A minus sign between two numbers indicates subtraction. In $3-5=-2$, the sign in front of 5 is a minus sign and the sign in front of 2 is a negative sign. A minus sign requires two quantities and a negative sign requires one quantity.

Practice

1. $28-30=$

2. $88-100=$

3. $25-110=$

4. $4-75=$

5. $5-90=$

Solutions

1. $28 - 30 = -2$

2. $88 - 100 = -12$

3. $25 - 110 = -85$

4. $4 - 75 = -71$

5. $5 - 90 = -85$

Finally, suppose you have gotten the first five problems incorrect. If you miss the next three problems, you move even further away from zero; you would now need to get five correct to bring the first five problems up to zero plus another three correct to bring the next three problems up to zero. In other words, you would need to get 8 more correct to bring your score up to zero: $-5 - 3 = -8$. To subtract a positive number from a negative number, add the two numbers. The sum will be negative.

Examples

$-30 - 15 = -45$ $\quad$ $-18 - 7 = -25$ $\quad$ $-500 - 81 = -581$

Practice

1. $-16 - 4 =$

2. $-70 - 19 =$

3. $-35 - 5 =$

4. $-100 - 8 =$

5. $-99 - 1 =$

Solutions

1. $-16 - 4 = -20$

2. $-70 - 19 = -89$

3. $-35 - 5 = -40$

4. $-100 - 8 = -108$

5. $-99 - 1 = -100$

Double Negatives

A negative sign in front of a quantity can be interpreted to mean "opposite." For instance -3 can be called "the opposite of 3." Viewed in this way, we can see that $-(-4)$ means "the opposite of -4." But the opposite of -4 is 4: $-(-4) = 4$.

Examples

$-(-25) = 25$ $\quad -(-x) = x$ $\quad -(-3y) = 3y$

Rewriting a Subtraction Problem as an Addition Problem

Sometimes in algebra it is easier to think of a subtraction problem as an addition problem. One advantage to this is that you can rearrange the terms in an addition problem but not a subtraction problem: $3 + 4 = 4 + 3$ but $4 - 3 \neq 3 - 4$. The minus sign can be replaced with a plus sign if you change the sign of the number following it: $4 - 3 = 4 + (-3)$. The parentheses are used to show that the sign in front of the 3 is a negative sign and not a minus sign.

Examples

$-82 - 14 = -82 + (-14)$ $\quad 20 - (-6) = 20 + 6$ $\quad x - y = x + (-y)$

Practice

Rewrite as an addition problem.

1. $8 - 5$
2. $-29 - 4$
3. $-6 - (-10)$
4. $15 - x$
5. $40 - 85$
6. $y - 37$
7. $-x - (-14)$
8. $-x - 9$

Solutions

1. $8 - 5 = 8 + (-5)$
2. $-29 - 4 = -29 + (-4)$
3. $-6 - (-10) = -6 + 10$
4. $15 - x = 15 + (-x)$
5. $40 - 85 = 40 + (-85)$
6. $y - 37 = y + (-37)$
7. $-x - (-14) = -x + 14$
8. $-x - 9 = -x + (-9)$

Adding and Subtracting Fractions (Again)

Remember to convert a mixed number to an improper fraction before subtracting.

Practice

1. $\frac{-4}{5} + \frac{2}{3} =$

2. $2\frac{1}{8} - 3\frac{1}{4} =$

3. $-4\frac{2}{9} - 1\frac{1}{2} =$

4. $\frac{5}{36} - 2 =$

5. $\frac{6}{25} + \frac{2}{3} - \frac{14}{15} =$

6. $\frac{-4}{3} + \frac{5}{6} - \frac{8}{21} =$

7. $1\frac{4}{5} - 3\frac{1}{2} - 1\frac{6}{7} =$

Solutions

1. $\frac{-4}{5} + \frac{2}{3} = \frac{-4}{5} \cdot \frac{3}{3} + \frac{2}{3} \cdot \frac{5}{5} = \frac{-12}{15} + \frac{10}{15} = \frac{-12 + 10}{15} = \frac{-2}{15}$

2. $2\frac{1}{8} - 3\frac{1}{4} = \frac{17}{8} - \frac{13}{4} = \frac{17}{8} - \frac{13}{4} \cdot \frac{2}{2} = \frac{17}{8} - \frac{26}{8} = \frac{17 - 26}{8} = \frac{-9}{8}$

3. $-4\frac{2}{9} - 1\frac{1}{2} = \frac{-38}{9} - \frac{3}{2} = \frac{-38}{9} \cdot \frac{2}{2} - \frac{3}{2} \cdot \frac{9}{9} = \frac{-76}{18} - \frac{27}{18}$
$= \frac{-76 - 27}{18} = \frac{-103}{18}$

4. $\frac{5}{36} - 2 = \frac{5}{36} - \frac{2}{1} \cdot \frac{36}{36} = \frac{5}{36} - \frac{72}{36} = \frac{5 - 72}{36} = \frac{-67}{36}$

5. $\frac{6}{25}+\frac{2}{3}-\frac{14}{15}=\frac{6}{25}\cdot\frac{3}{3}+\frac{2}{3}\cdot\frac{25}{25}-\frac{14}{15}\cdot\frac{5}{5}=\frac{18}{75}+\frac{50}{75}-\frac{70}{75}$
$=\frac{18+50-70}{75}=\frac{-2}{75}$

6. $\frac{-4}{3}+\frac{5}{6}-\frac{8}{21}=\frac{-4}{3}\cdot\frac{14}{14}+\frac{5}{6}\cdot\frac{7}{7}-\frac{8}{21}\cdot\frac{2}{2}=\frac{-56}{42}+\frac{35}{42}-\frac{16}{42}$
$=\frac{-56+35-16}{42}=\frac{-37}{42}$

7. $1\frac{4}{5}-3\frac{1}{2}-1\frac{6}{7}=\frac{9}{5}-\frac{7}{2}-\frac{13}{7}=\frac{9}{5}\cdot\frac{14}{14}-\frac{7}{2}\cdot\frac{35}{35}-\frac{13}{7}\cdot\frac{10}{10}$
$=\frac{126-245-130}{70}=\frac{-249}{70}$

Multiplication and Division with Negative Numbers

When taking the product of two or more quantities when one or more of them is negative, take the product as you ordinarily would as if the negative signs were not there. Count the number of negatives in the product. An even number of negative signs will yield a positive product and an odd number of negative signs will yield a negative product. Similarly, for a quotient (or fraction), two negatives yield a positive quotient and one negative and one positive yield a negative quotient.

Examples

$(4)(-3)(2)=-24$ $(-5)(-6)(-1)(3)=-90$ $8\div(-2)=-4$

$\frac{-55}{5}=-11$ $\frac{-2}{-3}=\frac{2}{3}$

Practice

1. $(15)(-2)=$

2. $-32 \div (-8) =$

3. $3(-3)(4) =$

4. $62 \div (-2) =$

5. $\frac{1}{2}\left(\frac{-3}{7}\right) =$

6. $(-4)(6)(-3) =$

7. $\frac{\frac{4}{3}}{-\frac{1}{2}} =$

8. $(-2)(5)(-6)(-8) =$

9. $\frac{-\frac{3}{5}}{-\frac{6}{5}} =$

10. $-28 \div (-4) =$

Solutions

1. $(15)(-2) = -30$

2. $-32 \div (-8) = 4$

3. $3(-3)(4) = -36$

4. $62 \div (-2) = -31$

5. $\frac{1}{2}\left(\frac{-3}{7}\right) = \frac{-3}{14}$

6. $(-4)(6)(-3) = 72$

7. $\frac{\frac{4}{3}}{-\frac{1}{2}} = \frac{4}{3} \div \frac{-1}{2} = \frac{4}{3} \cdot \frac{2}{-1} = \frac{8}{-3} = -\frac{8}{3}$

8. $(-2)(5)(-6)(-8) = -480$

9. $\frac{\frac{-3}{5}}{\frac{-6}{5}} = \frac{-3}{5} \div \frac{-6}{5} = \frac{-3}{5} \cdot \frac{5}{-6} = \frac{1}{2}$

10. $-28 \div (-4) = 7$

Negating a variable does not automatically mean that the quantity will be negative: $-x$ means "the opposite" of x. We cannot conclude that $-x$ is a negative number unless we have reason to believe x itself is a positive number. If x is a negative number, $-x$ is a positive number. (Although in practice we verbally say "negative x" for "$-x$" when we really mean "the opposite of x.")

The same rules apply when multiplying "negative" variables.

Examples

$-3(5x) = -15x$ $\qquad$ $5(-x) = -5x$

$-12(-4x) = 48x$ $\qquad$ $-x(-y) = xy$

$-2x(3y) = -6xy$ $\qquad$ $x(-y) = -xy$

$-16x(-4y) = 64xy$ $\qquad$ $4(-1.83x)(2.36y) = -17.2752xy$

$-3(-x) = 3x$

Practice

1. $18(-3x) =$
2. $-4(2x)(-9y) =$
3. $28(-3x) =$
4. $-5x(-7y) =$
5. $-1(-6)(-7x) =$
6. $1.1x(2.5y)-$
7. $-8.3(4.62x) =$

8. $-2.6(-13.14)(-6x) =$

9. $0.36(-8.1x)(-1.6y) =$

10. $4(-7)(2.1x)y =$

Solutions

1. $18(-3x) = -54x$

2. $-4(2x)(-9y) = 72xy$

3. $28(-3x) = -84x$

4. $-5x(-7y) = 35xy$

5. $-1(-6)(-7x) = -42x$

6. $1.1x(2.5y) = 2.75xy$

7. $-8.3(4.62x) = -38.346x$

8. $-2.6(-13.14)(-6x) = -204.984x$

9. $0.36(-8.1x)(-1.6y) = 4.6656xy$

10. $4(-7)(2.1x)y = -58.8xy$

Because a negative divided by a positive is negative and a positive divided by a negative is negative, a negative sign in a fraction can go wherever you want to put it.

$$\frac{\textit{negative}}{\textit{positive}} = \frac{\textit{positive}}{\textit{negative}} = -\frac{\textit{positive}}{\textit{positive}}$$

Examples

$$\frac{-2}{3} = \frac{2}{-3} = -\frac{2}{3} \qquad \frac{-x}{4} = \frac{x}{-4} = -\frac{x}{4}$$

Practice

Rewrite the fraction two different ways.

1. $\frac{-3}{5} =$

2. $\frac{-2x}{19} =$

3. $\frac{4}{-3x} =$

4. $-\frac{5}{9y} =$

Solutions

1. $\frac{-3}{5} = \frac{3}{-5} = -\frac{3}{5}$

2. $\frac{-2x}{19} = \frac{2x}{-19} = -\frac{2x}{19}$

3. $\frac{4}{-3x} = \frac{-4}{3x} = -\frac{4}{3x}$

4. $-\frac{5}{9y} = \frac{-5}{9y} = \frac{5}{-9y}$

Chapter Review

1. $-3 - 2 =$

(a) -1 (b) 1 (c) 5 (d) -5

2. $-8 + 10 =$

(a) -2 (b) 2 (c) 18 (d) -18

3. $5 - 6 =$

(a) -1 (b) 1 (c) 11 (d) -11

4. $-x - (-3) =$

(a) $-x - 3$ (b) $-x + 3$ (c) $x - 3$ (d) $x + 3$

5. $4 \times (-6) =$

(a) 24 (b) -24 (c) 2 (d) -2

6. $(-3) \times (-5) =$

(a) 15 (b) -15 (c) 2 (d) -2

7. $\dfrac{-4}{3x} =$

(a) $\dfrac{4}{3x}$ (b) $\dfrac{-4}{-3x}$ (c) $\dfrac{4}{-3x}$ (d) $\dfrac{4}{-3(-x)}$

8. $2\frac{1}{3} - 4 =$

(a) $\dfrac{4}{3}$ (b) 1 (c) $\dfrac{-5}{3}$ (d) $\dfrac{5}{3}$

9. $5 - (-2x) =$

(a) $5 - 2x$ (b) $5 + 2x$ (c) $10x$ (d) $-10x$

10. $(-4)(-1)(-3) =$

(a) 12 (b) -12

Solutions

1. (d) **2.** (b) **3.** (a) **4.** (b)
5. (b) **6.** (a) **7.** (c) **8.** (c)
9. (b) **10.** (b)

CHAPTER 5

Exponents and Roots

The expression $4x$ is shorthand for $x+x+x+x$, that is x added to itself four times. Likewise x^4 is shorthand for $x \cdot x \cdot x \cdot x$—$x$ multiplied by itself four times. In x^4, x is called the *base* and 4 is the *power* or *exponent*. We say "x to the fourth power" or simply "x to the fourth." There are many useful exponent properties. For the rest of the chapter, a is a nonzero number.

Property 1 $a^n a^m = a^{m+n}$

When multiplying two powers whose bases are the same, add the exponents.

Example

$2^3 \cdot 2^4 = (2 \cdot 2 \cdot 2)(2 \cdot 2 \cdot 2 \cdot 2) = 2^7 \qquad x^9 \cdot x^3 = x^{12}$

Property 2 $\dfrac{a^m}{a^n} = a^{m-n}$

When dividing two powers whose bases are the same, subtract the denominator's power from the numerator's power.

Example

$$\frac{3^4}{3^2} = \frac{3 \cdot 3 \cdot \cancel{3} \cdot \cancel{3}}{\cancel{3} \cdot \cancel{3}} = 3^2 \qquad \frac{y^7}{y^3} = y^{7-3} = y^4$$

Property 3 $(a^n)^m = a^{nm}$

If you have a quantity raised to a power then raised to another power, multiply the exponents.

Example

$$(5^3)^2 = (5 \cdot 5 \cdot 5)^2 = (5 \cdot 5 \cdot 5)(5 \cdot 5 \cdot 5) = 5^6 \qquad (x^6)^7 = x^{(6)(7)} = x^{42}$$

Be careful, Properties 1 and 3 are easily confused.

Property 4 $a^0 = 1$

Any nonzero number raised to the zero power is one. We will see that this is true by Property 2 and the fact that any nonzero number over itself is one.

$$1 = \frac{16}{16} = \frac{4^2}{4^2} = 4^{2-2} = 4^0$$ From this we can see that 4^0 must be 1.

Practice

Rewrite using a single exponent.

1. $\frac{x^8}{x^2} =$

2. $(x^3)^2 =$

3. $y^7y^3 =$

4. $x^{10}x =$

5. $\frac{x^3}{x^2} =$

6. $(y^5)^5 =$

Solutions

1. $\dfrac{x^8}{x^2} = x^{8-2} = x^6$

2. $(x^3)^2 = x^{(3)(2)} = x^6$

3. $y^7y^3 = y^{7+3} = y^{10}$

4. $x^{10}x = x^{10}x^1 = x^{10+1} = x^{11}$

5. $\dfrac{x^3}{x^2} = x^{3-2} = x^1 = x$

6. $(y^5)^5 = y^{(5)(5)} = y^{25}$

These properties also work with algebraic expressions.

Examples

$(x + 3)^2(x + 3)^4 = (x + 3)^6 \qquad [(x + 11)^3]^5 = (x + 11)^{15}$

$$\frac{(3x-4)^7}{(3x-4)^5} = (3x-4)^2$$

Be careful not to write $(3x - 4)^2$ as $(3x)^2 - 4^2$—we will see later that $(3x - 4)^2$ is $9x^2 - 24x + 16$.

Practice

Simplify.

1. $\dfrac{(5x^2 + x + 1)^3}{5x^2 + x + 1} =$

2. $\dfrac{(7x)^9}{(7x)^3} =$

3. $(2x - 5)^0 =$

4. $(x + 1)^{11}(x + 1)^6 =$

5. $(x^2 - 1)(x^2 - 1)^3 =$

6. $\left((16x - 4)^5\right)^2 =$

Solutions

1. $\dfrac{(5x^2 + x + 1)^3}{5x^2 + x + 1} = \dfrac{(5x^2 + x + 1)^3}{(5x^2 + x + 1)^1} = (5x^2 + x + 1)^{3-1}$
$= (5x^2 + x + 1)^2$

2. $\dfrac{(7x)^9}{(7x)^3} = (7x)^{9-3} = (7x)^6$

3. $(2x - 5)^0 = 1$

4. $(x + 1)^{11}(x + 1)^6 = (x + 1)^{11+6} = (x + 1)^{17}$

5. $(x^2 - 1)(x^2 - 1)^3 = (x^2 - 1)^1(x^2 - 1)^3 = (x^2 - 1)^{1+3} = (x^2 - 1)^4$

6. $((16x - 4)^5)^2 = (16x - 4)^{(5)(2)} = (16x - 4)^{10}$

Adding/Subtracting Fractions

When adding fractions with variables in one or more denominators, the LCD will have each variable (or algebraic expression) to its highest power as a factor. For example, the LCD for $\frac{1}{x^2} + \frac{1}{x} + \frac{1}{y^3} + \frac{1}{y^2}$ is x^2y^3.

Examples

$$\frac{4}{x^2} - \frac{3}{x} = \frac{4}{x^2} - \frac{3}{x} \cdot \frac{x}{x} = \frac{4}{x^2} - \frac{3x}{x^2} = \frac{4 - 3x}{x^2}$$

$$\frac{13}{xy^2} - \frac{6}{yz} = \frac{13}{xy^2} \cdot \frac{z}{z} - \frac{6}{yz} \cdot \frac{xy}{xy} = \frac{13z}{xy^2z} - \frac{6xy}{xy^2z} = \frac{13z - 6xy}{xy^2z}$$

$$\frac{2x}{(x+1)^2(4x+5)}+\frac{1}{x+1}=\frac{2x}{(x+1)^2(4x+5)}+\frac{1}{x+1}\cdot\frac{(x+1)(4x+5)}{(x+1)(4x+5)}$$
$$=\frac{2x+(x+1)(4x+5)}{(x+1)^2(4x+5)}$$

$$\frac{1}{x^2(x+y)}+\frac{1}{x}+\frac{1}{(x+y)^3}=\frac{1}{x^2(x+y)}\frac{(x+y)^2}{(x+y)^2}+\frac{1}{x}\cdot\frac{x(x+y)^3}{x(x+y)^3}$$
$$+\frac{1}{(x+y)^3}\cdot\frac{x^2}{x^2}$$
$$=\frac{(x+y)^2}{x^2(x+y)^3}+\frac{x(x+y)^3}{x^2(x+y)^3}+\frac{x^2}{x^2(x+y)^3}$$
$$=\frac{(x+y)^2+x(x+y)^3+x^2}{x^2(x+y)^3}$$

$$\frac{2}{xy}+\frac{1}{x^3y^2}+\frac{2}{xy^4}=\frac{2}{xy}\cdot\frac{x^2y^3}{x^2y^3}+\frac{1}{x^3y^2}\cdot\frac{y^2}{y^2}+\frac{2}{xy^4}\cdot\frac{x^2}{x^2}$$
$$=\frac{2x^2y^3}{x^3y^4}+\frac{y^2}{x^3y^4}+\frac{2x^2}{x^3y^4}=\frac{2x^2y^3+y^2+2x^2}{x^3y^4}$$

Practice

1. $\frac{6}{x}+\frac{2}{xy}=$

2. $\frac{3}{xy^2}-\frac{1}{y}=$

3. $\frac{1}{2x}+\frac{3}{10x^2}=$

4. $\frac{1}{x}+\frac{1}{xyz^2}+\frac{1}{x^2yz}=$

5. $2+\frac{x-1}{(x+4)^2}=$

6. $\frac{6}{2(x-1)(x+1)}+\frac{1}{6(x-1)^2}=$

7. $\dfrac{4}{3xy^2}+\dfrac{9}{2x^5y}-\dfrac{1}{x^3y^3}=$

Solutions

1. $\dfrac{6}{x}+\dfrac{2}{xy}=\dfrac{6}{x}\cdot\dfrac{y}{y}+\dfrac{2}{xy}=\dfrac{6y}{xy}+\dfrac{2}{xy}=\dfrac{6y+2}{xy}$

2. $\dfrac{3}{xy^2}-\dfrac{1}{y}=\dfrac{3}{xy^2}-\dfrac{1}{y}\cdot\dfrac{xy}{xy}=\dfrac{3}{xy^2}-\dfrac{xy}{xy^2}=\dfrac{3-xy}{xy^2}$

3. $\dfrac{1}{2x}+\dfrac{3}{10x^2}=\dfrac{1}{2x}\cdot\dfrac{5x}{5x}+\dfrac{3}{10x^2}=\dfrac{5x}{10x^2}+\dfrac{3}{10x^2}=\dfrac{5x+3}{10x^2}$

4. $\dfrac{1}{x}+\dfrac{1}{xyz^2}+\dfrac{1}{x^2yz}=\dfrac{1}{x}\dfrac{xyz^2}{xyz^2}+\dfrac{1}{xyz^2}\cdot\dfrac{x}{x}+\dfrac{1}{x^2yz}\cdot\dfrac{z}{z}$

$$=\frac{xyz^2}{x^2yz^2}+\frac{x}{x^2yz^2}+\frac{z}{x^2yz^2}=\frac{xyz^2+x+z}{x^2yz^2}$$

5. $2+\dfrac{x-1}{(x+4)^2}=\dfrac{2}{1}+\dfrac{x-1}{(x+4)^2}=\dfrac{2}{1}\cdot\dfrac{(x+4)^2}{(x+4)^2}+\dfrac{x-1}{(x+4)^2}$

$$=\frac{2(x+4)^2+x-1}{(x+4)^2}$$

6. $\dfrac{6}{2(x-1)(x+1)}+\dfrac{1}{6(x-1)^2}=\dfrac{6}{2(x-1)(x+1)}\cdot\dfrac{3(x-1)}{3(x-1)}$

$$+\frac{1}{6(x-1)^2}\cdot\frac{x+1}{x+1}=\frac{18(x-1)}{6(x-1)^2(x+1)}$$

$$+\frac{x+1}{6(x-1)^2(x+1)}=\frac{18(x-1)+x+1}{6(x-1)^2(x+1)}$$

7. $\dfrac{4}{3xy^2}+\dfrac{9}{2x^5y}-\dfrac{1}{x^3y^3}=\dfrac{4}{3xy^2}\cdot\dfrac{2x^4y}{2x^4y}+\dfrac{9}{2x^5y}\cdot\dfrac{3y^2}{3y^2}-\dfrac{1}{x^3y^3}\cdot\dfrac{6x^2}{6x^2}$

$$=\frac{8x^4y}{6x^5y^3}+\frac{27y^2}{6x^5y^3}-\frac{6x^2}{6x^5y^3}=\frac{8x^4y+27y^2-6x^2}{6x^5y^3}$$

Property 5 $a^{-1}=\dfrac{1}{a}$

This property says that a^{-1} is the reciprocal of a. In other words, a^{-1} means "invert a."

Examples

$$2^{-1} = \frac{1}{2} \qquad x^{-1} = \frac{1}{x}$$

$$\left(\frac{x}{y}\right)^{-1} = \frac{1}{\frac{x}{y}} = 1 \div \frac{x}{y} = 1 \cdot \frac{y}{x} = \frac{y}{x} \qquad \left(\frac{2}{3}\right)^{-1} = \frac{3}{2} \qquad \left(\frac{1}{4}\right)^{-1} = 4$$

Property 6 $a^{-n} = \dfrac{1}{a^n}$

This is a combination of Properties 3 and 5: $\frac{1}{a^n} = \left(\frac{1}{a}\right)^n = (a^{-1})^n = a^{-n}$.

Examples

$$5^{-6} = \frac{1}{5^6}$$

$$x^{-10} = \frac{1}{x^{10}}$$

$$\left(\frac{x}{y}\right)^{-3} = \left(\left(\frac{x}{y}\right)^{-1}\right)^3 = \left(\frac{y}{x}\right)^3$$

$$\left(\frac{2}{5}\right)^{-4} = \left(\frac{5}{2}\right)^4$$

Often a combination of exponent properties is needed. In the following examples the goal is to rewrite the expression without using a negative exponent.

Examples

$$(x^6)^{-2} = x^{-12} = \frac{1}{x^{12}} \qquad x^{-7}x^6 = x^{-1} = \frac{1}{x}$$

$$\frac{y^3}{y^{-2}} = y^{3-(-2)} = y^{3+2} = y^5 \qquad \frac{x^{-4}}{x^2} = x^{-4-2} = x^{-6} = \frac{1}{x^6}$$

Practice

Use Properties 1–6 to rewrite without a negative exponent.

1. $6^{-1} =$
2. $(x^2y)^{-1} =$
3. $\left(\frac{5}{8}\right)^{-1} =$
4. $(x^3)^{-1} =$
5. $\frac{x^2}{x^{-1}} =$
6. $x^4x^{-3} =$
7. $x^8x^{-11} =$
8. $(x^{-4})^2 =$
9. $\frac{y^{-7}}{y^{-2}} =$
10. $\frac{x^{-5}}{x^3} =$
11. $(12x - 5)^{-2} =$
12. $(6x)^{-1} =$
13. $\frac{(3x-2)^4}{(3x-2)^{-1}} =$
14. $(2x^3 + 4)^{-6}(2x^3 + 4)^4 =$
15. $((x-8)^3)^{-1} =$

16. $\left(\frac{x+7}{2x-3}\right)^{-1} =$

Solutions

1. $6^{-1} = \frac{1}{6}$

2. $(x^2y)^{-1} = \frac{1}{x^2y}$

3. $\left(\frac{5}{8}\right)^{-1} = \frac{8}{5}$

4. $(x^3)^{-1} = x^{-3} = \frac{1}{x^3}$

5. $\frac{x^2}{x^{-1}} = x^{2-(-1)} = x^{2+1} = x^3$

6. $x^4x^{-3} = x^{4+(-3)} = x^{4-3} = x^1 = x$

7. $x^8x^{-11} = x^{8+(-11)} = x^{-3} = \frac{1}{x^3}$

8. $(x^{-4})^2 = x^{(-4)(2)} = x^{-8} = \frac{1}{x^8}$

9. $\frac{y^{-7}}{y^{-2}} = y^{-7-(-2)} = y^{-7+2} = y^{-5} = \frac{1}{y^5}$

10. $\frac{x^{-5}}{x^3} = x^{-5-3} = x^{-8} = \frac{1}{x^8}$

11. $(12x-5)^{-2} = \frac{1}{(12x-5)^2}$

12. $(6x)^{-1} = \frac{1}{6x}$

13. $\frac{(3x-2)^4}{(3x-2)^{-1}} = (3x-2)^{4-(-1)} = (3x-2)^{4+1} = (3x-2)^5$

14. $(2x^3+4)^{-6}(2x^3+4)^4 = (2x^3+4)^{-6+4} = (2x^3+4)^{-2} = \dfrac{1}{(2x^3+4)^2}$

15. $((x-8)^3)^{-1} = \dfrac{1}{(x-8)^3}$

16. $\left(\dfrac{x+7}{2x-3}\right)^{-1} = \dfrac{2x-3}{x+7}$

In expressions such as $(2x)^{-1}$ the exponent "-1" applies to $2x$, but in $2x^{-1}$ the exponent "-1" applies only to x:

$$(2x)^{-1} = \frac{1}{2x} \quad \text{and} \quad 2x^{-1} = 2 \cdot \frac{1}{x} = \frac{2}{x}.$$

Property 7 $(ab)^n = a^n b^n$

By Property 7 we can take a product then the power or take the powers then the product.

Examples

$$(4x)^3 = (4x)(4x)(4x) = (4 \cdot 4 \cdot 4)(x \cdot x \cdot x) = 4^3x^3 = 64x^3$$

$$[4(x+1)]^2 = 4^2(x+1)^2 = 16(x+1)^2$$

$$(x^2y)^4 = (x^2)^4y^4 = x^8y^4$$

$$(2x)^{-3} = \frac{1}{(2x)^3} = \frac{1}{2^3x^3} = \frac{1}{8x^3}$$

$$(2x^{-1})^{-3} = 2^{-3}(x^{-1})^{-3} = \frac{1}{2^3}x^{(-1)(-3)} = \frac{1}{8}x^3$$

$$[(5x+8)^2(x+6)]^4 = [(5x+8)^2]^4(x+6)^4$$
$$= (5x+8)^{(2)(4)}(x+6)^4 = (5x+8)^8(x+6)^4$$

$$(4x^3y)^2 = 4^2(x^3)^2y^2 = 16x^{(3)(2)}y^2 = 16x^6y^2$$

$$4(3x)^3 = 4(3^3x^3) = 4(27x^3) = 108x^3$$

It is *not* true that $(a+b)^n = a^n + b^n$. This mistake is very common.

Property 8 $\left(\frac{a}{b}\right)^n = \frac{a^n}{b^n}$

Property 8 says that we can take the quotient first then the power or each power followed by the quotient.

Examples

$$\left(\frac{2}{5}\right)^3 = \frac{2}{5}\cdot\frac{2}{5}\cdot\frac{2}{5} = \frac{2^3}{5^3} = \frac{8}{125} \qquad \left(\frac{x}{y}\right)^4 = \frac{x^4}{y^4}$$

$$\left(\frac{x^2}{y^5}\right)^4 = \frac{(x^2)^4}{(y^5)^4} = \frac{x^8}{y^{20}}$$

$$\left(\frac{x}{y}\right)^{-n} = \left(\frac{x}{y}\right)^{(-1)(n)} = \left[\left(\frac{x}{y}\right)^{-1}\right]^n = \left(\frac{y}{x}\right)^n = \frac{y^n}{x^n}$$

This example will be used for the rest of the examples and practice problems.

$$\left(\frac{2}{3}\right)^{-2} = \frac{3^2}{2^2} = \frac{9}{4} \qquad \left(\frac{1}{2}\right)^{-3} = \frac{2^3}{1^3} = \frac{8}{1} = 8$$

$$\left(\frac{6x+5}{x-1}\right)^{-3} = \frac{(x-1)^3}{(6x+5)^3} \qquad \left(\frac{1}{(x+2)^3}\right)^{-5} = \frac{[(x+2)^3]^5}{1^5} = (x+2)^{15}$$

$$\left(\frac{y^{-3}}{x^{-4}}\right)^{-6} = \frac{(x^{-4})^6}{(y^{-3})^6} = \frac{x^{-24}}{y^{-18}} = \frac{\frac{1}{x^{24}}}{\frac{1}{y^{18}}} = \frac{1}{x^{24}} \div \frac{1}{y^{18}} = \frac{1}{x^{24}}\cdot\frac{y^{18}}{1} = \frac{y^{18}}{x^{24}}$$

This expression can be simplified more quickly using Property 8 and Property 3.

$$\left(\frac{y^{-3}}{x^{-4}}\right)^{-6} = \frac{\left(y^{-3}\right)^{-6}}{\left(x^{-4}\right)^{-6}} = \frac{y^{(-3)(-6)}}{x^{(-4)(-6)}} = \frac{y^{18}}{x^{24}}$$

$$\left(\frac{(x+7)^3}{x^{-2}}\right)^{-4} = \frac{(x^{-2})^4}{[(x+7)^3]^4} = \frac{x^{-8}}{(x+7)^{12}}$$

$$= x^{-8}\cdot\frac{1}{(x+7)^{12}} = \frac{1}{x^8}\cdot\frac{1}{(x+7)^{12}} = \frac{1}{x^8(x+7)^{12}}$$

Practice

Simplify and eliminate any negative exponents.

1. $(xy^3)^2 =$

2. $(3x)^{-3} =$

3. $(2x)^4 =$

4. $(3(x-4))^2 =$

5. $6(2x)^3 =$

6. $6y^2(3y^4)^2 =$

7. $(5x^2y^4z^6)^2 =$

8. $\left(\frac{4}{y^2}\right)^3 =$

9. $\left(\frac{2}{x-9}\right)^{-3} =$

10. $\left(\frac{(x+8)^2}{x^4}\right)^{-3} =$

11. $\left(\frac{(x+3)^{-2}}{y^{-4}}\right)^{-3} =$

Solutions

1. $(xy^3)^2 = x^2(y^3)^2 = x^2y^6$

2. $(3x)^{-3} = \frac{1}{(3x)^3} = \frac{1}{3^3x^3} = \frac{1}{27x^3}$

3. $(2x)^4 = 2^4x^4 = 16x^4$

4. $(3(x-4))^2 = 3^2(x-4)^2 = 9(x-4)^2$

5. $6(2x)^3 = 6(2^3x^3) = 6(8x^3) = 48x^3$

6. $6y^2(3y^4)^2 = 6y^2(3^2(y^4)^2) = 6y^2(9y^8) = 54y^{10}$

7. $(5x^2y^4z^6)^2 = 5^2(x^2)^2(y^4)^2(z^6)^2 = 25x^4y^8z^{12}$

8. $\left(\frac{4}{y^2}\right)^3 = \frac{4^3}{(y^2)^3} = \frac{64}{y^6}$

9. $\left(\frac{2}{x-9}\right)^{-3} = \frac{(x-9)^3}{2^3} = \frac{(x-9)^3}{8}$

10. $\left(\frac{(x+8)^2}{x^4}\right)^{-3} = \frac{(x^4)^3}{[(x+8)^2]^3} = \frac{x^{12}}{(x+8)^6}$

11. $\left(\frac{(x+3)^{-2}}{y^{-4}}\right)^{-3} = \frac{[(x+3)^{-2}]^{-3}}{(y^{-4})^{-3}} = \frac{(x+3)^{(-2)(-3)}}{y^{(-4)(-3)}} = \frac{(x+3)^6}{y^{12}}$

Multiplying/Dividing with Exponents

When multiplying (or dividing) quantities that have exponents, use the exponent properties to simplify each factor (or numerator and denominator) then multiply (or divide).

Examples

$$3x^3(4xy^5)^2 = 3x^3(4^2x^2(y^5)^2) = 3x^3(16x^2y^{10}) = 3 \cdot 16x^3x^2y^{10} = 48x^5y^{10}$$

$$(2x)^3(3x^3y)^2 = (2^3x^3)(3^2(x^3)^2y^2) = (8x^3)(9x^6y^2) = 8 \cdot 9x^3x^6y^2 = 72x^9y^2$$

$$\frac{(5x^3y^2)^3}{(10x)^2} = \frac{5^3(x^3)^3(y^2)^3}{10^2x^2} = \frac{125x^9y^6}{100x^2} = \frac{125}{100}x^{9-2}y^6 = \frac{5}{4}x^7y^6 = \frac{5x^7y^6}{4}$$

$$(6xy^4)^2(4xy^8)^{-3} = (6^2x^2(y^4)^2)(4^{-3}x^{-3}(y^8)^{-3}) = (36x^2y^8)\left(\frac{1}{64}x^{-3}y^{-24}\right)$$
$$= \frac{36}{64}x^2x^{-3}y^8y^{-24} = \frac{9}{16}x^{-1}y^{-16} = \frac{9}{16}\cdot\frac{1}{x}\cdot\frac{1}{y^{16}} = \frac{9}{16xy^{16}}$$

$$\left(\frac{8x^5y^2}{9x^2y}\right)^{-2} = \frac{(9x^2y)^2}{(8x^5y^2)^2} = \frac{9^2(x^2)^2y^2}{8^2(x^5)^2(y^2)^2} = \frac{81x^4y^2}{64x^{10}y^4} = \frac{81}{64}x^{4-10}y^{2-4}$$
$$= \frac{81}{64}x^{-6}y^{-2} = \frac{81}{64}\cdot\frac{1}{x^6}\cdot\frac{1}{y^2} = \frac{81}{64x^6y^2}$$

Practice

1. $\left(\frac{x^3}{y^2}\right)^5\left(\frac{x}{y}\right)^{-2} =$
2. $\frac{(2x^3y^5)^4}{(6x^5y^3)^2} =$
3. $(2x^3)^2(3x^{-1})^3 =$
4. $(3xy^4)^{-2}(12x^2y)^2 =$
5. $(4x^{-1}y^{-2})^2(2x^4y^5)^3 =$
6. $\frac{(5x^4y^3)^3}{(15xy^5)^2} =$
7. $\frac{(9x^{-2}y^3)^2}{(6xy^2)^3} =$
8. $[9(x+3)^2]^2[2(x+3)]^3 =$
9. $(2xy^2z^4)^4(3x^{-1}z^2)^3(xy^5z^{-4}) =$
10. $\frac{2(xy^4)^3(yz^2)^4}{(3xyz)^4} =$

Solutions

1. $$\left(\frac{x^3}{y^2}\right)^5\left(\frac{x}{y}\right)^{-2} = \frac{(x^3)^5}{(y^2)^5}\cdot\frac{y^2}{x^2} = \frac{x^{15}}{y^{10}}\cdot\frac{y^2}{x^2} = x^{15-2}y^{2-10} = x^{13}y^{-8} = \frac{x^{13}}{y^8}$$

2. $$\frac{(2x^3y^5)^4}{(6x^5y^3)^2} = \frac{2^4(x^3)^4(y^5)^4}{6^2(x^5)^2(y^3)^2} = \frac{16x^{12}y^{20}}{36x^{10}y^6} = \frac{16}{36}x^{12-10}y^{20-6}$$
$$= \frac{4}{9}x^2y^{14} = \frac{4x^2y^{14}}{9}$$

3. $$(2x^3)^2(3x^{-1})^3 = [2^2(x^3)^2][3^3(x^{-1})^3] = (4x^6)(27x^{-3})$$
$$= 108x^{6+(-3)} = 108x^3$$

4. $$(3xy^4)^{-2}(12x^2y)^2 = (3^{-2}x^{-2}(y^4)^{-2})(12^2(x^2)^2y^2$$
$$= \left(\frac{1}{9}x^{-2}y^{-8}\right)(144x^4y^2) = \frac{144}{9}x^{-2+4}y^{-8+2}$$
$$= 16x^2y^{-6} = 16x^2\frac{1}{y^6} = \frac{16x^2}{y^6}$$

5. $$(4x^{-1}y^{-2})^2(2x^4y^5)^3 = [4^2(x^{-1})^2(y^{-2})^2][2^3(x^4)^3(y^5)^3]$$
$$= (16x^{-2}y^{-4})(8x^{12}y^{15}) = 128x^{-2+12}y^{-4+15}$$
$$= 128x^{10}y^{11}$$

6. $$\frac{(5x^4y^3)^3}{(15xy^5)^2} = \frac{5^3(x^4)^3(y^3)^3}{15^2x^2(y^5)^2} = \frac{125x^{12}y^9}{225x^2y^{10}} = \frac{125}{225}x^{12-2}y^{9-10} = \frac{5}{9}x^{10}y^{-1}$$
$$= \frac{5x^{10}}{9}\cdot\frac{1}{y} = \frac{5x^{10}}{9y}$$

7. $$\frac{(9x^{-2}y^3)^2}{(6xy^2)^3} = \frac{9^2(x^{-2})^2(y^3)^2}{6^3x^3(y^2)^3} = \frac{81x^{-4}y^6}{216x^3y^6} = \frac{81}{216}x^{-4-3}y^{6-6}$$
$$= \frac{3}{8}x^{-7}y^0 = \frac{3}{8}\cdot\frac{1}{x^7}\cdot 1 = \frac{3}{8x^7}$$

8. $$[9(x+3)^2]^2[2(x+3)]^3 = [9^2((x+3)^2)^2][2^3(x+3)^3]$$
$$= [81(x+3)^4][8(x+3)^3] = 648(x+3)^{4+3}$$
$$= 648(x+3)^7$$

9. $$(2xy^2z^4)^4(3x^{-1}z^2)^3(xy^5z^4) = [2^4x^4(y^2)^4(z^4)^4][3^3(x^{-1})^3(z^2)^3](xy^5z^4)$$
$$= (16x^4y^8z^{16})(27x^{-3}z^6)(x^1y^5z^4)$$
$$= 432x^{4+(-3)+1}y^{8+5}z^{16+6+4} = 432x^2y^{13}z^{26}$$

10. $$\frac{2(xy^4)^3(yz^2)^4}{(3xyz)^4} = \frac{2x^3(y^4)^3y^4(z^2)^4}{3^4x^4y^4z^4} = \frac{2x^3y^{12}y^4z^8}{81x^4y^4z^4} = \frac{2x^3y^{16}z^8}{81x^4y^4z^4}$$
$$= \frac{2}{81}x^{3-4}y^{16-4}z^{8-4} = \frac{2}{81}x^{-1}y^{12}z^4 = \frac{2}{81}\frac{1}{x}y^{12}z^4 = \frac{2y^{12}z^4}{81x}$$

There are times in algebra, and especially in calculus, when you will need to convert a fraction into a product. Using the fact that $\frac{1}{a} = a^{-1}$, we can rewrite a fraction as a product of the numerator and denominator raised to the -1 power. Here is the idea:

$$\frac{\textit{numerator}}{\textit{denominator}} = (\textit{numerator})(\textit{denominator})^{-1}.$$

Examples

$$\frac{3}{x} = 3x^{-1} \qquad \frac{4}{x+3} = 4(x+3)^{-1} \qquad \frac{x^n}{y^m} = x^n(y^m)^{-1} = x^ny^{-m}$$

$$\frac{5x-8}{(2x+3)^3} = (5x-8)(2x+3)^{-3}$$

Practice

1. $\frac{4x^2}{y^5} =$

2. $\frac{2x(x-3)}{(x+1)^2} =$

3. $\frac{x}{y} =$

4. $\dfrac{2x}{(3y)^2} =$

5. $\dfrac{2x-3}{2x+5} =$

Solutions

1. $\dfrac{4x^2}{y^5} = 4x^2y^{-5}$

2. $\dfrac{2x(x-3)}{(x+1)^2} = 2x(x-3)(x+1)^{-2}$

3. $\dfrac{x}{y} = xy^{-1}$

4. $\dfrac{2x}{(3y)^2} = 2x(3y)^{-2}$

5. $\dfrac{2x-3}{2x+5} = (2x-3)(2x+5)^{-1}$

Roots

The square root of a number is the nonnegative number whose square is the root. For example 3 is the square root of 9 because $3^2 = 9$.

Examples

$\sqrt{16} = 4$ because $4^2 = 16$ $\sqrt{81} = 9$ because $9^2 = 81$

$\sqrt{625} = 25$ because $25^2 = 625$

It may seem that negative numbers could be square roots. It is true that $(-3)^2 = 9$. But $\sqrt{9}$ is the symbol for the nonnegative number whose square is 9. Sometimes we say that 3 is the *principal* square root of 9. When we speak of an even root, we mean the nonnegative root. In general, $\sqrt[n]{a} = b$ if $b^n = a$. There is no problem with odd roots being negative numbers:

$$\sqrt[3]{-64} = -4 \text{ because } (-4)^3 = (-4)(-4)(-4) = -64.$$

If n is even, b is assumed to be the nonnegative root. Also even roots of negative numbers do not exist in the real number system. In this book, it is assumed that even roots will be taken only of nonnegative numbers. For instance in $\sqrt{x}$, it is assumed that x is not negative.

Root properties are similar to exponent properties.

Property 1 $\sqrt[n]{ab} = \sqrt[n]{a}\sqrt[n]{b}$

We can take the product then the root or take the individual roots then the product.

Examples

$$\sqrt{64} = \sqrt{4 \cdot 16} = \sqrt{4} \cdot \sqrt{16} = 2 \cdot 4 = 8$$

$$\sqrt[5]{3}\sqrt[5]{4x} = \sqrt[5]{12x} \qquad \sqrt[4]{6x}\sqrt[4]{4y} = \sqrt[4]{24xy}$$

Property 1 only applies to multiplication. There is no similar property for addition (nor subtraction). A common mistake is to "simplify" the sum of two squares. For example $\sqrt{x^2 + 9} = x + 3$ is incorrect. The following example should give you an idea of why these two expressions are not equal. If there were the property $\sqrt[n]{a+b} = \sqrt[n]{a} + \sqrt[n]{b}$, then we would have

$$\sqrt{58} = \sqrt{49 + 9} = \sqrt{49} + \sqrt{9} = 7 + 3 = 10.$$

This could only be true if $10^2 = 58$.

Property 2 $\sqrt[n]{\dfrac{a}{b}} = \dfrac{\sqrt[n]{a}}{\sqrt[n]{b}}$

We can take the quotient then the root or the individual roots then the quotient.

$$\sqrt{\frac{4}{9}} = \frac{\sqrt{4}}{\sqrt{9}} = \frac{2}{3}$$

Property 3 $\left(\sqrt[n]{a}\right)^m = \sqrt[n]{a^m}$ (Remember that if n is even, then a must not be negative.)

We can take the root then the power or the power then take the root.

Property 4 $(\sqrt[n]{a})^n = \sqrt[n]{a^n} = a$

Property 4 can be thought of as a root-power cancellation law.

Example

$\sqrt[3]{27} = \sqrt[3]{3^3} = 3 \qquad (\sqrt{5})^2 = 5 \qquad \sqrt[3]{8x^3} = \sqrt[3]{(2x)^3} = 2x$

Practice

1. $\sqrt{25x^2} =$

2. $\sqrt[3]{8y^3} =$

3. $\sqrt{(4-x)^2} =$

4. $\sqrt[3]{[5(x-1)]^3} =$

Solutions

1. $\sqrt{25x^2} = \sqrt{(5x)^2} = 5x$

2. $\sqrt[3]{8y^3} = \sqrt[3]{(2y)^3} = 2y$

3. $\sqrt{(4-x)^2} = 4 - x$

4. $\sqrt[3]{[5(x-1)]^3} = 5(x-1)$

These properties can be used to simplify roots in the same way canceling is used to simplify fractions. For instance you normally would not leave $\sqrt{25}$ without simplifying it as 5 any more than you would leave $\frac{12}{4}$ without reducing it to 3. In $\sqrt[n]{a^m}$ if m is at least as large as n, then $\sqrt[n]{a^m}$ can be simplified using Property 1 ($\sqrt[n]{ab} = \sqrt[n]{a}\sqrt[n]{b}$) and Property 4 ($\sqrt[n]{a^n} = a$).

Examples

$$\sqrt{27} = \sqrt{3^2 3^1} = \sqrt{3^2}\sqrt{3} = 3\sqrt{3}$$

$$\sqrt{32x^3} = \sqrt{2^2 2^2 2^1 x^2 x^1} = \sqrt{2^2}\sqrt{2^2}\sqrt{x^2}\sqrt{2x} = 2 \cdot 2x\sqrt{2x} = 4x\sqrt{2x}$$

$$\sqrt[3]{625x^5y^4} = \sqrt[3]{5^3 5^1 x^3 x^2 y^3 y^1} = \sqrt[3]{5^3}\sqrt[3]{x^3}\sqrt[3]{y^3}\sqrt[3]{5x^2y} = 5xy\sqrt[3]{5x^2y}$$

$$\sqrt[4]{(x-6)^9} = \sqrt[4]{(x-6)^4(x-6)^4(x-6)^1} = \sqrt[4]{(x-6)^4}\sqrt[4]{(x-6)^4}\sqrt[4]{x-6}$$
$$= (x-6)(x-6)\sqrt[4]{x-6} = (x-6)^2\sqrt[4]{x-6}$$

$$\sqrt{\frac{8}{9}} = \frac{\sqrt{8}}{\sqrt{9}} = \frac{\sqrt{2^2 2^1}}{\sqrt{9}} = \frac{\sqrt{2^2}\sqrt{2}}{3} = \frac{2\sqrt{2}}{3}$$

Practice

1. $\sqrt[3]{x^7} =$

2. $\sqrt{x^{10}} =$

3. $\sqrt[3]{16x^7y^5} =$

4. $\sqrt[5]{(4x-1)^8} =$

5. $\sqrt{25(x+4)^2} =$

6. $\sqrt[4]{x^9y^6} =$

7. $\sqrt[50]{\dfrac{x^{100}}{y^{200}}} =$

Solutions

1. $\sqrt[3]{x^7} = \sqrt[3]{x^3x^3x^1} = \sqrt[3]{x^3}\sqrt[3]{x^3}\sqrt[3]{x^1} = xx\sqrt[3]{x} = x^2\sqrt[3]{x}$

2. $\sqrt{x^{10}} = \sqrt{x^5 x^5} = \sqrt{(x^5)^2} = x^5$

3. $\sqrt[3]{16x^7y^5} = \sqrt[3]{2^3 2x^3x^3xy^3y^2} = \sqrt[3]{2^3}\sqrt[3]{x^3}\sqrt[3]{x^3}\sqrt[3]{y^3}\sqrt[3]{2xy^2}$
$= 2xxy\sqrt[3]{2xy^2} = 2x^2y\sqrt[3]{2xy^2}$

4. $\sqrt[5]{(4x-1)^8} = \sqrt[5]{(4x-1)^5(4x-1)^3} = \sqrt[5]{(4x-1)^5}\sqrt[5]{(4x-1)^3}$
$= (4x-1)\sqrt[5]{(4x-1)^3}$

5. $\sqrt{25(x+4)^2} = \sqrt{5^2(x+4)^2} = \sqrt{5^2}\sqrt{(x+4)^2} = 5(x+4)$

6. $\sqrt[4]{x^9y^6} = \sqrt[4]{x^4x^4x^1y^4y^2} = \sqrt[4]{x^4}\sqrt[4]{x^4}\sqrt[4]{y^4}\sqrt[4]{xy^2} = xxy\sqrt[4]{x^1y^2} = x^2y\sqrt[4]{xy^2}$

7. $\sqrt[50]{\dfrac{x^{100}}{y^{200}}} = \dfrac{\sqrt[50]{x^{100}}}{\sqrt[50]{y^{200}}} = \dfrac{\sqrt[50]{x^{50}x^{50}}}{\sqrt[50]{y^{50}y^{50}y^{50}y^{50}}} = \dfrac{\sqrt[50]{x^{50}}\ \sqrt[50]{x^{50}}}{\sqrt[50]{y^{50}}\ \sqrt[50]{y^{50}}\ \sqrt[50]{y^{50}}\ \sqrt[50]{y^{50}}}$
$= \dfrac{xx}{yyyy} = \dfrac{x^2}{y^4}$

Numbers like 18, 48, and 50 are not perfect squares but they do have perfect squares as factors. Using the same properties, $\sqrt[n]{ab} = \sqrt[n]{a}\sqrt[n]{b}$ and $\sqrt[n]{a^n} = a$, we can simplify quantities like $\sqrt{18}$.

Examples

$\sqrt{18} = \sqrt{3^2 2} = \sqrt{3^2}\sqrt{2} = 3\sqrt{2}$ $\quad\sqrt{48} = \sqrt{4^2 3} = \sqrt{4^2}\sqrt{3} = 4\sqrt{3}$

$\sqrt{50} = \sqrt{5^2 2} = \sqrt{5^2}\sqrt{2} = 5\sqrt{2}$ $\quad\sqrt[3]{162} = \sqrt[3]{3^3 \cdot 3 \cdot 2} = \sqrt[3]{3^3}\sqrt[3]{3 \cdot 2}$
$= 3\sqrt[3]{6}$

$\sqrt[5]{64x^6y^3} = \sqrt[5]{2^5 \cdot 2x^5xy^3} = \sqrt[5]{2^5}\sqrt[5]{x^5}\sqrt[5]{2xy^3} = 2x\sqrt[5]{2xy^3}$

$\sqrt[3]{(2x-7)^5} = \sqrt[3]{(2x-7)^3(2x-7)^2} = \sqrt[3]{(2x-7)^3}\sqrt[3]{(2x-7)^2}$
$= (2x-7)\sqrt[3]{(2x-7)^2}$

$$\sqrt{\frac{48x^3}{25}} = \frac{\sqrt{48x^3}}{\sqrt{25}} = \frac{\sqrt{4^2 3x^2 x}}{\sqrt{5^2}} = \frac{\sqrt{4^2}\sqrt{x^2}\sqrt{3x}}{5} = \frac{4x\sqrt{3x}}{5}$$

Practice

1. $\sqrt[3]{54x^5} =$

2. $\sqrt{50x^3 y} =$

3. $\sqrt{\frac{8}{9}} =$

4. $\sqrt[4]{32x^7 y^5} =$

5. $\sqrt[3]{\frac{40(3x-1)^4}{x^6}} =$

Solutions

1. $\sqrt[3]{54x^5} = \sqrt[3]{3^3 2x^3 x^2} = \sqrt[3]{3^3}\sqrt[3]{x^3}\sqrt[3]{2x^2} = 3x\sqrt[3]{2x^2}$

2. $\sqrt{50x^3 y} = \sqrt{5^2 2x^2 xy} = \sqrt{5^2}\sqrt{x^2}\sqrt{2xy} = 5x\sqrt{2xy}$

3. $\sqrt{\frac{8}{9}} = \frac{\sqrt{8}}{\sqrt{9}} = \frac{\sqrt{2^2 2}}{3} = \frac{\sqrt{2^2}\sqrt{2}}{3} = \frac{2\sqrt{2}}{3}$

4. $\sqrt[4]{32x^7 y^5} = \sqrt[4]{2^4 2x^4 x^3 y^4 y} = \sqrt[4]{2^4}\sqrt[4]{x^4}\sqrt[4]{y^4}\sqrt[4]{2x^3 y} = 2xy\sqrt[4]{2x^3 y}$

5. $$\sqrt[3]{\frac{40(3x-1)^4}{x^6}} = \frac{\sqrt[3]{40(3x-1)^4}}{\sqrt[3]{x^6}} = \frac{\sqrt[3]{2^3 5(3x-1)^3(3x-1)}}{\sqrt[3]{x^3 x^3}}$$
$$= \frac{\sqrt[3]{2^3}\sqrt[3]{(3x-1)^3}\sqrt[3]{5(3x-1)}}{\sqrt[3]{x^3}\sqrt[3]{x^3}} = \frac{2(3x-1)\sqrt[3]{5(3x-1)}}{xx}$$
$$= \frac{2(3x-1)\sqrt[3]{5(3x-1)}}{x^2}$$

Roots of fractions or fractions with a root in the denominator are not simplified. To eliminate roots in denominators, use the fact that $\sqrt[n]{a^n} = a$ and that any nonzero number over itself is one. We will begin with square roots. If the denominator is a square root, multiply the fraction by the denominator over itself. This will force the new denominator to be a perfect square.

Examples

$$\frac{1}{\sqrt{2}} = \frac{1}{\sqrt{2}} \cdot \frac{\sqrt{2}}{\sqrt{2}} = \frac{\sqrt{2}}{\sqrt{2^2}} = \frac{\sqrt{2}}{2} \qquad \frac{4}{\sqrt{x}} = \frac{4}{\sqrt{x}} \cdot \frac{\sqrt{x}}{\sqrt{x}} = \frac{4\sqrt{x}}{\sqrt{x^2}} = \frac{4\sqrt{x}}{x}$$

$$\sqrt{\frac{2}{3}} = \frac{\sqrt{2}}{\sqrt{3}} \cdot \frac{\sqrt{3}}{\sqrt{3}} = \frac{\sqrt{6}}{\sqrt{3^2}} = \frac{\sqrt{6}}{3}$$

Practice

1. $\dfrac{3}{\sqrt{5}} =$

2. $\dfrac{7}{\sqrt{y}} =$

3. $\sqrt{\dfrac{6}{7}} =$

4. $\dfrac{8x}{\sqrt{3}} =$

5. $\sqrt{\dfrac{7xy}{11}} =$

Solutions

1. $\dfrac{3}{\sqrt{5}} = \dfrac{3}{\sqrt{5}} \cdot \dfrac{\sqrt{5}}{\sqrt{5}} = \dfrac{3\sqrt{5}}{\sqrt{5^2}} = \dfrac{3\sqrt{5}}{5}$

2. $\dfrac{7}{\sqrt{y}} = \dfrac{7}{\sqrt{y}} \cdot \dfrac{\sqrt{y}}{\sqrt{y}} = \dfrac{7\sqrt{y}}{\sqrt{y^2}} = \dfrac{7\sqrt{y}}{y}$

3. $\sqrt{\frac{6}{7}} = \frac{\sqrt{6}}{\sqrt{7}} \cdot \frac{\sqrt{7}}{\sqrt{7}} = \frac{\sqrt{42}}{\sqrt{7^2}} = \frac{\sqrt{42}}{7}$

4. $\frac{8x}{\sqrt{3}} = \frac{8x}{\sqrt{3}} \cdot \frac{\sqrt{3}}{\sqrt{3}} = \frac{8x\sqrt{3}}{\sqrt{3^2}} = \frac{8x\sqrt{3}}{3}$

5. $\sqrt{\frac{7xy}{11}} = \frac{\sqrt{7xy}}{\sqrt{11}} \cdot \frac{\sqrt{11}}{\sqrt{11}} = \frac{\sqrt{77xy}}{\sqrt{11^2}} = \frac{\sqrt{77xy}}{11}$

In the case of a cube (or higher) root, multiplying the fraction by the denominator over itself usually does not work. To eliminate the nth root in the denominator, we need to write the denominator as the nth root of some quantity to the nth power. For example, to simplify $\frac{1}{\sqrt[3]{5}}$ we need a 5^3 under the cube root sign. There is only one 5 under the cube root. We need a total of three 5s, so we need two more 5s. Multiply 5 by 5^2 to get 5^3:

$$\frac{1}{\sqrt[3]{5}} \cdot \frac{\sqrt[3]{5^2}}{\sqrt[3]{5^2}} = \frac{\sqrt[3]{5^2}}{\sqrt[3]{5^3}} = \frac{\sqrt[3]{25}}{5}.$$

When the denominator is written as a power (often the power is 1) subtract this power from the root. The factor will have this number as a power.

Examples

$\frac{8}{\sqrt[4]{x^3}}$ The root minus the power is $4 - 3 = 1$. We need another x^1 under the root.

$$\frac{8}{\sqrt[4]{x^3}} \cdot \frac{\sqrt[4]{x^1}}{\sqrt[4]{x^1}} = \frac{8\sqrt[4]{x}}{\sqrt[4]{x^4}} = \frac{8\sqrt[4]{x}}{x}$$

$\frac{4x}{\sqrt[5]{x^2}}$ The root minus the power is $5 - 2 = 3$. We need another x^3 under the root.

$$\frac{4x}{\sqrt[5]{x^2}} \cdot \frac{\sqrt[5]{x^3}}{\sqrt[5]{x^3}} = \frac{4x\sqrt[5]{x^3}}{\sqrt[5]{x^5}} = \frac{4x\sqrt[5]{x^3}}{x} = 4\sqrt[5]{x^3}$$

$$\sqrt[7]{\frac{2}{x^3}} = \frac{\sqrt[7]{2}}{\sqrt[7]{x^3}} \cdot \frac{\sqrt[7]{x^4}}{\sqrt[7]{x^4}} = \frac{\sqrt[7]{2x^4}}{\sqrt[7]{x^7}} = \frac{\sqrt[7]{2x^4}}{x}$$

$$\frac{1}{\sqrt[3]{2x}} = \frac{1}{\sqrt[3]{2x}} \cdot \frac{\sqrt[3]{(2x)^2}}{\sqrt[3]{(2x)^2}} = \frac{\sqrt[3]{2^2x^2}}{\sqrt[3]{(2x)^3}} = \frac{\sqrt[3]{4x^2}}{2x}$$

$$\frac{y}{\sqrt[5]{9}} = \frac{y}{\sqrt[5]{3^2}} \cdot \frac{\sqrt[5]{3^3}}{\sqrt[5]{3^3}} = \frac{y\sqrt[5]{27}}{\sqrt[5]{3^5}} = \frac{y\sqrt[5]{27}}{3}$$

$$\frac{2}{\sqrt[4]{xy^2}} = \frac{2}{\sqrt[4]{xy^2}} \cdot \frac{\sqrt[4]{x^3y^2}}{\sqrt[4]{x^3y^2}} = \frac{2\sqrt[4]{x^3y^2}}{\sqrt[4]{x^4y^4}} = \frac{2\sqrt[4]{x^3y^2}}{\sqrt[4]{(xy)^4}} = \frac{2\sqrt[4]{x^3y^2}}{xy}$$

Practice

1. $\dfrac{6}{\sqrt[3]{x^2}}$

2. $\sqrt[5]{\dfrac{3}{2}}$

3. $\dfrac{1}{\sqrt[5]{(x-4)^2}}$

4. $\dfrac{x}{\sqrt[4]{x}}$

5. $\dfrac{9}{\sqrt[5]{8}}$

6. $\dfrac{x}{\sqrt[4]{9}}$

7. $\dfrac{1}{\sqrt[5]{x^2y^4}}$

8. $\sqrt[8]{\dfrac{12}{x^5y^6}}$

9. $\dfrac{1}{\sqrt[4]{8xy^2}}$

10. $\sqrt[5]{\dfrac{4}{27x^3y}}$

Solutions

1. $\dfrac{6}{\sqrt[3]{x^2}} = \dfrac{6}{\sqrt[3]{x^2}} \cdot \dfrac{\sqrt[3]{x^1}}{\sqrt[3]{x^1}} = \dfrac{6\sqrt[3]{x}}{\sqrt[3]{x^3}} = \dfrac{6\sqrt[3]{x}}{x}$

2. $\sqrt[5]{\dfrac{3}{2}} = \dfrac{\sqrt[5]{3}}{\sqrt[5]{2^1}} \cdot \dfrac{\sqrt[5]{2^4}}{\sqrt[5]{2^4}} = \dfrac{\sqrt[5]{3 \cdot 2^4}}{\sqrt[5]{2^5}} = \dfrac{\sqrt[5]{48}}{2}$

3. $\dfrac{1}{\sqrt[5]{(x-4)^2}} = \dfrac{1}{\sqrt[5]{(x-4)^2}} \cdot \dfrac{\sqrt[5]{(x-4)^3}}{\sqrt[5]{(x-4)^3}} = \dfrac{\sqrt[5]{(x-4)^3}}{\sqrt[5]{(x-4)^5}} = \dfrac{\sqrt[5]{(x-4)^3}}{x-4}$

4. $\dfrac{x}{\sqrt[4]{x}} = \dfrac{x}{\sqrt[4]{x^1}} \cdot \dfrac{\sqrt[4]{x^3}}{\sqrt[4]{x^3}} = \dfrac{x\sqrt[4]{x^3}}{\sqrt[4]{x^4}} = \dfrac{x\sqrt[4]{x^3}}{x} = \sqrt[4]{x^3}$

5. $\dfrac{9}{\sqrt[5]{8}} = \dfrac{9}{\sqrt[5]{2^3}} = \dfrac{9}{\sqrt[5]{2^3}} \cdot \dfrac{\sqrt[5]{2^2}}{\sqrt[5]{2^2}} = \dfrac{9\sqrt[5]{4}}{\sqrt[5]{2^5}} = \dfrac{9\sqrt[5]{4}}{2}$

6. $\dfrac{x}{\sqrt[4]{9}} = \dfrac{x}{\sqrt[4]{3^2}} = \dfrac{x}{\sqrt[4]{3^2}} \cdot \dfrac{\sqrt[4]{3^2}}{\sqrt[4]{3^2}} = \dfrac{x\sqrt[4]{9}}{\sqrt[4]{3^4}} = \dfrac{x\sqrt[4]{9}}{3}$

7. $\dfrac{1}{\sqrt[5]{x^2y^4}} = \dfrac{1}{\sqrt[5]{x^2y^4}} \cdot \dfrac{\sqrt[5]{x^3y^1}}{\sqrt[5]{x^3y^1}} = \dfrac{\sqrt[5]{x^3y}}{\sqrt[5]{x^5y^5}} = \dfrac{\sqrt[5]{x^3y}}{\sqrt[5]{(xy)^5}} = \dfrac{\sqrt[5]{x^3y}}{xy}$

8. $\sqrt[8]{\dfrac{12}{x^5y^6}} = \dfrac{\sqrt[8]{12}}{\sqrt[8]{x^5y^6}} \cdot \dfrac{\sqrt[8]{x^3y^2}}{\sqrt[8]{x^3y^2}} = \dfrac{\sqrt[8]{12x^3y^2}}{\sqrt[8]{x^8y^8}} = \dfrac{\sqrt[8]{12x^3y^2}}{\sqrt[8]{(xy)^8}} = \dfrac{\sqrt[8]{12x^3y^2}}{xy}$

9. $\dfrac{1}{\sqrt[4]{8xy^2}} = \dfrac{1}{\sqrt[4]{2^3x^1y^2}} \cdot \dfrac{\sqrt[4]{2x^3y^2}}{\sqrt[4]{2x^3y^2}} = \dfrac{\sqrt[4]{2x^3y^2}}{\sqrt[4]{2^4x^4y^4}} = \dfrac{\sqrt[4]{2x^3y^2}}{\sqrt[4]{(2xy)^4}} = \dfrac{\sqrt[4]{2x^3y^2}}{2xy}$

10. $\sqrt[5]{\dfrac{4}{27x^3y}} = \dfrac{\sqrt[5]{4}}{\sqrt[5]{3^3x^3y^1}} \cdot \dfrac{\sqrt[5]{3^2x^2y^4}}{\sqrt[5]{3^2x^2y^4}} = \dfrac{\sqrt[5]{4 \cdot 9x^2y^4}}{\sqrt[5]{3^5x^5y^5}} = \dfrac{\sqrt[5]{36x^2y^4}}{\sqrt[5]{(3xy)^5}} = \dfrac{\sqrt[5]{36x^2y^4}}{3xy}$

Roots Expressed as Exponents

Roots can be written as exponents by using the following two properties. This ability is useful in algebra and calculus.

Property 1 $\sqrt[n]{a} = a^{1/n}$

The exponent is a fraction whose numerator is 1 and whose denominator is the root.

Examples

$\sqrt{x} = x^{1/2}$ $\quad \sqrt[3]{2x+1} = (2x+1)^{1/3}$ $\quad \dfrac{1}{\sqrt{x}} = \dfrac{1}{x^{1/2}} = x^{-1/2}$

Property 2 $(\sqrt[n]{a})^m = \sqrt[n]{a^m} = a^{m/n}$ (If n is even, a must be nonnegative.)

The exponent is a fraction whose numerator is the power and whose denominator is the root.

Examples

$\sqrt[5]{x^3} = x^{3/5}$ $\quad \sqrt[5]{x^6} = x^{6/5}$ $\quad \sqrt{(2x^2-1)^7} = (2x^2-1)^{7/2}$

$\sqrt[3]{(12x)^2} = (12x)^{2/3}$ $\quad \dfrac{15}{\sqrt{x^3}} = \dfrac{15}{x^{3/2}} = 15x^{-3/2}$

Practice

1. $\sqrt{14x} =$

2. $\dfrac{3}{\sqrt{2x}} =$

3. $\sqrt[6]{(x+4)^5} =$

4. $\dfrac{3x-5}{\sqrt{x-5}} =$

5. $\dfrac{1}{\sqrt[3]{(10x)^4}} =$

6. $2x^2\sqrt{(x-y)^3} =$

7. $\dfrac{3x+8}{\sqrt[7]{(12x+5)^3}} =$

8. $\sqrt{\dfrac{x-3}{y^5}} =$

9. $\sqrt[4]{\dfrac{16x^3}{3x+1}} =$

10. $\sqrt[5]{\dfrac{(x-1)^4}{(x+1)^3}} =$

Solutions

1. $\sqrt{14x} = (14x)^{1/2}$

2. $\dfrac{3}{\sqrt{2x}} = \dfrac{3}{(2x)^{1/2}} = 3(2x)^{-1/2}$

3. $\sqrt[6]{(x+4)^5} = (x+4)^{5/6}$

4. $\dfrac{3x-5}{\sqrt{x-5}} = \dfrac{3x-5}{(x-5)^{1/2}} = (3x-5)(x-5)^{-1/2}$

5. $\dfrac{1}{\sqrt[3]{(10x)^4}} = \dfrac{1}{(10x)^{4/3}} = (10x)^{-4/3}$

6. $2x^2\sqrt{(x-y)^3} = 2x^2(x-y)^{3/2}$

7. $\dfrac{3x+8}{\sqrt[7]{(12x+5)^3}} = \dfrac{3x+8}{(12x+5)^{3/7}} = (3x+8)(12x+5)^{-3/7}$

8. $\sqrt{\dfrac{x-3}{y^5}} = \dfrac{\sqrt{x-3}}{\sqrt{y^5}} = \dfrac{(x-3)^{1/2}}{y^{5/2}} = (x-3)^{1/2}y^{-5/2}$

9. $\sqrt[4]{\dfrac{16x^3}{3x+1}} = \dfrac{\sqrt[4]{16x^3}}{\sqrt[4]{3x+1}} = \dfrac{(16x^3)^{1/4}}{(3x+1)^{1/4}} = (16x^3)^{1/4}(3x+1)^{-1/4}$

10. $\sqrt[5]{\dfrac{(x-1)^4}{(x+1)^3}} = \dfrac{\sqrt[5]{(x-1)^4}}{\sqrt[5]{(x+1)^3}} = \dfrac{(x-1)^{4/5}}{(x+1)^{3/5}} = (x-1)^{4/5}(x+1)^{-3/5}$

One of the uses of these exponent-root properties is to simplify multiple roots. Using the properties $\sqrt[n]{a^m} = a^{m/n}$ and $(a^m)^n = a^{mn}$, gradually rewrite the multiple roots as an exponent then as a single root.

Examples

$$\sqrt[4]{\sqrt[5]{x}} = \sqrt[4]{x^{1/5}} = (x^{1/5})^{1/4} = x^{(1/5)(1/4)} = x^{1/20} = \sqrt[20]{x}$$

$$\sqrt[6]{\sqrt[3]{y^5}} = \sqrt[6]{y^{5/3}} = (y^{5/3})^{1/6} = y^{(5/3)(1/6)} = y^{5/18} = \sqrt[18]{y^5}$$

Practice

1. $\sqrt{\sqrt{10}} =$

2. $\sqrt{\sqrt[4]{x^3}} =$

3. $\sqrt[5]{\sqrt[7]{2x^4}} =$

4. $\sqrt[2]{\sqrt[15]{(x-8)^4}} =$

5. $\sqrt{\sqrt{\sqrt[3]{y}}} =$

Solutions

1. $\sqrt{\sqrt{10}} = \sqrt{10^{1/2}} = (10^{1/2})^{1/2} = 10^{1/4} = \sqrt[4]{10}$

2. $\sqrt{\sqrt[4]{x^3}} = \sqrt{x^{3/4}} = (x^{3/4})^{1/2} = x^{3/8} = \sqrt[8]{x^3}$

3. $\sqrt[5]{\sqrt[7]{2x^4}} = \sqrt[5]{(2x^4)^{1/7}} = ((2x^4)^{1/7})^{1/5} = (2x^4)^{1/35} = \sqrt[35]{2x^4}$

4. $\sqrt[2]{\sqrt[15]{(x-8)^4}} = \sqrt[2]{(x-8)^{4/15}} = ((x-8)^{4/15})^{1/2} = (x-8)^{4/30}$
$= (x-8)^{2/15} = \sqrt[15]{(x-8)^2}$

5. $\sqrt{\sqrt{\sqrt[3]{y}}} = \sqrt{\sqrt{y^{1/3}}} = \sqrt{(y^{1/3})^{1/2}} = \sqrt{y^{1/6}} = (y^{1/6})^{1/2} = y^{1/12} = \sqrt[12]{y}$

Chapter Review

1. $\dfrac{(6x+5)^3}{(6x+5)^2} =$

(a) $(6x+5)^5$ (b) $6x+5$ (c) $(6x+5)^{-1}$ (d) $(6x+5)^6$

2. $((10x)^4)^7 =$

(a) $(10x)^{28}$ (b) $10x^{28}$ (c) $(10x)^{11}$ (d) $10x^{11}$

3. $(xy)^2(xy)^9 =$

(a) xy^{11} (b) $(xy)^{18}$ (c) $(xy)^{11}$ (d) xy^{18}

4. $6x^2\sqrt{x} =$

(a) $6x$ (b) $\sqrt{36x^3}$ (c) $6x^{5/2}$ (d) $(6x)^{5/2}$

5. $\dfrac{1}{(x+100)^4} =$

(a) $(x+100)^4$ (b) $x^{-4} + 100^{-4}$ (c) $-(x+100)^4$
(d) $(x+100)^{-4}$

6. $\left(\dfrac{2}{x}\right)^{-1} =$

(a) $\dfrac{x}{2}$ (b) $\dfrac{x^{-1}}{2^{-1}}$ (c) $\dfrac{-x}{2}$ (d) $\dfrac{-2}{x}$

7. $\left(\dfrac{x-5}{y}\right)^{-2} =$

(a) $-\left(\dfrac{x-5}{y}\right)^2$ (b) $\left(\dfrac{y}{x-5}\right)^2$ (c) $\dfrac{y^2}{x^2-25}$
(d) $\dfrac{-y^2}{x^2-25}$

8. $\dfrac{x^3}{x^{-4}} =$

(a) x^{-12} (b) x^{-1} (c) x^7 (d) x^{-7}

9. $\dfrac{y^3}{\sqrt[4]{y}} =$

(a) $x^{11/4}$ (b) $x^{3/4}$ (c) x^{12} (d) $x^{13/4}$

10. $[x(y-x)]^{-2} =$

(a) $\dfrac{x^2}{(y-x)^2}$ (b) $\dfrac{-x^2}{(y-x)^2}$ (c) $\dfrac{1}{x^2(y-x)^2}$
(d) $\dfrac{x^2}{y^2-x^2}$

11. $(y^{-3})^7 =$

(a) $\frac{1}{y^{21}}$ (b) $\frac{1}{y^{10}}$ (c) y^4 (d) $-y^4$

12. $\frac{1}{\sqrt[5]{8xy^2}} =$

(a) $(8xy)^{2/5}$ (b) $(8xy^2)^{-1/5}$ (c) $(8xy)^{-2/5}$ (d) $(8xy^2)^{1/5}$

13. $\sqrt{5}\sqrt{2x} =$

(a) $\sqrt{10x}$ (b) $\sqrt{50x}$ (c) $\sqrt{100x}$ (d) $\sqrt{10}x$

14. $(3x^2y^4)^3 =$

(a) $3x^5y^7$ (b) $3x^6y^{12}$ (c) $9x^5y^7$ (d) $27x^6y^{12}$

15. $(-6x)^2 =$

(a) $36x^2$ (b) $-36x^2$ (c) $6x^2$ (d) $-6x^2$

16. $(-x^5) =$

(a) $-x^5$ (b) $x^{-1/5}$ (c) x^{-5} (d) $x^{1/5}$

17. $\sqrt{\frac{3}{2}} =$

(a) $\frac{\sqrt{3}}{2}$ (b) $\frac{\sqrt{6}}{2}$ (c) $\sqrt{\frac{2}{3}}$ (d) $\frac{\sqrt{6}}{3}$

18. $\sqrt[4]{16x^8} =$

(a) $2x^2$ (b) $4x^2$ (c) $4x^4$ (d) $16x^2$

19. $\sqrt[3]{125x^6y^{21}} =$

(a) $5x^3y^{18}$ (b) $25x^2y^7$ (c) $5x^2y^7$ (d) $25x^3y^{18}$

20. $\sqrt{27x^5} =$

(a) $3x\sqrt{3x}$ (b) $3x^2\sqrt{3x}$ (c) $3x^2\sqrt{x}$ (d) $3x\sqrt{3x}$

21. $\sqrt[5]{\dfrac{2x-7}{x^3y}} =$

(a) $\dfrac{\sqrt[5]{x^2y^4(2x-7)}}{xy}$ (b) $\dfrac{\sqrt[5]{x^2(2x-7)}}{x}$ (c) $\dfrac{\sqrt[5]{y^4(2x-7)}}{x}$

(d) $\dfrac{\sqrt[5]{x^3y(2x-7)}}{x^3y}$

22. $x^4x^5 =$

(a) x^{-1} (b) x (c) x^9 (d) x^{20}

23. $\left(\dfrac{y}{x}\right)^{-3} =$

(a) $\dfrac{-y^3}{x^3}$ (b) $\dfrac{-x^3}{y^3}$ (c) $\dfrac{y^3}{x^3}$ (d) $\dfrac{x^3}{y^3}$

24. $\dfrac{1}{\sqrt[4]{x^3}} =$

(a) $x^{3/4}$ (b) $x^{-3/4}$ (c) $x^{4/3}$ (d) $x^{-4/3}$

25. $\sqrt[3]{\sqrt[4]{y}} =$

(a) $y^{1/7}$ (b) y^7 (c) $y^{1/12}$ (d) y^{12}

26. $5y^4(3y^3)^2 =$

(a) $15y^{10}$ (b) $45y^{10}$ (c) $15y^9$ (d) $45y^9$

27. $\left(\dfrac{x^{-2}y^2}{2}\right)^{-3} =$

(a) $\dfrac{8x^6}{y^6}$ (b) $\dfrac{2x^6}{y^6}$ (c) $\dfrac{x^6}{8y^6}$ (d) $\dfrac{x^6}{2y^6}$

Solutions

1. (b)	**2.** (a)	**3.** (c)	**4.** (c)
5. (d)	**6.** (a)	**7.** (b)	**8.** (c)
9. (a)	**10.** (c)	**11.** (a)	**12.** (b)
13. (a)	**14.** (d)	**15.** (a)	**16.** (a)
17. (b)	**18.** (a)	**19.** (c)	**20.** (b)
21. (a)	**22.** (c)	**23.** (d)	**24.** (b)
25. (c)	**26.** (b)	**27.** (a)	

Factoring

Distributing Multiplication over Addition and Subtraction

Distributing multiplication over addition (and subtraction) and factoring (the opposite of distributing) are extremely important in algebra. The distributive law of multiplication over addition, $a(b+c) = ab + ac$, says that you can first take the sum $(b+c)$ then the product (a times the sum of b and c) or the individual products (ab and ac) then the sum (the sum of ab and ac). For instance, $12(6+4)$ could be computed as $12(6+4) = 12(6) + 12(4) = 72 + 48 = 120$ or as $12(6+4) = 12(10) = 120$. The distributive law of multiplication over subtraction, $a(b-c) = ab - ac$, says the same about a product and difference.

Examples

$7(x-y) = 7x - 7y$

$4(3x+1) = 12x + 4$

$x^2(3x - 5y) = 3x^3 - 5x^2y$

$8xy(x^3 + 4y) = 8x^4y + 32xy^2$

$6x^2y^3(5x - 2y^2) = 30x^3y^3 - 12x^2y^5$

$\sqrt{x}(x^2 + 12) = x^2\sqrt{x} + 12\sqrt{x}$

$y^{-2}(y^4 + 6) = y^{-2}y^4 + 6y^{-2} = y^2 + 6y^{-2}$

Practice

1. $3(14 - 2) =$
2. $\frac{1}{2}(6 + 8) =$
3. $4(6 - 2x) =$
4. $9x(4y + x) =$
5. $3xy^4(9x^3 + 2y) =$
6. $3\sqrt[3]{x}(6y - 2x) =$
7. $\sqrt{x}(1 + \sqrt{x}) =$
8. $10y^{-3}(xy^4 - 8) =$
9. $4x^2(2y - 5x + 6) =$

Solutions

1. $3(14 - 2) = 3(14) - 3(2) = 42 - 6 = 36$
2. $\frac{1}{2}(6 + 8) = \frac{1}{2}(6) + \frac{1}{2}(8) = 3 + 4 = 7$
3. $4(6 - 2x) = 4(6) - 4(2x) = 24 - 8x$
4. $9x(4y + x) = 36xy + 9x^2$
5. $3xy^4(9x^3 + 2y) = 27x^4y^4 + 6xy^5$
6. $3\sqrt[3]{x}(6y - 2x) = 18\sqrt[3]{x}y - 6x\sqrt[3]{x}$
7. $\sqrt{x}(1 + \sqrt{x}) = \sqrt{x} + (\sqrt{x})(\sqrt{x}) = \sqrt{x} + (\sqrt{x})^2 = \sqrt{x} + x$
8. $10y^{-3}(xy^4 - 8) = 10xy^{-3}y^4 - 80y^{-3} = 10xy - 80y^{-3}$
9. $4x^2(2y - 5x + 6) = 8x^2y - 20x^3 + 24x^2$

Sometimes you will need to "distribute" a minus sign or negative sign: $-(a+b) = -a-b$ and $-(a-b) = -a+b$. You can use the distributive properties and think of $-(a+b)$ as $(-1)(a+b)$ and $-(a-b)$ as $(-1)(a-b)$:

$$-(a+b) = (-1)(a+b) = (-1)a + (-1)b = -a + -b = -a - b$$

and

$$-(a-b) = (-1)(a-b) = (-1)a - (-1)b = -a - (-1)b$$
$$= -a - (-b) = -a + b.$$

A common mistake is to write $-(a+b) = -a+b$ and $-(a-b) = -a-b$. The minus sign and negative sign in front of the parentheses changes the signs of *every* term (a quantity separated by a plus or minus sign) inside the parentheses.

Examples

$-(3+x) = -3-x$

$-(y-x^2) = -y+x^2$

$-(-2+y) = 2-y$

$-(-9-y) = 9+y$

$-(2+x-3y) = -2-x+3y$

$-(x^2-x-2) = -x^2+x+2$

$-(-4x-7y-2) = 4x+7y+2$

Practice

1. $-(4+x) =$
2. $-(-x-y) =$
3. $-(2x^2-5) =$
4. $-(-18+xy^2) =$
5. $-(2x-16y+5) =$
6. $-(x^2-5x-6) =$

Solutions

1. $-(4+x) = -4 - x$
2. $-(-x-y) = x + y$
3. $-(2x^2 - 5) = -2x^2 + 5$
4. $-(-18 + xy^2) = 18 - xy^2$
5. $-(2x - 16y + 5) = -2x + 16y - 5$
6. $-(x^2 - 5x - 6) = -x^2 + 5x + 6$

Distributing negative quantities has the same effect on signs as distributing a minus sign: every sign in the parentheses changes.

Examples

$-8(4+5x) = -32 - 40x$

$-xy(1-x) = -xy + x^2y$

$-3x^2(-2y+9x) = 6x^2y - 27x^3$

$-100(-4-x) = 400 + 100x$

Practice

1. $-2(16+y) =$
2. $-50(3-x) =$
3. $-12xy(-2x+y) =$
4. $-7x^2(-x-4y) =$
5. $-6y(-3x-y+4) =$

Solutions

1. $-2(16+y) = -32 - 2y$

2. $-50(3-x) = -150 + 50x$

3. $-12xy(-2x+y) = 24x^2y - 12xy^2$

4. $-7x^2(-x-4y) = 7x^3 + 28x^2y$

5. $-6y(-3x-y+4) = 18xy + 6y^2 - 24y$

Combining Like Terms

Two or more terms are alike if they have the same variables and the exponents (or roots) on those variables are the same: $3x^2y$ and $5x^2y$ are like terms but $6xy$ and $4xy^2$ are not. Constants are terms with no variables. The number in front of the variable(s) is the *coefficient*—in $4x^2y^3$, 4 is the coefficient. If no number appears in front of the variable, then the coefficient is 1. Add or subtract like terms by adding or subtracting their coefficients.

Examples

$3x^2y + 5x^2y = (3+5)x^2y = 8x^2y$

$14\sqrt{x} - 10\sqrt{x} = (14-10)\sqrt{x} = 4\sqrt{x}$

$8xyz + 9xyz - 6xyz = (8+9-6)xyz = 11xyz$

$3x + x = 3x + 1x = (3+1)x = 4x$

$7x\sqrt{y} - x\sqrt{y} = 7x\sqrt{y} - 1x\sqrt{y} = (7-1)x\sqrt{y} = 6x\sqrt{y}$

$$\frac{2}{3}x^2 - 4xy + \frac{3}{4}x^2 + \frac{5}{2}xy = \left(\frac{2}{3}+\frac{3}{4}\right)x^2 + \left(-4+\frac{5}{2}\right)xy$$
$$= \left(\frac{8}{12}+\frac{9}{12}\right)x^2 + \left(\frac{-8}{2}+\frac{5}{2}\right)xy = \frac{17}{12}x^2 - \frac{3}{2}xy$$

$$\begin{aligned}3x^2 + 4xy - 8xy^2 - (2x^2 - 3xy - 4xy^2 + 6) &= 3x^2 + 4xy - 8xy^2 - 2x^2 \\ &\quad + 3xy + 4xy^2 - 6 \\ &= 3x^2 - 2x^2 - 8xy^2 + 4xy^2 \\ &\quad + 4xy + 3xy - 6 \\ &= (3-2)x^2 + (-8+4)xy^2 \\ &\quad + (4+3)xy - 6 \\ &= x^2 - 4xy^2 + 7xy - 6\end{aligned}$$

Practice

1. $3xy + 7xy =$
2. $4x^2 - 6x^2 =$
3. $-\frac{3}{5}xy^2 + 2xy^2 =$
4. $8\sqrt{x} - \sqrt{x} =$
5. $2xy^2 - 4x^2y - 7xy^2 + 17x^2y =$
6. $14x + 8 - (2x - 4) =$
7. $16x^{-4} + 3x^{-2} - 4x + 9x^{-4} - x^{-2} + 5x - 6 =$
8. $5x\sqrt{y} + 7\sqrt{xy} + 1 - (3x\sqrt{y} - 7\sqrt{xy} + 4) =$
9. $x^2y + xy^2 + 6x + 4 - (4x^2y + 3xy^2 - 2x + 5) =$

Solutions

1. $3xy + 7xy = (3+7)xy = 10xy$
2. $4x^2 - 6x^2 = (4-6)x^2 = -2x^2$
3. $\frac{-3}{5}xy^2 + 2xy^2 = \left(\frac{-3}{5} + 2\right)xy^2 = \left(\frac{-3}{5} + \frac{10}{5}\right)xy^2 = \frac{7}{5}xy^2$

4. $8\sqrt{x} - \sqrt{x} = (8-1)\sqrt{x} = 7\sqrt{x}$

5. $$\begin{aligned} 2xy^2 - 4x^2y - 7xy^2 + 17x^2y &= 2xy^2 - 7xy^2 - 4x^2y + 17x^2y \\ &= (2-7)xy^2 + (-4+17)x^2y \\ &= -5xy^2 + 13x^2y \end{aligned}$$

6. $$\begin{aligned} 14x + 8 - (2x-4) &= 14x + 8 - 2x + 4 = 14x - 2x + 8 + 4 \\ &= (14-2)x + 12 = 12x + 12 \end{aligned}$$

7. $$\begin{aligned} &16x^{-4} + 3x^{-2} - 4x + 9x^{-4} - x^{-2} + 5x - 6 \\ &= 16x^{-4} + 9x^{-4} + 3x^{-2} - x^{-2} - 4x + 5x - 6 \\ &= (16+9)x^{-4} + (3-1)x^{-2} + (-4+5)x - 6 \\ &= 25x^{-4} + 2x^{-2} + x - 6 \end{aligned}$$

8. $$\begin{aligned} &5x\sqrt{y} + 7\sqrt{xy} + 1 - (3x\sqrt{y} - 7\sqrt{xy} + 4) \\ &= 5x\sqrt{y} + 7\sqrt{xy} + 1 - 3x\sqrt{y} + 7\sqrt{xy} - 4 \\ &= 5x\sqrt{y} - 3x\sqrt{y} + 7\sqrt{xy} + 7\sqrt{xy} + 1 - 4 \\ &= (5-3)x\sqrt{y} + (7+7)\sqrt{xy} - 3 \\ &= 2x\sqrt{y} + 14\sqrt{xy} - 3 \end{aligned}$$

9. $$\begin{aligned} &x^2y + xy^2 + 6x + 4 - (4x^2y + 3xy^2 - 2x + 5) \\ &= x^2y + xy^2 + 6x + 4 - 4x^2y - 3xy^2 + 2x - 5 \\ &= x^2y - 4x^2y + xy^2 - 3xy^2 + 6x + 2x + 4 - 5 \\ &= (1-4)x^2y + (1-3)xy^2 + (6+2)x - 1 \\ &= -3x^2y - 2xy^2 + 8x - 1 \end{aligned}$$

Adding/Subtracting Fractions

With the distributive property and the ability to combine like terms, the numerator of fraction sums/differences can be simplified. For now, we will leave the denominators factored.

Examples

$$\frac{2}{x-4}+\frac{x}{x+1}=\frac{2}{x-4}\cdot\frac{x+1}{x+1}+\frac{x}{x+1}\cdot\frac{x-4}{x-4}=\frac{2(x+1)+x(x-4)}{(x+1)(x-4)}$$

$$=\frac{2x+2+x^2-4x}{(x+1)(x-4)}$$

$$=\frac{x^2-2x+2}{(x+1)(x-4)}$$

$$4-\frac{2x+1}{x+3}=\frac{4}{1}-\frac{2x+1}{x+3}=\frac{4}{1}\cdot\frac{x+3}{x+3}-\frac{2x+1}{x+3}=\frac{4(x+3)-(2x+1)}{x+3}$$

$$=\frac{4x+12-2x-1}{x+3}=\frac{2x+11}{x+3}$$

$$\frac{x}{x-5}-\frac{x}{x+2}=\frac{x}{x-5}\cdot\frac{x+2}{x+2}-\frac{x}{x+2}\cdot\frac{x-5}{x-5}=\frac{x(x+2)-x(x-5)}{(x+2)(x-5)}$$

$$=\frac{x^2+2x-x^2+5x}{(x+2)(x-5)}=\frac{7x}{(x+2)(x-5)}$$

Practice

1. $\frac{7}{2x+3}+\frac{4}{x-2}=$

2. $\frac{1}{x-1}+\frac{x}{x+2}=$

3. $\frac{3x-4}{x+5}-2=$

4. $\frac{x}{2x+y}+\frac{y}{3x-4y}=$

5. $\frac{x}{6x+3}+\frac{x}{6x-3}=$

Solutions

1. $\frac{7}{2x+3}+\frac{4}{x-2}=\frac{7}{2x+3}\cdot\frac{x-2}{x-2}+\frac{4}{x-2}\cdot\frac{2x+3}{2x+3}$

Solution 1 (*continued*)

$$= \frac{7(x-2)+4(2x+3)}{(x-2)(2x+3)} = \frac{7x-14+8x+12}{(x-2)(2x+3)}$$

$$= \frac{15x-2}{(x-2)(2x+3)}$$

2. $$\frac{1}{x-1}+\frac{x}{x+2} = \frac{1}{x-1}\cdot\frac{x+2}{x+2}+\frac{x}{x+2}\cdot\frac{x-1}{x-1} = \frac{1(x+2)+x(x-1)}{(x+2)(x-1)}$$

$$= \frac{x+2+x^2-x}{(x+2)(x-1)} = \frac{x^2+2}{(x+2)(x-1)}$$

3. $$\frac{3x-4}{x+5}-2 = \frac{3x-4}{x+5}-\frac{2}{1} = \frac{3x-4}{x+5}-\frac{2}{1}\cdot\frac{x+5}{x+5} = \frac{3x-4-2(x+5)}{x+5}$$

$$= \frac{3x-4-2x-10}{x+5} = \frac{x-14}{x+5}$$

4. $$\frac{x}{2x+y}+\frac{y}{3x-4y} = \frac{x}{2x+y}\cdot\frac{3x-4y}{3x-4y}+\frac{y}{3x-4y}\cdot\frac{2x+y}{2x+y}$$

$$= \frac{x(3x-4y)+y(2x+y)}{(3x-4y)(2x+y)} = \frac{3x^2-4xy+2xy+y^2}{(3x-4y)(2x+y)}$$

$$= \frac{3x^2-2xy+y^2}{(3x-4y)(2x+y)}$$

5. $$\frac{x}{6x+3}+\frac{x}{6x-3} = \frac{x}{6x+3}\cdot\frac{6x-3}{6x-3}+\frac{x}{6x-3}\cdot\frac{6x+3}{6x+3}$$

$$= \frac{x(6x-3)+x(6x+3)}{(6x-3)(6x+3)} = \frac{6x^2-3x+6x^2+3x}{(6x-3)(6x+3)}$$

$$= \frac{12x^2}{(6x-3)(6x+3)}$$

Factoring

The distributive property, $a(b+c) = ab + ac$, can be used to factor a quantity from two or more terms. In the formula $ab + ac = a(b+c)$, a is factored from (or divided into) ab and ac. The first step in factoring is to decide what quantity you want to factor from each term. Second write each term as a product of the factor and something else (this step will become unnecessary once you are experienced). Third apply the distributive property in reverse.

Examples

$4 + 6x$

Each term is divisible by 2, so factor 2 from 4 and $6x$: $4 + 6x = 2 \cdot \mathbf{2} + 2 \cdot \mathbf{3x} = 2(2 + 3x)$.

$2x + 5x^2 = x \cdot \mathbf{2} + x \cdot \mathbf{5x} = x(2 + 5x)$

$3x^2 + 6x = 3x \cdot \mathbf{x} + 3x \cdot \mathbf{2} = 3x(x + 2)$

$8x + 8 = 8 \cdot \mathbf{x} + 8 \cdot \mathbf{1} = 8(x + 1)$

$4xy + 6x^2 + 2xy^2 = 2x \cdot \mathbf{2y} + 2x \cdot \mathbf{3x} + 2x \cdot \mathbf{y^2} = 2x(2y + 3x + y^2)$

Complicated expressions can be factored in several steps. Take for example $48x^5y^3z^6 + 60x^4yz^3 + 36x^6y^2z$, each term is divisible by $12xyz$. Start with this.

$$48x^5y^3z^6 + 60x^4yz^3 + 36x^6y^2z = 12xyz \cdot \mathbf{4x^4y^2z^5} + 12xyz \cdot \mathbf{5x^3z^2} + 12xyz \cdot \mathbf{3x^5y} = 12xyz(4x^4y^2z^5 + 5x^3z^2 + 3x^5y)$$

Each term in the parentheses is divisible by x^2:

$$4x^4y^2z^5 + 5x^3z^2 + 3x^5y = x^2 \cdot \mathbf{4x^2y^2z^5} + x^2 \cdot \mathbf{5xz^2} + x^2 \cdot \mathbf{3x^3y} = x^2(4x^2y^2z^5 + 5xz^2 + 3x^3y)$$

$$48x^5y^3z^6 + 60x^4yz^3 + 36x^6y^2z = 12xyz \cdot x^2(4x^2y^2z^5 + 5xz^2 + 3x^3y) = 12x^3yz(4x^2y^2z^5 + 5xz^2 + 3x^3y)$$

Practice

1. $4x - 10y =$
2. $3x + 6y - 12 =$
3. $5x^2 + 15 =$

4. $4x^2 + 4x =$

5. $4x^3 - 6x^2 + 12x =$

6. $-24xy^2 + 6x^2 + 18x =$

7. $30x^4 - 6x^2 =$

8. $15x^3y^2z^7 - 30xy^2z^4 + 6x^4y^2z^6 =$

Solutions

1. $4x - 10y = 2 \cdot 2x - 2 \cdot 5y = 2(2x - 5y)$

2. $3x + 6y - 12 = 3 \cdot x + 3 \cdot 2y - 3 \cdot 4 = 3(x + 2y - 4)$

3. $5x^2 + 15 = 5 \cdot x^2 + 5 \cdot 3 = 5(x^2 + 3)$

4. $4x^2 + 4x = 4x \cdot x + 4x \cdot 1 = 4x(x + 1)$

5. $4x^3 - 6x^2 + 12x = 2x \cdot 2x^2 - 2x \cdot 3x + 2x \cdot 6 = 2x(2x^2 - 3x + 6)$

6. $$\begin{aligned} -24xy^2 + 6x^2 + 18x &= 6x \cdot (-4y^2) + 6x \cdot x + 6x \cdot 3 \\ &= 6x(-4y^2 + x + 3) \end{aligned}$$

7. $30x^4 - 6x^2 = 6x^2 \cdot 5x^2 - 6x^2 \cdot 1 = 6x^2(5x^2 - 1)$

8. $$\begin{aligned} 15x^3y^2z^7 - 30xy^2z^4 + 6x^4y^2z^6 &= 3xy^2z^4 \cdot 5x^2z^3 - 3xy^2z^4 \cdot 10 \\ &\quad + 3xy^2z^4 \cdot 2x^3z^2 \\ &= 3xy^2z^4(5x^2z^3 - 10 + 2x^3z^2) \end{aligned}$$

Factoring a negative quantity has the same effect on signs within parentheses as distributing a negative quantity does—every sign changes. Negative quantities are factored in the next examples and practice problems.

Examples

$x + y = -(-x - y)$ $\qquad$ $-4 + x = -(4 - x)$

$-2 - 3x = -(2 + 3x)$ $\qquad$ $2x^2 + 4x = -2x(-x - 2)$

$-14xy + 21x^2y = -7xy(2 - 3x)$ $\qquad$ $12xy - 25x = -x(-12y + 25)$

$16y^2 - 1 = -(-16y^2 + 1) = -(1 - 16y^2)$

$-4x + 3y = -(4x - 3y)$ $\qquad$ $x - y - z + 5 = -(-x + y + z - 5)$

Practice

Factor a negative quantity from the expression.

1. $28xy^2 - 14x =$
2. $4x + 16xy =$
3. $-18y^2 + 6xy =$
4. $25 + 15y =$
5. $-8x^2y^2 - 4xy^2 =$
6. $-18x^2y^2 - 24xy^3 =$
7. $20xyz^2 - 5yz =$

Solutions

1. $28xy^2 - 14x = -7x(-4y^2 + 2)$
2. $4x + 16xy = -4x(-1 - 4y)$
3. $-18y^2 + 6xy = -6y(3y - x)$
4. $25 + 15y = -5(-5 - 3y)$
5. $-8x^2y^2 - 4xy^2 = -4xy^2(2x + 1)$
6. $-18x^2y^2 - 24xy^3 = -6xy^2(3x + 4y)$
7. $20xyz^2 - 5yz = -5yz(-4xz + 1)$

The associative and distributive properties can be confusing. The associative property states $(ab)c = a(bc)$. This property says that when multiplying three (or more) quantities you can multiply the first two then the third or multiply the second two then the first. For example, it might be tempting to write $5(x+1)(y-3) = (5x+5)(5y-15)$. But $(5x+5)(5y-15) = [5(x+1)][5(y-3)] = 25(x+1)(y-3)$. The "5" can be grouped either with "$x+1$" or with "$y-3$" but not both: $[5(x+1)](y-3) = (5x+5)(y-3)$ or $(x+1)[5(y-3)] = (x+1)(5y-15)$.

Factors themselves can have more than one term. For instance $\mathbf{3}(x+4) - \boldsymbol{x}(x+4)$ has $x+4$ as a factor in each term, so $x+4$ can be factored from $3(x+4)$ and $x(x+4)$:

$$\mathbf{3}(x+4) - \boldsymbol{x}(x+4) = (\mathbf{3} - \boldsymbol{x})(x+4).$$

Examples

$$\mathbf{2}\boldsymbol{x}(3x+y) + \mathbf{5}\boldsymbol{y}(3x+y) = (\mathbf{2}\boldsymbol{x} + \mathbf{5}\boldsymbol{y})(3x+y)$$

$$10y(x-y) + x - y = \mathbf{10}\boldsymbol{y}(x-y) + \mathbf{1}(x-y) = (\mathbf{10}\boldsymbol{y} + \mathbf{1})(x-y)$$

$$\mathbf{8}(2x-1) + \mathbf{2}\boldsymbol{x}(2x-1) - \mathbf{3}\boldsymbol{y}(2x-1) = (\mathbf{8} + \mathbf{2}\boldsymbol{x} - \mathbf{3}\boldsymbol{y})(2x-1)$$

Practice

1. $2(x-y) + 3y(x-y) =$
2. $4(2+7x) - x(2+7x) =$
3. $3(3+x) + x(3+x) =$
4. $6x(4-3x) - 2y(4-3x) - 5(4-3x) =$
5. $2x + 1 + 9x(2x+1) =$
6. $3(x-2y)^4 + 2x(x-2y)^4 =$

Solutions

1. $2(x-y) + 3y(x-y) = (2+3y)(x-y)$

2. $4(2+7x) - x(2+7x) = (4-x)(2+7x)$

3. $3(3+x) + x(3+x) = (3+x)(3+x) = (3+x)^2$

4. $6x(4-3x) - 2y(4-3x) - 5(4-3x) = (6x-2y-5)(4-3x)$

5. $2x+1+9x(2x+1) = 1(2x+1) + 9x(2x+1) = (1+9x)(2x+1)$

6. $3(x-2y)^4 + 2x(x-2y)^4 = (3+2x)(x-2y)^4$

More Factoring

An algebraic expression raised to different powers might appear in different terms. Factor out this expression raised to the lowest power.

Examples

$$\begin{aligned} 6(x+1)^2 - 5(x+1) &= [\mathbf{6(x+1)}](x+1) - \mathbf{5}(x+1) \\ &= [\mathbf{6(x+1) - 5}](x+1) = (6x+6-5)(x+1) \\ &= (6x+1)(x+1) \end{aligned}$$

$$\begin{aligned} 10(2x-3)^3 + 3(2x-3)^2 &= [\mathbf{10(2x-3)}](2x-3)^2 + \mathbf{3}(2x-3)^2 \\ &= [\mathbf{10(2x-3) + 3}](2x-3)^2 \\ &= (20x-30+3)(2x-3)^2 = (20x-27)(2x-3)^2 \end{aligned}$$

$$\begin{aligned} 9(14x+5)^4 + 6x(14x+5) - (14x+5) &= [\mathbf{9(14x+5)^3}](14x+5) \\ &\quad + \mathbf{6x}(14x+5) - \mathbf{1}(14x+5) \\ &= [\mathbf{9(14x+5)^3 + 6x - 1}](14x+5) \end{aligned}$$

Practice

1. $8(x+2)^3 + 5(x+2)^2 =$

2. $-4(x+16)^4 + 9(x+16)^2 + x + 16 =$

3. $(x+2y)^3 - 4(x+2y) =$

4. $2(x^2-6)^9 + (x^2-6)^4 + 4(x^2-6)^3 + (x^2-6)^2 =$

5. $(15xy-1)(2x-1)^3 - 8(2x-1)^2 =$

Solutions

1. $$\begin{aligned} 8(x+2)^3 + 5(x+2)^2 &= [\mathbf{8(x+2)}](x+2)^2 + \mathbf{5}(x+2)^2 \\ &= [\mathbf{8(x+2)+5}](x+2)^2 \\ &= (8x+16+5)(x+2)^2 = (8x+21)(x+2)^2 \end{aligned}$$

2. $$\begin{aligned} &-4(x+16)^4 + 9(x+16)^2 + x + 16 \\ &= [\mathbf{-4(x+16)^3}](x+16) + \mathbf{9(x+16)}(x+16) + \mathbf{1}(x+16) \\ &= [\mathbf{-4(x+16)^3 + 9(x+16) + 1}](x+16) \\ &= [-4(x+16)^3 + 9x + 144 + 1](x+16) \\ &= [-4(x+16)^3 + 9x + 145)(x+16) \end{aligned}$$

3. $$\begin{aligned} (x+2y)^3 - 4(x+2y) &= \mathbf{(x+2y)^2}(x+2y) - \mathbf{4}(x+2y) \\ &= [\mathbf{(x+2y)^2 - 4}](x+2y) \end{aligned}$$

4. $$\begin{aligned} &2(x^2-6)^9 + (x^2-6)^4 + 4(x^2-6)^3 + (x^2-6)^2 \\ &= \mathbf{2(x^2-6)^7}(x^2-6)^2 + \mathbf{(x^2-6)^2}(x^2-6)^2 \\ &\quad + \mathbf{4(x^2-6)}(x^2-6)^2 + \mathbf{1}(x^2-6)^2 \\ &= [\mathbf{2(x^2-6)^7 + (x^2-6)^2 + 4(x^2-6) + 1}](x^2-6)^2 \\ &= [2(x^2-6)^7 + (x^2-6)^2 + 4x^2 - 24 + 1](x^2-6)^2 \\ &= [2(x^2-6)^7 + (x^2-6)^2 + 4x^2 - 23](x^2-6)^2 \end{aligned}$$

5. $$\begin{aligned} (15xy-1)(2x-1)^3 - 8(2x-1)^2 &= \mathbf{(15xy-1)(2x-1)}(2x-1)^2 - \mathbf{8}(2x-1)^2 \\ &= [\mathbf{(15xy-1)(2x-1) - 8}](2x-1)^2 \end{aligned}$$

Factoring by Grouping

Sometimes you can combine two or more terms at a time in such a way that each term has an algebraic expression as a common factor.

Examples

$3x^2 - 3 + x^3 - x$

If 3 is factored from the first two terms and x is factored from the last two terms, we would have two terms with a factor of $x^2 - 1$.

$3x^2 - 3 + x^3 - x = 3(x^2 - 1) + x(x^2 - 1) = (3 + x)(x^2 - 1)$

You could also combine the first and third terms and the second and fourth terms.

$$3x^2 - 3 + x^3 - x = 3x^2 + x^3 - 3 - x = x^2(3 + x) - (3 + x)$$
$$= (x^2 - 1)(3 + x)$$

$3xy - 2y + 3x^2 - 2x = y(3x - 2) + x(3x - 2) = (y + x)(3x - 2)$

$5x^2 - 25 - x^2y + 5y = 5(x^2 - 5) - y(x^2 - 5) = (5 - y)(x^2 - 5)$

$4x^4 + x^3 - 4x - 1 = x^3(4x + 1) - (4x + 1) = (x^3 - 1)(4x + 1)$

Practice

1. $6xy^2 + 4xy + 9xy + 6x =$
2. $x^3 + x^2 - x - 1 =$
3. $15xy + 5x + 6y + 2 =$
4. $2x^4 - 6x - x^3y + 3y =$
5. $9x^3 + 18x^2 - x - 2 =$

Solutions

1. $6xy^2 + 4xy + 9xy + 6x = 2xy(3y + 2) + 3x(3y + 2)$
$= (2xy + 3x)(3y + 2) = x(2y + 3)(3y + 2)$

2. $x^3 + x^2 - x - 1 = x^2(x + 1) - 1(x + 1) = (x^2 - 1)(x + 1)$

3. $15xy + 5x + 6y + 2 = 5x(3y + 1) + 2(3y + 1)$
$= (5x + 2)(3y + 1)$

4. $2x^4 - 6x - x^3y + 3y = 2x(x^3 - 3) - y(x^3 - 3) = (2x - y)(x^3 - 3)$

5. $9x^3 + 18x^2 - x - 2 = 9x^2(x + 2) - 1(x + 2) = (9x^2 - 1)(x + 2)$

Factoring to Reduce Fractions

Among factoring's many uses is in reducing fractions. If the numerator's terms and the denominator's terms have common factors, factor them then cancel. It might not be necessary to factor the numerator and denominator completely.

Examples

$\dfrac{6x^2 + 2xy}{4x^2y - 10xy}$ Each term in the numerator and denominator has a factor of $2x$.

$$\frac{6x^2 + 2xy}{4x^2y - 10xy} = \frac{2x(3x + y)}{2x(2xy - 5y)} = \frac{3x + y}{2xy - 5y}$$

$$\frac{xy - x}{16x^2 - xy} = \frac{x(y - 1)}{x(16x - y)} = \frac{y - 1}{16x - y}$$

Practice

1. $\dfrac{18x - 24y}{6} =$

2. $\dfrac{8xy - 9x^2}{2x} =$

3. $\dfrac{14x^2y^2 + 21xy}{3x^2} =$

4. $\dfrac{28x - 14y}{7x} =$

5. $\dfrac{16x^3y^2 + 4xy}{12xy^2 - 8x^2y} =$

6. $\dfrac{15xyz^2 + 5x^2z}{30x^2y + 25x} =$

7. $\dfrac{24xyz^4 + 6x^2yz^3 - 18xz^2}{54xy^3z^3 + 48x^3y^2z^5} =$

Solutions

1. $\dfrac{18x - 24y}{6} = \dfrac{6(3x - 4y)}{6} = 3x - 4y$

2. $\dfrac{8xy - 9x^2}{2x} = \dfrac{x(8y - 9x)}{2x} = \dfrac{8y - 9x}{2}$

3. $\dfrac{14x^2y^2 + 21xy}{3x^2} = \dfrac{x(14xy^2 + 21y)}{3x^2} = \dfrac{14xy^2 + 21y}{3x}$

4. $\dfrac{28x - 14y}{7x} = \dfrac{7(4x - 2y)}{7x} = \dfrac{4x - 2y}{x}$

5. $\dfrac{16x^3y^2 + 4xy}{12xy^2 - 8x^2y} = \dfrac{4xy(4x^2y + 1)}{4xy(3y - 2x)} = \dfrac{4x^2y + 1}{3y - 2x}$

6. $\dfrac{15xyz^2 + 5x^2z}{30x^2y + 25x} = \dfrac{5x(3yz^2 + xz)}{5x(6xy + 5)} = \dfrac{3yz^2 + xz}{6xy + 5}$

7. $\dfrac{24xyz^4 + 6x^2yz^3 - 18xz^2}{54xy^3z^3 + 48x^3y^2z^5} = \dfrac{6xz^2(4yz^2 + xyz - 3)}{6xz^2(9y^3z + 8x^2y^2z^3)} = \dfrac{4yz^2 + xyz - 3}{9y^3z + 8x^2y^2z^3}$

Reducing a fraction or adding two fractions sometimes only requires that -1 be factored from one or more denominators. For instance in $\frac{y-x}{x-y}$ the numerator and denominator are only off by a factor of -1. To reduce this fraction, factor -1 from the numerator or denominator:

$$\frac{y-x}{x-y} = \frac{-(-y+x)}{x-y} = \frac{-(x-y)}{x-y} = \frac{-1}{1} = -1 \text{ or}$$

$$\frac{y-x}{x-y} = \frac{y-x}{-(-x+y)} = \frac{y-x}{-(y-x)} = \frac{1}{-1} = -1.$$

In the sum $\frac{3}{y-x} + \frac{x}{x-y}$ the denominators are off by a factor of -1. Factor -1 from one of the denominators and use the fact that $\frac{a}{-b} = \frac{-a}{b}$ to write both terms with the same denominator.

$$\frac{3}{y-x} + \frac{x}{x-y} = \frac{3}{-(-y+x)} + \frac{x}{x-y} = \frac{3}{-(x-y)} + \frac{x}{x-y}$$

$$= \frac{-3}{x-y} + \frac{x}{x-y} = \frac{-3+x}{x-y}$$

In the next examples and practice problems a "-1" is factored from the denominator and moved to the numerator.

Examples

$$\frac{1}{1-x} = \frac{1}{-(-1+x)} = \frac{1}{-(x-1)} = \frac{-1}{x-1} \qquad \frac{3}{-x-6} = \frac{3}{-(x+6)} = \frac{-3}{x+6}$$

$$\frac{-3x}{14+9x} = \frac{-3x}{-(-14-9x)} = \frac{-(-3x)}{-14-9x} = \frac{3x}{-14-9x}$$

$$\frac{16x-5}{7x-3} = \frac{16x-5}{-(-7x+3)} = \frac{16x-5}{-(3-7x)} = \frac{-(16x-5)}{3-7x} = \frac{-16x+5}{3-7x}$$

Practice

1. $\frac{1}{y-x} =$

2. $\frac{16}{4-x} =$

3. $\frac{-10x^2}{7-3x} =$

4. $\frac{9+8y}{-6-x} =$

5. $\frac{8xy-5}{5-8xy} =$

6. $\frac{5xy-4+3x}{9x-16} =$

Solutions

1. $\frac{1}{y-x} = \frac{1}{-(-y+x)} = \frac{1}{-(x-y)} = \frac{-1}{x-y}$

2. $\frac{16}{4-x} = \frac{16}{-(-4+x)} = \frac{16}{-(x-4)} = \frac{-16}{x-4}$

3. $\frac{-10x^2}{7-3x} = \frac{-10x^2}{-(-7+3x)} = \frac{-10x^2}{-(3x-7)} = \frac{-(-10x^2)}{3x-7} = \frac{10x^2}{3x-7}$

4. $\frac{9+8y}{-6-x} = \frac{9+8y}{-(6+x)} = \frac{-(9+8y)}{6+x} = \frac{-9-8y}{6+x}$

5. $\frac{8xy-5}{5-8xy} = \frac{8xy-5}{-(-5+8xy)} = \frac{8xy-5}{-(8xy-5)} = \frac{-(8xy-5)}{8xy-5} = \frac{-1}{1} = -1$

6. $\frac{5xy-4+3x}{9x-16} = \frac{5xy-4+3x}{-(-9x+16)} = \frac{5xy-4+3x}{-(16-9x)} = \frac{-(5xy-4+3x)}{16-9x}$

$$= \frac{-5xy+4-3x}{16-9x}$$

More on the Distribution Property—the FOIL Method

The FOIL method helps us to use the distribution property to help expand expressions like $(x+4)(2x-1)$. The letters in "FOIL" describe the sums and products.

F	First × First	$(x+4)(2x-1)$: $x(2x) = 2x^2$ (F on x and $2x$)
+ O	Outer × Outer	$(x+4)(2x-1)$: $x(-1) = -x$ (O on x and -1)
+ I	Inner × Inner	$(x+4)(2x-1)$: $4(2x) = 8x$ (I on 4 and $2x$)
+ L	Last × Last	$(x+4)(2x-1)$: $4(-1) = -4$ (L on 4 and -1)

$$(x+4)(2x-1) = 2x^2 - x + 8x - 4 = 2x^2 + 7x - 4$$

Examples

$$(x+16)(x-4) = x \cdot x + x(-4) + 16x + 16(-4) = x^2 - 4x + 16x - 64$$
$$= x^2 + 12x - 64$$

$$(2x+3)(7x-6) = 2x(7x) + 2x(-6) + 3(7x) + 3(-6)$$
$$= 14x^2 - 12x + 21x - 18 = 14x^2 + 9x - 18$$

$$(2x+1)^2 = (2x+1)(2x+1) = 2x(2x) + 2x(1) + 1(2x) + 1(1)$$
$$= 4x^2 + 2x + 2x + 1 = 4x^2 + 4x + 1$$

$$(x-7)(x+7) = x \cdot x + 7x + (-7)x + (-7)(7) = x^2 + 7x - 7x - 49$$
$$= x^2 - 49$$

Practice

1. $(5x - 1)(2x + 3) =$
2. $(4x + 2)(x - 6) =$
3. $(2x + 1)(9x + 4) =$
4. $(12x - 1)(2x - 5) =$
5. $(x^2 + 2)(x - 1) =$
6. $(x^2 - y)(x + 2y) =$
7. $(\sqrt{x} - 3)(\sqrt{x} + 4) =$
8. $(x - 5)(x + 5) =$
9. $(x - 6)(x + 6) =$
10. $(\sqrt{x} + 2)(\sqrt{x} - 2) =$
11. $(x + 8)^2 =$
12. $(x - y)^2 =$
13. $(2x + 3y)^2 =$
14. $(\sqrt{x} + \sqrt{y})^2 =$
15. $(\sqrt{x} + \sqrt{y})(\sqrt{x} - \sqrt{y}) =$

Solutions

1. $(5x - 1)(2x + 3) = 5x(2x) + 5x(3) + (-1)(2x) + (-1)(3)$
$= 10x^2 + 15x - 2x - 3 = 10x^2 + 13x - 3$

2. $(4x + 2)(x - 6) = 4x(x) + 4x(-6) + 2x + 2(-6)$
$= 4x^2 - 24x + 2x - 12 = 4x^2 - 22x - 12$

3. $(2x+1)(9x+4) = 2x(9x) + 2x(4) + 1(9x) + 1(4)$
$= 18x^2 + 8x + 9x + 4 = 18x^2 + 17x + 4$

4. $(12x-1)(2x-5) = 12x(2x) + 12x(-5) + (-1)(2x) + (-1)(-5)$
$= 24x^2 - 60x - 2x + 5 = 24x^2 - 62x + 5$

5. $(x^2+2)(x-1) = x^2(x) + x^2(-1) + 2x + 2(-1) = x^3 - x^2 + 2x - 2$

6. $(x^2-y)(x+2y) = x^2(x) + x^2(2y) + (-y)x + (-y)(2y)$
$= x^3 + 2x^2y - xy - 2y^2$

7. $(\sqrt{x}-3)(\sqrt{x}+4) = \sqrt{x}\cdot\sqrt{x} + 4\sqrt{x} + (-3)\sqrt{x} + (-3)(4)$
$= (\sqrt{x})^2 + 1\sqrt{x} - 12 = x + \sqrt{x} - 12$

8. $(x-5)(x+5) = x(x) + 5x + (-5)x + (-5)(5)$
$= x^2 + 5x - 5x - 25 = x^2 - 25$

9. $(x-6)(x+6) = x(x) + 6x + (-6)x + (-6)(6)$
$= x^2 + 6x - 6x - 36 = x^2 - 36$

10. $(\sqrt{x}+2)(\sqrt{x}-2) = (\sqrt{x})(\sqrt{x}) + (-2)\sqrt{x} + 2\sqrt{x} + 2(-2)$
$= (\sqrt{x})^2 - 2\sqrt{x} + 2\sqrt{x} - 4 = x - 4$

11. $(x+8)^2 = (x+8)(x+8) = x(x) + 8x + 8x + 8(8)$
$= x^2 + 16x + 64$

12. $(x-y)^2 = (x-y)(x-y) = x(x) + x(-y) + x(-y) + (-y)(-y)$
$= x^2 - xy - xy + y^2 = x^2 - 2xy + y^2$

13. $(2x+3y)^2 = (2x+3y)(2x+3y)$
$= 2x(2x) + 2x(3y) + 3y(2x) + (3y)(3y)$
$= 4x^2 + 6xy + 6xy + 9y^2 = 4x^2 + 12xy + 9y^2$

14. $(\sqrt{x}+\sqrt{y})^2 = (\sqrt{x}+\sqrt{y})(\sqrt{x}+\sqrt{y})$
$= \sqrt{x}(\sqrt{x}) + \sqrt{x}(\sqrt{y}) + \sqrt{x}(\sqrt{y}) + \sqrt{y}(\sqrt{y})$
$= (\sqrt{x})^2 + 2\sqrt{x}\sqrt{y} + (\sqrt{y})^2$
$= x + 2\sqrt{xy} + y$

15. $(\sqrt{x}+\sqrt{y})(\sqrt{x}-\sqrt{y}) = \sqrt{x}(\sqrt{x}) + \sqrt{x}(-\sqrt{y}) + \sqrt{x}\sqrt{y} + \sqrt{y}(\sqrt{y})$
$= (\sqrt{x})^2 + \sqrt{x}\sqrt{y} - \sqrt{x}\sqrt{y} + (\sqrt{y})^2 = x - y$

Factoring Quadratic Polynomials

We will now work in the opposite direction—factoring. First we will factor *quadratic polynomials*, expressions of the form $ax^2 + bx + c$ (where a is not 0). For example $x^2 + 5x + 6$ is factored as $(x+2)(x+3)$. Quadratic polynomials whose first factors are x^2 are the easiest to factor. Their factorization always begins as $(x \pm \_\_)(x \pm \_\_)$. This forces the first factor to be x^2 when the FOIL method is used. All you need to do is fill in the two blanks and decide when to use plus and minus signs. All quadratic polynomials factor though some do not factor "nicely." We will only concern ourselves with "nicely" factorable polynomials in this chapter.

If the second sign is minus, then the signs in the factors will be different (one plus and one minus). If the second sign is plus then both of the signs will be the same. In this case, if the first sign in the trinomial is a plus sign, both signs in the factors will be plus; and if the first sign in the trinomial is a minus sign, both signs in the factors will be minus.

Examples

$x^2 - 4x - 5 = (x - \_\_)(x + \_\_)$ or $(x + \_\_)(x - \_\_)$

$x^2 + x - 12 = (x + \_\_)(x - \_\_)$ or $(x - \_\_)(x + \_\_)$

$x^2 - 6x + 8 = (x - \_\_)(x - \_\_)$

$x^2 + 4x + 3 = (x + \_\_)(x + \_\_)$

Practice

Determine whether to begin the factoring as $(x + \_\_)(x + \_\_)$, $(x - \_\_)(x - \_\_)$, or $(x - \_\_)(x + \_\_)$.

1. $x^2 - 5x - 6 =$

2. $x^2 + 2x + 1 =$

3. $x^2 + 3x - 10 =$

4. $x^2 - 6x + 8 =$

5. $x^2 - 11x - 12 =$

6. $x^2 - 9x + 14 =$

7. $x^2 + 7x + 10 =$

8. $x^2 + 4x - 21 =$

Solutions

1. $x^2 - 5x - 6 = (x - \_\_)(x + \_\_)$

2. $x^2 + 2x + 1 = (x + \_\_)(x + \_\_)$

3. $x^2 + 3x - 10 = (x - \_\_)(x + \_\_)$

4. $x^2 - 6x + 8 = (x - \_\_)(x - \_\_)$

5. $x^2 - 11x - 12 = (x - \_\_)(x + \_\_)$

6. $x^2 - 9x + 14 = (x - \_\_)(x - \_\_)$

7. $x^2 + 7x + 10 = (x + \_\_)(x + \_\_)$

8. $x^2 + 4x - 21 = (x - \_\_)(x + \_\_)$

Once the signs are determined all that remains is to fill in the two blanks. Look at all of the pairs of factors of the constant term. These pairs will be the candidates for the blanks. For example, if the constant term is 12, you will need to consider 1 and 12, 2 and 6, and 3 and 4. If both signs in the factors are the same, these will be the only ones you need to try. If the signs are different, you will need to reverse the order: 1 and 12 as well as 12 and 1; 2 and 6 as well as 6 and 2; 3 and 4 as well as 4 and 3. Try the FOIL method on these pairs. (Not every trinomial can be factored in this way.)

Examples

$x^2 + x - 12$

Factors to check: $(x + 1)(x - 12)$, $(x - 1)(x + 12)$, $(x + 2)(x - 6)$, $(x - 2)(x + 6)$, $(x - 4)(x + 3)$, and $(x + 4)(x - 3)$.

$(x + 1)(x - 12) = x^2 - 11x - 12$

$(x - 1)(x + 12) = x^2 + 11x - 12$

$(x + 2)(x - 6) = x^2 - 4x - 12$

$(x - 2)(x + 6) = x^2 + 4x - 12$

$(x - 4)(x + 3) = x^2 - x - 12$

$(x + 4)(x - 3) = x^2 + x - 12$ (This works.)

Examples

$x^2 - 2x - 15$

Factors to check: $(x + 15)(x - 1)$, $(x - 15)(x + 1)$, $(x + 5)(x - 3)$, and $(x - 5)(x + 3)$ (works).

$x^2 - 11x + 18$

Factors to check: $(x - 1)(x - 18)$, $(x - 3)(x - 6)$, and $(x - 2)(x - 9)$ (works).

$x^2 + 8x + 7$

Factors to check: $(x + 1)(x + 7)$ (works).

Practice

Factor the quadratic polynomial.

1. $x^2 - 5x - 6 =$

2. $x^2 + 2x + 1 =$

3. $x^2 + 3x - 10 =$

4. $x^2 - 6x + 8 =$

5. $x^2 - 11x - 12 =$

6. $x^2 - 9x + 14 =$

7. $x^2 + 7x + 10 =$

8. $x^2 + 4x - 21 =$

9. $x^2 + 13x + 36 =$

10. $x^2 + 5x - 24 =$

Solutions

1. $x^2 - 5x - 6 = (x - 6)(x + 1)$

2. $x^2 + 2x + 1 = (x + 1)(x + 1) = (x + 1)^2$

3. $x^2 + 3x - 10 = (x + 5)(x - 2)$

4. $x^2 - 6x + 8 = (x - 4)(x - 2)$

5. $x^2 - 11x - 12 = (x - 12)(x + 1)$

6. $x^2 - 9x + 14 = (x - 7)(x - 2)$

7. $x^2 + 7x + 10 = (x + 5)(x + 2)$

8. $x^2 + 4x - 21 = (x + 7)(x - 3)$

9. $x^2 + 13x + 36 = (x + 4)(x + 9)$

10. $x^2 + 5x - 24 = (x + 8)(x - 3)$

There is a factoring shortcut when the first term is x^2. If the second sign is plus, choose the factors whose *sum* is the coefficient of the second term. For example the factors of 6 we need for $x^2 - 7x + 6$ need to sum to 7:

$x^2 - 7x + 6 = (x - 1)(x - 6)$. The factors of 6 we need for $x^2 + 5x + 6$ need to sum to 5: $x^2 + 5x + 6 = (x + 2)(x + 3)$.

If the second sign is minus, the *difference* of the factors needs to be the coefficient of the middle term. If the first sign is plus, the bigger factor will have the plus sign. If the first sign is minus, the bigger factor will have the minus sign.

Examples

$x^2 + 3x - 10$: The factors of 10 whose difference is 3 are 2 and 5. The first sign is plus, so the plus sign goes with 5, the bigger factor: $x^2 + 3x - 10 = (x + 5)(x - 2)$.

$x^2 - 5x - 14$: The factors of 14 whose difference is 5 are 2 and 7. The first sign is minus, so the minus sign goes with 7, the bigger factor: $x^2 - 5x - 14 = (x - 7)(x + 2)$.

$x^2 + 11x + 24$: $3 \cdot 8 = 24$ and $3 + 8 = 11$

$x^2 + 11x + 24 = (x + 3)(x + 8)$

$x^2 - 9x + 18$: $3 \cdot 6 = 18$ and $3 + 6 = 9$

$x^2 - 9x + 18 = (x - 3)(x - 6)$

$x^2 + 9x - 36$: $3 \cdot 12 = 36$ and $12 - 3 = 9$

$x^2 + 9x - 36 = (x + 12)(x - 3)$

$x^2 - 2x - 8$: $2 \cdot 4 = 8$ and $4 - 2 = 2$

$x^2 - 2x - 8 = (x + 2)(x - 4)$

Practice

1. $x^2 - 6x + 9 =$
2. $x^2 - x - 12 =$
3. $x^2 + 9x - 22 =$

4. $x^2 + x - 20 =$

5. $x^2 + 13x + 36 =$

6. $x^2 - 19x + 34 =$

7. $x^2 - 18x + 17 =$

8. $x^2 + 24x - 25 =$

9. $x^2 - 14x + 48 =$

10. $x^2 + 16x + 64 =$

11. $x^2 - 49 =$

(*Hint*: $x^2 - 49 = x^2 + 0x - 49$)

Solutions

1. $x^2 - 6x + 9 = (x - 3)(x - 3) = (x - 3)^2$

2. $x^2 - x - 12 = (x - 4)(x + 3)$

3. $x^2 + 9x - 22 = (x + 11)(x - 2)$

4. $x^2 + x - 20 = (x + 5)(x - 4)$

5. $x^2 + 13x + 36 = (x + 4)(x + 9)$

6. $x^2 - 19x + 34 = (x - 2)(x - 17)$

7. $x^2 - 18x + 17 = (x - 1)(x - 17)$

8. $x^2 + 24x - 25 = (x + 25)(x - 1)$

9. $x^2 - 14x + 48 = (x - 6)(x - 8)$

10. $x^2 + 16x + 64 = (x + 8)(x + 8) = (x + 8)^2$

11. $x^2 - 49 = (x - 7)(x + 7)$

This shortcut can help you identify quadratic polynomials that do not factor "nicely" without spending too much time on them. The next three examples are quadratic polynomials that do not factor "nicely."

$$x^2 + x + 1 \qquad x^2 + 14x + 19 \qquad x^2 - 5x + 10$$

Quadratic polynomials of the form $x^2 - c^2$ are called the *difference of two squares*. We can use the shortcut on $x^2 - c^2 = x^2 + 0x - c^2$. The factors of c^2 must have a difference of 0. This can only happen if they are the same, so the factors of c^2 we want are c and c.

Examples

$x^2 - 9 = (x - 3)(x + 3)$ $\qquad$ $x^2 - 100 = (x - 10)(x + 10)$

$x^2 - 49 = (x - 7)(x + 7)$ $\qquad$ $16 - x^2 = (4 - x)(4 + x)$

When the sign between x^2 and c^2 is plus, the quadratic cannot be factored using real numbers.

Practice

1. $x^2 - 4 =$
2. $x^2 - 81 =$
3. $x^2 - 25 =$
4. $x^2 - 64 =$
5. $x^2 - 1 =$
6. $x^2 - 15 =$
7. $25 - x^2 =$

Solutions

1. $x^2 - 4 = (x - 2)(x + 2)$
2. $x^2 - 81 = (x - 9)(x + 9)$

3. $x^2 - 25 = (x-5)(x+5)$

4. $x^2 - 64 = (x-8)(x+8)$

5. $x^2 - 1 = (x-1)(x+1)$

6. $x^2 - 15 = (x-\sqrt{15})(x+\sqrt{15})$

7. $25 - x^2 = (5-x)(5+x)$

The difference of two squares can come in the form $x^n - c^n$ where n is any even number. The factorization is $x^n - c^n = (x^{n/2} - c^{n/2})(x^{n/2} + c^{n/2})$. [When n is odd, $x^n - c^n$ can be factored also but this factorization will not be covered here.]

Examples

$$x^6 - 1 = x^6 - 1^6 = (x^3-1)(x^3+1)$$

$$16 - x^4 = 2^4 - x^4 = (2^2 - x^2)(2^2 + x^2) = (4 - x^2)(4 + x^2)$$
$$= (2-x)(2+x)(4+x^2)$$

$$16x^4 - 1 = (2x)^4 - 1^4 = (4x^2-1)(4x^2+1) = (2x-1)(2x+1)(4x^2+1)$$

$$x^6 - \frac{1}{64} = x^6 - \left(\frac{1}{2}\right)^6 = \left(x^3 - \frac{1}{2}\right)\left(x^3 + \frac{1}{2}\right)$$

$$x^{10} - 1 = x^{10} - 1^{10} = (x^5-1)(x^5+1)$$

$$x^8 - 1 = x^8 - 1^8 = (x^4-1)(x^4+1) = (x^2-1)(x^2+1)(x^4+1)$$
$$= (x-1)(x+1)(x^2+1)(x^4+1)$$

Practice

1. $x^4 - 1 =$

2. $x^8 - 16 =$

3. $x^8 - \frac{1}{16} =$

4. $256x^4 - 1 =$

5. $x^4 - 81 =$

6. $81x^4 - 1 =$

7. $\frac{1}{64}x^6 - 1 =$

8. $16x^4 - 81 =$

9. $\frac{16}{81}x^4 - 16 =$

10. $x^{12} - 1 =$

Solutions

1. $x^4 - 1 = (x^2 - 1)(x^2 + 1) = (x - 1)(x + 1)(x^2 + 1)$

2. $x^8 - 16 = (x^4 - 4)(x^4 + 4) = (x^2 - 2)(x^2 + 2)(x^4 + 4)$

3. $x^8 - \frac{1}{16} = \left(x^4 - \frac{1}{4}\right)\left(x^4 + \frac{1}{4}\right) = \left(x^2 - \frac{1}{2}\right)\left(x^2 + \frac{1}{2}\right)\left(x^4 + \frac{1}{4}\right)$

4. $256x^4 - 1 = (16x^2 - 1)(16x^2 + 1) = (4x - 1)(4x + 1)(16x^2 + 1)$

5. $x^4 - 81 = (x^2 - 9)(x^2 + 9) = (x - 3)(x + 3)(x^2 + 9)$

6. $81x^4 - 1 = (9x^2 - 1)(9x^2 + 1) = (3x - 1)(3x + 1)(9x^2 + 1)$

7. $\frac{1}{64}x^6 - 1 = \left(\frac{1}{8}x^3 - 1\right)\left(\frac{1}{8}x^3 + 1\right)$

8. $16x^4 - 81 = (4x^2 - 9)(4x^2 + 9) = (2x - 3)(2x + 3)(4x^2 + 9)$

9. $\frac{16}{81}x^4 - 16 = \left(\frac{4}{9}x^2 - 4\right)\left(\frac{4}{9}x^2 + 4\right)$
$= \left(\frac{2}{3}x - 2\right)\left(\frac{2}{3}x + 2\right)\left(\frac{4}{9}x^2 + 4\right)$

10. $x^{12} - 1 = (x^6 - 1)(x^6 + 1) = (x^3 - 1)(x^3 + 1)(x^6 + 1)$

When the first term is not x^2, see if you can factor out the coefficient of x^2. If you can, then you are left with a quadratic whose first term is x^2. For example each term in $2x^2 + 16x - 18$ is divisible by 2:

$$2x^2 + 16x - 18 = 2(x^2 + 8x - 9) = 2(x + 9)(x - 1).$$

Practice

1. $4x^2 + 28x + 48 =$
2. $3x^2 - 9x - 54 =$
3. $9x^2 - 9x - 18 =$
4. $15x^2 - 60 =$
5. $6x^2 + 24x + 24 =$

Solutions

1. $4x^2 + 28x + 48 = 4(x^2 + 7x + 12) = 4(x + 4)(x + 3)$
2. $3x^2 - 9x - 54 = 3(x^2 - 3x - 18) = 3(x - 6)(x + 3)$
3. $9x^2 - 9x - 18 = 9(x^2 - x - 2) = 9(x - 2)(x + 1)$
4. $15x^2 - 60 = 15(x^2 - 4) = 15(x - 2)(x + 2)$
5. $6x^2 + 24x + 24 = 6(x^2 + 4x + 4) = 6(x + 2)(x + 2) = 6(x + 2)^2$

The coefficient of the x^2 term will not always factor away. In order to factor quadratics such as $4x^2 + 8x + 3$ you will need to try all combinations of factors of 4 and of 3: $(4x + \_\_)(x + \_\_)$ and $(2x + \_\_)(2x + \_\_)$. The blanks will be filled in with the factors of 3. You will need to check all of the possibilities: $(4x + 1)(x + 3)$, $(4x + 3)(x + 1)$, and $(2x + 1)(2x + 3)$.

Example

$4x^2 - 4x - 15$

The possibilities to check are

(a) $(4x+15)(x-1)$ (b) $(4x-15)(x+1)$

(c) $(4x-1)(x+15)$ (d) $(4x+1)(x-15)$

(e) $(4x+5)(x-3)$ (f) $(4x-5)(x+3)$

(g) $(4x+3)(x-5)$ (h) $(4x-3)(x+5)$

(i) $(2x+15)(2x-1)$ (j) $(2x-15)(2x+1)$

(k) $(2x+5)(2x-3)$ (l) $(2x-5)(2x+3)$

We have chosen these combinations to force the first and last terms of the quadratic to be $4x^2$ and -15, respectively, we only need to check the combination that will give a middle term of $-4x$ (if there is one).

(a) $-4x+15x=11x$ (b) $4x-15x=-11x$

(c) $60x-x=59x$ (d) $-60x+x=-59x$

(e) $-12x+5x=-7x$ (f) $12x-5x=7x$

(g) $-20x+3x=-17x$ (h) $20x-3x=17x$

(i) $-2x+30x=28x$ (j) $2x-30x=-28x$

(k) $-6x+10x=4x$ (l) $6x-10x=-4x$

Combination (l) is the correct factorization:

$$4x^2-4x-15=(2x-5)(2x+3).$$

You can see that when the constant term and x^2's coefficient have many factors, this list of factorizations to check can grow rather long. Fortunately there is a way around this problem as we shall see in a later chapter.

Practice

1. $6x^2+25x-9=$
2. $18x^2+21x+5=$
3. $8x^2-35x+12=$

4. $25x^2 + 25x - 14 =$

5. $4x^2 - 9 =$

6. $4x^2 + 20x + 25 =$

7. $12x^2 + 32x - 35 =$

Solutions

1. $6x^2 + 25x - 9 = (2x + 9)(3x - 1)$

2. $18x^2 + 21x + 5 = (3x + 1)(6x + 5)$

3. $8x^2 - 35x + 12 = (8x - 3)(x - 4)$

4. $25x^2 + 25x - 14 = (5x - 2)(5x + 7)$

5. $4x^2 - 9 = (2x - 3)(2x + 3)$

6. $4x^2 + 20x + 25 = (2x + 5)(2x + 5) = (2x + 5)^2$

7. $12x^2 + 32x - 35 = (6x - 5)(2x + 7)$

Quadratic Type Expressions

An expression with three terms where the power of the first term is twice that of the second and the third term is a constant is called a *quadratic type* expression. They factor in the same way as quadratic polynomials. The power on x in the factorization will be the power on x in the middle term. To see the effect of changing the exponents, let us look at $x^2 - 2x - 3 = (x - 3)(x + 1)$.

$$x^4 - 2x^2 - 3 = (x^2 - 3)(x^2 + 1)$$

$$x^6 - 2x^3 - 3 = (x^3 - 3)(x^3 + 1)$$

$$x^{10} - 2x^5 - 3 = (x^5 - 3)(x^5 + 1)$$

$$x^{-4} - 2x^{-2} - 3 = (x^{-2} - 3)(x^{-2} + 1)$$

$$x^{2/3} - 2x^{1/3} - 3 = (x^{1/3} - 3)(x^{1/3} + 1)$$

$$x^{1} - 2x^{1/2} - 3 = (x^{1/2} - 3)(x^{1/2} + 1)$$

Examples

$$4x^6 + 20x^3 + 21 = (2x^3 + 3)(2x^3 + 7)$$

$$x^{2/3} - 5x^{1/3} + 6 = (x^{1/3} - 2)(x^{1/3} - 3)$$

$$x^4 + x^2 - 2 = (x^2 + 2)(x^2 - 1) = (x^2 + 2)(x - 1)(x + 1)$$

$$x - 2\sqrt{x} - 8 = x^1 - 2x^{1/2} - 8 = (x^{1/2} - 4)(x^{1/2} + 2)$$
$$= (\sqrt{x} - 4)(\sqrt{x} + 2)$$

$$\sqrt{x} - 2\sqrt[4]{x} - 15 = x^{1/2} - 2x^{1/4} - 15 = (x^{1/4} - 5)(x^{1/4} + 3)$$
$$= (\sqrt[4]{x} - 5)(\sqrt[4]{x} + 3)$$

Practice

1. $x^4 - 3x^2 + 2 =$
2. $x^{10} - 3x^5 + 2 =$
3. $x^{2/5} - 3x^{1/5} + 2 =$
4. $x^{-6} - 3x^{-3} + 2 =$
5. $x^{1/2} - 3x^{1/4} + 2 =$
6. $x^4 + 10x^2 + 9 =$
7. $x^6 - 4x^3 - 21 =$
8. $4x^6 + 4x^3 - 35 =$

9. $10x^{10} + 23x^5 + 6 =$

10. $9x^4 - 6x^2 + 1 =$

11. $x^{2/7} - 3x^{1/7} - 18 =$

12. $6x^{2/3} - 7x^{1/3} - 3 =$

13. $x^{1/3} + 11x^{1/6} + 10 =$

14. $15x^{1/2} - 8x^{1/4} + 1 =$

15. $14x - \sqrt{x} - 3 =$

16. $25x^6 + 20x^3 + 4 =$

17. $x + 6\sqrt{x} + 9 =$

Solutions

1. $x^4 - 3x^2 + 2 = (x^2 - 2)(x^2 - 1)$

2. $x^{10} - 3x^5 + 2 = (x^5 - 2)(x^5 - 1)$

3. $x^{2/5} - 3x^{1/5} + 2 = (x^{1/5} - 2)(x^{1/5} - 1)$

4. $x^{-6} - 3x^{-3} + 2 = (x^{-3} - 2)(x^{-3} - 1)$

5. $x^{1/2} - 3x^{1/4} + 2 = (x^{1/4} - 2)(x^{1/4} - 1)$

6. $x^4 + 10x^2 + 9 = (x^2 + 9)(x^2 + 1)$

7. $x^6 - 4x^3 - 21 = (x^3 - 7)(x^3 + 3)$

8. $4x^6 + 4x^3 - 35 = (2x^3 + 7)(2x^3 - 5)$

9. $10x^{10} + 23x^5 + 6 = (10x^5 + 3)(x^5 + 2)$

10. $9x^4 - 6x^2 + 1 = (3x^2 - 1)(3x^2 - 1) = (3x^2 - 1)^2$

11. $x^{2/7} - 3x^{1/7} - 18 = (x^{1/7} - 6)(x^{1/7} + 3)$

12. $6x^{2/3} - 7x^{1/3} - 3 = (2x^{1/3} - 3)(3x^{1/3} + 1)$

13. $x^{1/3} + 11x^{1/6} + 10 = (x^{1/6} + 10)(x^{1/6} + 1)$

14. $15x^{1/2} - 8x^{1/4} + 1 = (3x^{1/4} - 1)(5x^{1/4} - 1)$

15. $14x - \sqrt{x} - 3 = 14x^1 - x^{1/2} - 3 = (2x^{1/2} - 1)(7x^{1/2} + 3)$
$= (2\sqrt{x} - 1)(7\sqrt{x} + 3)$

16. $25x^6 + 20x^3 + 4 = (5x^3 + 2)(5x^3 + 2) = (5x^3 + 2)^2$

17. $x + 6\sqrt{x} + 9 = x^1 + 6x^{1/2} + 9 = (x^{1/2} + 3)(x^{1/2} + 3)$
$= (x^{1/2} + 3)^2 = (\sqrt{x} + 3)^2$

Factoring To Reduce Fractions

To reduce a fraction to its lowest terms, factor the numerator and denominator. Cancel any like factors.

Examples

$$\frac{x^2 - 1}{x + 1} = \frac{(x + 1)(x - 1)}{x + 1} = \frac{x - 1}{1} = x - 1$$

$$\frac{y - x}{x^2 - y^2} = \frac{y - x}{(x - y)(x + y)} = \frac{-(x - y)}{(x - y)(x + y)} = \frac{-1}{x + y}$$

$$\frac{x^2 - 5x + 6}{x^2 - 2x - 3} = \frac{(x - 3)(x - 2)}{(x - 3)(x + 1)} = \frac{x - 2}{x + 1}$$

$$\frac{3x^3 - 3x^2 - 6x}{6x^3 - 12x^2} = \frac{3x(x^2 - x - 2)}{6x^2(x - 2)} = \frac{3x(x + 1)(x - 2)}{6x^2(x - 2)} = \frac{x + 1}{2x}$$

$$\frac{x^2 + 10x + 25}{x^2 + 6x + 5} = \frac{(x + 5)(x + 5)}{(x + 5)(x + 1)} = \frac{x + 5}{x + 1}$$

Practice

1. $\dfrac{16x^3 - 24x^2}{8x^4 - 12x^3} =$

2. $\dfrac{x^2 + 2x - 8}{x^2 + 7x + 12} =$

3. $\dfrac{x^2 - 7x + 6}{x^2 + 4x - 5} =$

4. $\dfrac{2x^2 - 5x - 12}{2x^2 - x - 6} =$

5. $\dfrac{3x^2 - 7x + 2}{6x^2 + x - 1} =$

6. $\dfrac{2x^3 + 6x^2 + 4x}{3x^3 + 3x^2 - 36x} =$

7. $\dfrac{4x^3y + 28x^2y + 40xy}{6x^3y^2 - 6x^2y^2 - 36xy^2} =$

8. $\dfrac{x - 4}{16 - x^2} =$

9. $\dfrac{2x^2 - 5x + 2}{1 - 2x} =$

10. $\dfrac{x^2 - y^2}{x^4 - y^4} =$

Solutions

1. $\dfrac{16x^3 - 24x^2}{8x^4 - 12x^3} = \dfrac{8x^2(2x - 3)}{4x^3(2x - 3)} = \dfrac{2}{x}$

2. $\dfrac{x^2 + 2x - 8}{x^2 + 7x + 12} = \dfrac{(x + 4)(x - 2)}{(x + 4)(x + 3)} = \dfrac{x - 2}{x + 3}$

3. $$\frac{x^2-7x+6}{x^2+4x-5}=\frac{(x-1)(x-6)}{(x-1)(x+5)}=\frac{x-6}{x+5}$$

4. $$\frac{2x^2-5x-12}{2x^2-x-6}=\frac{(2x+3)(x-4)}{(2x+3)(x-2)}=\frac{x-4}{x-2}$$

5. $$\frac{3x^2-7x+2}{6x^2+x-1}=\frac{(3x-1)(x-2)}{(3x-1)(2x+1)}=\frac{x-2}{2x+1}$$

6. $$\frac{2x^3+6x^2+4x}{3x^3+3x^2-36x}=\frac{2x(x^2+3x+2)}{3x(x^2+x-12)}$$
$$=\frac{2x(x+1)(x+2)}{3x(x+4)(x-3)}=\frac{2(x+1)(x+2)}{3(x+4)(x-3)}$$

7. $$\frac{4x^3y+28x^2y+40xy}{6x^3y^2-6x^2y^2-36xy^2}=\frac{4xy(x^2+7x+10)}{6xy^2(x^2-x-6)}$$
$$=\frac{4xy(x+2)(x+5)}{6xy^2(x+2)(x-3)}=\frac{2(x+5)}{3y(x-3)}$$

8. $$\frac{x-4}{16-x^2}=\frac{x-4}{(4-x)(4+x)}=\frac{-(4-x)}{(4-x)(4+x)}=\frac{-1}{4+x}$$

9. $$\frac{2x^2-5x+2}{1-2x}=\frac{(2x-1)(x-2)}{1-2x}=\frac{(2x-1)(x-2)}{-(2x-1)}=\frac{x-2}{-1}$$
$$=-(x-2)$$

10. $$\frac{x^2-y^2}{x^4-y^4}=\frac{x^2-y^2}{(x^2-y^2)(x^2+y^2)}=\frac{1}{x^2+y^2}$$

Before adding or subtracting fractions factor the denominator. Once the denominator is factored you can determine the LCD.

Examples

$$\frac{4}{x^2-3x-4}+\frac{2x}{x^2-1}=\frac{4}{(x-4)(x+1)}+\frac{2x}{(x-1)(x+1)}$$

From the first fraction we see that the LCD needs $x-4$ and $x+1$ as factors. From the second fraction we see that the LCD needs $x-1$ and $x+1$, but $x+1$ has been accounted for by the first fraction. The LCD is $(x-4)(x-1)(x+1)$.

$$\frac{7x+5}{2x^2-6x-36}-\frac{10x-1}{x^2+x-6}=\frac{7x+5}{2(x+3)(x-6)}-\frac{10x-1}{(x+3)(x-2)}$$

LCD $= 2(x+3)(x-6)(x-2)$

$$\frac{x-2}{x^2+6x+5}+\frac{1}{3x^2+18x+15}=\frac{x-2}{(x+5)(x+1)}+\frac{1}{3(x+5)(x+1)}$$

LCD $= 3(x+5)(x+1)$

$$\frac{1}{x-1}+\frac{3}{1-x}=\frac{1}{x-1}+\frac{3}{-(x-1)}=\frac{1}{x-1}+\frac{-3}{x-1}$$

LCD $= x-1$

$$4-\frac{2x+9}{x-5}=\frac{4}{1}-\frac{2x+9}{x-5}$$

LCD $= x-5$

$$\frac{3x}{x^2+8x+16}+\frac{2}{x^2+6x+8}=\frac{3x}{(x+4)(x+4)}+\frac{2}{(x+4)(x+2)}$$

LCD $= (x+4)(x+4)(x+2) = (x+4)^2(x+2)$

Practice

Find the LCD.

1. $\dfrac{x-10}{x^2+8x+7}+\dfrac{5}{2x^2-2}$

2. $\dfrac{3x+22}{x^2-5x-24}-\dfrac{x+14}{x^2+x-6}$

3. $\dfrac{7}{x-3}+\dfrac{1}{3-x}$

4. $\dfrac{x+1}{6x^2+21x-12}+\dfrac{4}{9x^2+27x+18}$

5. $\dfrac{3}{2x^2+4x-48}+\dfrac{7}{6x-24}-\dfrac{1}{4x^2+20x-24}$

6. $\dfrac{2x-4}{x^2-7x+12}-\dfrac{x}{x^2-6x+9}$

7. $\dfrac{6x-7}{x^2-5}+3$

Solutions

1. $\dfrac{x-10}{x^2+8x+7}+\dfrac{5}{2x^2-2}=\dfrac{x-10}{(x+7)(x+1)}+\dfrac{5}{2(x-1)(x+1)}$

LCD $= 2(x+7)(x+1)(x-1)$

2. $\dfrac{3x+22}{x^2-5x-24}-\dfrac{x+14}{x^2+x-6}=\dfrac{3x+22}{(x-8)(x+3)}-\dfrac{x+14}{(x+3)(x-2)}$

LCD $= (x-8)(x+3)(x-2)$

3. $\dfrac{7}{x-3}+\dfrac{1}{3-x}=\dfrac{7}{x-3}+\dfrac{1}{-(x-3)}=\dfrac{7}{x-3}+\dfrac{-1}{x-3}$

LCD $= x-3$

4. $$\dfrac{x+1}{6x^2+21x-12}+\dfrac{4}{9x^2+27x+18}=\dfrac{x+1}{3(2x-1)(x+4)}+\dfrac{4}{9(x+1)(x+2)}$$

LCD $= 9(2x-1)(x+4)(x+1)(x+2)$

5. $$\dfrac{3}{2x^2+4x-48}+\dfrac{7}{6x-24}-\dfrac{1}{4x^2+20x-24}=\dfrac{3}{2(x-4)(x+6)}+\dfrac{7}{6(x-4)}-\dfrac{1}{4(x+6)(x-1)}$$

LCD $= 12(x-4)(x+6)(x-1)$

6. $\dfrac{2x-4}{x^2-7x+12}-\dfrac{x}{x^2-6x+9}=\dfrac{2x-4}{(x-3)(x-4)}-\dfrac{x}{(x-3)(x-3)}$

LCD $= (x-3)(x-3)(x-4)=(x-3)^2(x-4)$

7. $\dfrac{6x-7}{x^2-5}+3=\dfrac{6x-7}{x^2-5}+\dfrac{3}{1}$

LCD $= x^2-5$

Once the LCD is found rewrite each fraction in terms of the LCD—multiply each fraction by the "missing" factors over themselves. Then add or subtract the numerators.

Examples

$$\frac{1}{x^2+2x-3}+\frac{x}{x^2-9}=\frac{1}{(x+3)(x-1)}+\frac{x}{(x-3)(x+3)}$$

LCD $=(x+3)(x-1)(x-3)$

The factor $x-3$ is "missing" in the first denominator so multiply the first fraction by $\frac{x-3}{x-3}$. An $x-1$ is "missing" from the second denominator so multiply the second fraction by $\frac{x-1}{x-1}$.

$$\frac{1}{(x+3)(x-1)}\cdot\frac{x-3}{x-3}+\frac{x}{(x-3)(x+3)}\cdot\frac{x-1}{x-1}=\frac{x-3}{(x+3)(x-1)(x-3)}+\frac{x(x-1)}{(x+3)(x-1)(x-3)}=$$

$$\frac{x-3+x(x-1)}{(x+3)(x-1)(x-3)}=\frac{x-3+x^2-x}{(x+3)(x-1)(x-3)}=\frac{x^2-3}{(x+3)(x-1)(x-3)}$$

$$\begin{aligned}\frac{6x}{x^2+2x+1}-\frac{2}{x^2+4x+3}&=\frac{6x}{(x+1)(x+1)}-\frac{2}{(x+1)(x+3)}\\&=\frac{6x}{(x+1)^2}\cdot\frac{x+3}{x+3}-\frac{2}{(x+1)(x+3)}\cdot\frac{x+1}{x+1}\\&=\frac{6x(x+3)-2(x+1)}{(x+1)^2(x+3)}\\&=\frac{6x^2+18x-2x-2}{(x+1)^2(x+3)}=\frac{6x^2+16x-2}{(x+1)^2(x+3)}\end{aligned}$$

$$6+\frac{1}{2x+5}=\frac{6}{1}+\frac{1}{2x+5}=\frac{6}{1}\cdot\frac{2x+5}{2x+5}+\frac{1}{2x+5}=\frac{6(2x+5)+1}{2x+5}$$
$$=\frac{12x+30+1}{2x+5}=\frac{12x+31}{2x+5}$$

$$\frac{3}{x^2+4x-5}+\frac{x-2}{x+5}=\frac{3}{(x+5)(x-1)}+\frac{x-2}{x+5}$$
$$=\frac{3}{(x+5)(x-1)}+\frac{x-2}{x+5}\cdot\frac{x-1}{x-1}$$
$$=\frac{3+(x-2)(x-1)}{(x+5)(x-1)}=\frac{3+x^2-3x+2}{(x+5)(x-1)}$$
$$=\frac{x^2-3x+5}{(x+5)(x-1)}$$

Practice

1. $\dfrac{1}{x^2+5x+6}+\dfrac{1}{x^2+2x-3}=$

2. $\dfrac{5}{2x^2-5x-12}+\dfrac{2x}{3x-12}=$

3. $\dfrac{1}{x^2+x-20}-\dfrac{2}{x^2+x-12}=$

4. $\dfrac{1}{6x^2+24x-30}+\dfrac{5}{2x^2-2x-60}=$

5. $\dfrac{2}{x^2-4x}+\dfrac{1}{2x^2-32}-\dfrac{3}{2x^2+10x+8}=$

6. $1-\dfrac{2x-3}{x+4}=$

7. $\dfrac{1}{x-3}+\dfrac{1}{x^2-2x-3}=$

Solutions

1. $$\frac{1}{x^2+5x+6}+\frac{1}{x^2+2x-3}=\frac{1}{(x+2)(x+3)}+\frac{1}{(x-1)(x+3)}$$
$$=\frac{1}{(x+2)(x+3)}\cdot\frac{x-1}{x-1}+\frac{1}{(x-1)(x+3)}$$
$$\cdot\frac{x+2}{x+2}=\frac{1(x-1)+1(x+2)}{(x+2)(x+3)(x-1)}$$
$$=\frac{2x+1}{(x+2)(x+3)(x-1)}$$

2. $$\frac{5}{2x^2-5x-12}+\frac{2x}{3x-12}=\frac{5}{(2x+3)(x-4)}+\frac{2x}{3(x-4)}$$
$$=\frac{5}{(2x+3)(x-4)}\cdot\frac{3}{3}+\frac{2x}{3(x-4)}\cdot\frac{2x+3}{2x+3}$$
$$=\frac{5\cdot 3+2x(2x+3)}{3(2x+3)(x-4)}=\frac{15+4x^2+6x}{3(2x+3)(x-4)}$$
$$=\frac{4x^2+6x+15}{3(2x+3)(x-4)}$$

3. $$\frac{1}{x^2+x-20}-\frac{2}{x^2+x-12}=\frac{1}{(x+5)(x-4)}-\frac{2}{(x+4)(x-3)}$$
$$=\frac{1}{(x+5)(x-4)}\cdot\frac{(x+4)(x-3)}{(x+4)(x-3)}$$
$$-\frac{2}{(x+4)(x-3)}\cdot\frac{(x+5)(x-4)}{(x+5)(x-4)}$$
$$=\frac{1(x+4)(x-3)-2(x+5)(x-4)}{(x+5)(x-4)(x+4)(x-3)}$$
$$=\frac{x^2+x-12-2(x^2+x-20)}{(x+5)(x-4)(x+4)(x-3)}$$
$$=\frac{x^2+x-12-2x^2-2x+40}{(x+5)(x-4)(x+4)(x-3)}$$
$$=\frac{-x^2-x+28}{(x+5)(x-4)(x+4)(x-3)}$$

4. $$\frac{1}{6x^2+24x-30}+\frac{5}{2x^2-2x-60}=\frac{1}{6(x-1)(x+5)}+\frac{5}{2(x+5)(x-6)}$$
$$=\frac{1}{6(x-1)(x+5)}\cdot\frac{x-6}{x-6}$$
$$+\frac{5}{2(x+5)(x-6)}\cdot\frac{3(x-1)}{3(x-1)}$$
$$=\frac{1(x-6)+5(3)(x-1)}{6(x-1)(x+5)(x-6)}$$
$$=\frac{x-6+15(x-1)}{6(x-1)(x+5)(x-6)}$$
$$=\frac{x-6+15x-15}{6(x-1)(x+5)(x-6)}$$
$$=\frac{16x-21}{6(x-1)(x+5)(x-6)}$$

5. $$\frac{2}{x^2-4x}+\frac{1}{2x^2-32}-\frac{3}{2x^2+10x+8}$$
$$=\frac{2}{x(x-4)}+\frac{1}{2(x-4)(x+4)}-\frac{3}{2(x+4)(x+1)}$$
$$=\frac{2}{x(x-4)}\cdot\frac{2(x+4)(x+1)}{2(x+4)(x+1)}+\frac{1}{2(x-4)(x+4)}\cdot\frac{x(x+1)}{x(x+1)}$$
$$-\frac{3}{2(x+4)(x+1)}\cdot\frac{x(x-4)}{x(x-4)}$$
$$=\frac{2(2)(x+4)(x+1)+x(x+1)-3x(x-4)}{2x(x-4)(x+4)(x+1)}$$
$$=\frac{4(x^2+5x+4)+x^2+x-3x^2+12x}{2x(x-4)(x+4)(x+1)}$$
$$=\frac{4x^2+20x+16+x^2+x-3x^2+12x}{2x(x-4)(x+4)(x+1)}$$
$$=\frac{2x^2+33x+16}{2x(x-4)(x+4)(x+1)}$$

6. $$1-\frac{2x-3}{x+4}=\frac{1}{1}-\frac{2x-3}{x+4}=\frac{1}{1}\cdot\frac{x+4}{x+4}-\frac{2x-3}{x+4}=\frac{x+4-(2x-3)}{x+4}$$
$$=\frac{x+4-2x+3}{x+4}=\frac{-x+7}{x+4}$$

7. $$\frac{1}{x-3}+\frac{1}{x^2-2x-3}=\frac{1}{x-3}+\frac{1}{(x-3)(x+1)}$$
$$=\frac{1}{x-3}\cdot\frac{x+1}{x+1}+\frac{1}{(x-3)(x+1)}$$
$$=\frac{x+1+1}{(x-3)(x+1)}=\frac{x+2}{(x-3)(x+1)}$$

Chapter Review

1. $10xy^2 + 5x^2 =$

(a) $5(2y^2 + x)$ (b) $5x(2y^2 + 1)$ (c) $5x(2y^2 + x)$
(d) $5x^2(2y^2 + 1)$

2. $30x^2 + 6xy - 6x =$

(a) $6x(5x + y - 1)$ (b) $6x(5x + y)$ (c) $6(5x^2 + y - x)$
(d) $6(5x^2 + y)$

3. $x^2 - x - 30 =$

(a) $(x - 5)(x + 6)$ (b) $(x + 5)(x - 6)$ (c) $(x - 3)(x + 10)$
(d) $(x + 3)(x - 10)$

4. $-2x^2 - 16x - 30 =$

(a) $-2(x + 3)(x + 5)$ (b) $2(x - 3)(x - 5)$
(c) $-2(x - 3)(x - 5)$ (d) $2(x + 3)(x + 5)$

5. $7x^2 - 7 =$

(a) $7(x^2 - 7)$ (b) $7(x - 1)(x - 1)$ (c) $7(x - 1)(x + 1)$
(d) $7(x + 1)(x + 1)$

6. $8x^2 - 2x - 3 =$

(a) $(4x - 1)(2x + 3)$ (b) $(4x + 1)(2x - 3)$
(c) $(4x + 3)(2x - 1)$ (d) $(4x - 3)(2x + 1)$

7. $9x^2 - 3x + 6xy =$

(a) $-3x(-3x + 1 + 2y)$ (b) $-3x(-3x - 1 + 2y)$
(c) $-3x(-3x + 1 - 2y)$ (d) $-3x(-3x - 1 - 2y)$

8. $x^2 - \frac{1}{9} =$

(a) $\left(x - \frac{1}{3}\right)^2$ (b) $\left(x + \frac{1}{3}\right)^2$ (c) $\left(x - \frac{1}{6}\right)\left(x + \frac{1}{3}\right)$
(d) $\left(x - \frac{1}{3}\right)\left(x + \frac{1}{3}\right)$

9. Completely factor $16x^4 - 81$.

(a) $(2x - 3)(2x + 3)(4x^2 + 9)$ (b) $(4x^2 - 9)(4x^2 + 9)$
(c) $(2x - 3)(2x + 3)(2x - 3)$ (d) $(4x^2 - 9)(2x - 3)(2x + 3)$

10. $\frac{8}{4x^2 + 12y} =$

(a) $\frac{2}{x^2 + 12}$ (b) $\frac{2}{4x^2 + 3y}$ (c) $\frac{2}{x^2 + 3y}$ (d) $\frac{2}{x^2 + 12y}$

11. $\frac{15}{45xy^2 + 30x + 15} =$

(a) $\frac{1}{3xy^2 + 30x + 15}$ (b) $\frac{1}{3xy^2 + 2x + 1}$
(c) $\frac{1}{3x^2y + 2x + 15}$ (d) $\frac{1}{45xy^2 + 30x + 1}$

12. $\frac{x^2 - 2x - 3}{3x^2 + x - 2} =$

(a) $\frac{x - 3}{3x - 2}$ (b) $\frac{-2x - 3}{3x - 2}$ (c) $\frac{-2x - 2}{x + 1}$ (d) $\frac{x - 3}{3x + 2}$

13. $\frac{x^2 - 1}{x + 1} =$

(a) $\frac{x - 1}{2}$ (b) $x - 1$ (c) $x + 1$ (d) $\frac{x + 1}{2}$

14. $\dfrac{x-3}{3-x} =$

(a) $\dfrac{x-3}{x+3}$ (b) $\dfrac{-x-3}{3-x}$ (c) 1 (d) -1

15. $\dfrac{10x-2y}{y-5x} =$

(a) 2 (b) -2 (c) $\dfrac{10x-2y}{-5x-y}$ (d) $\dfrac{-10x-2y}{5x-y}$

16. $\dfrac{16x^2+4x-4}{2} =$

(a) $8x^2+4x-4$ (b) $8x^2+2x-4$

(c) $8x^2+2x-2$ (d) $8x^2+4x-2$

17. $\dfrac{5}{x^2-x-2} + \dfrac{3}{x^2+4x+3} =$

(a) $\dfrac{8x+21}{(x-1)(x+2)(x+3)}$ (b) $\dfrac{8x-3}{(x+1)(x-2)(x+3)}$

(c) $\dfrac{8x+6}{(x-1)(x+2)(x+3)}$ (d) $\dfrac{8x+9}{(x+1)(x-2)(x+3)}$

18. $\dfrac{7}{x^2+3x-18} - \dfrac{1}{x^2+x-12} =$

(a) $\dfrac{6x+22}{(x+6)(x-3)(x+4)}$ (b) $\dfrac{6x+34}{(x-6)(x+3)(x+4)}$

(c) $\dfrac{6x+34}{(x+6)(x-3)(x+4)}$ (d) $\dfrac{6x+22}{(x-6)(x+3)(x+4)}$

19. $\dfrac{8}{x^2+4x-21} + \dfrac{3}{x+7} =$

(a) $\dfrac{11}{(x+7)(x-3)}$ (b) $\dfrac{3x+2}{(x+7)(x-3)}$ (c) $\dfrac{3x-1}{(x+7)(x-3)}$

(d) $\dfrac{3x+5}{(x+7)(x-3)}$

20. $(3x-4)(2x+3) =$

(a) $15x-12$ (b) $5x-1$ (c) $6x^2-x-12$

(d) $6x^2+x-12$

21. $-2x(4x - 3y - 5) =$

(a) $-8x^2 - 6xy - 10x$ (b) $-8x^2 + 6xy + 10x$

(c) $-8x^2 + 6xy - 10x$ (d) $-8x^2 - 6xy + 10x$

22. $7(x - 5)(x + 3) =$

(a) $49x^2 - 98x - 735$ (b) $49x^2 + 98x - 735$

(c) $7x^2 + 14x - 105$ (d) $7x^2 - 14x - 105$

23. $4xy^2 - 3x + x^2y - (5x^2y - 6x - 2xy^2) =$

(a) $6xy^2 + 3x - 4x^2y$ (b) $-xy^2 + 3x + 3x^2y$

(c) $-xy^2 - x^2y - 9x$ (d) $2xy^2 - 4x^2y - 9x$

24. $x^3 - x^2 - 2x =$

(a) $x(x - 2)(x + 1)$ (b) $x(x + 2)(x - 1)$

(c) $-x(x - 2)(x + 1)$ (d) $-x(x + 2)(x - 1)$

25. $x^{2/3} - x^{1/3} - 2 =$

(a) $(x^{2/3} - 2)(x^{2/3} + 1)$ (b) $(x^{1/3} - 2)(x^{1/3} + 1)$

(c) $(x^{1/6} - 2)(x^{1/6} + 1)$ (d) $(x^{2/3} + 2)(x^{2/3} - 1)$

26. $2xy + 10y - 3x - 15 =$

(a) $(2y + 3)(x + 5)$ (b) $(2y + 3)(x - 5)$ (c) $(2y - 3)(x + 5)$

(d) Cannot be factored

Solutions

1.	(c)	**2.**	(a)	**3.**	(b)	**4.**	(a)
5.	(c)	**6.**	(d)	**7.**	(c)	**8.**	(d)
9.	(a)	**10.**	(c)	**11.**	(b)	**12.**	(a)
13.	(b)	**14.**	(d)	**15.**	(b)	**16.**	(c)
17.	(d)	**18.**	(a)	**19.**	(c)	**20.**	(d)
21.	(b)	**22.**	(d)	**23.**	(a)	**24.**	(a)
25.	(b)	**26.**	(c)				

Linear Equations

Now we can use the tools we have developed to solve equations. Up to now, we have rewritten expressions and added fractions. This chapter is mostly concerned with linear equations. In a linear equation, the variables are raised to the first power—there are no variables in denominators, no variables to any power (other than one), and no variables under root signs.

In solving for linear equations, there will be an unknown, usually only one but possibly several. What is meant by "solve for x" is to isolate x on one side of the equation and to move everything else on the other side. Usually, although not always, the last line is the sentence

"$x =$ (number)"

where the number satisfies the original equation. That is, when the number is substituted for x, the equation is true.

In the equation $3x + 7 = 1$; $x = -2$ is the solution because $3(-2) + 7 = 1$ is a true statement. For any other number, the statement would be false. For instance, if we were to say that $x = 4$, the sentence would be $3(4) + 7 = 1$, which is false.

Not every equation will have a solution. For example, $x + 3 = x + 10$ has no solution. Why not? There is no number that can be added to three and be the same quantity as when it is added to 10. If you were to try to solve for x, you would end up with the false statement $3 = 10$.

In order to solve equations and to verify solutions, you must know the order of operations. For example, in the formula

$$s = \sqrt{\frac{(x-y)^2 + (z-y)^2}{n-1}}$$

what is done first? Second? Third?

A pneumonic for remembering operation order is "**P**lease **e**xcuse **m**y **d**ear **A**unt **S**ally."

P—parentheses first
E—exponents (and roots) second
M—multiplication third
D—division third (multiplication and division should be done together, working from left to right)
A—addition fourth
S—subtraction fourth (addition and subtraction should be done together, working from left to right)

When working with fractions, think of numerators and denominators as being in parentheses.

Examples

$$[6(3^2) - 8 + 2]^{3/2} = [6(9) - 8 + 2]^{3/2} = (54 - 8 + 2)^{3/2} = 48^{3/2}$$
$$= \sqrt{48^3} = 48\sqrt{48} = 48(4)\sqrt{3} = 192\sqrt{3}$$

$$3^2 - 2(4) = 9 - 2(4) = 9 - 8 = 1$$

$$2(3+1)^2 = 2(4)^2 = 2(16) = 32$$

$$\frac{5(6-2)}{3(3+1)^2} = \frac{5(4)}{3(4)^2} = \frac{5(4)}{3(16)} = \frac{20}{48} = \frac{5}{12}$$

$$\sqrt{\frac{4(10+6)}{10+3(5)}} = \sqrt{\frac{4(16)}{10+15}} = \sqrt{\frac{64}{25}} = \frac{8}{5}$$

Practice

1. $\dfrac{4(3^2) - 6(2)}{6 + 2^2}$

2. $\sqrt{\dfrac{8^2 - 3^2 - 5(6)}{2(5) + 6}}$

3. $5^2 - [2(3) + 1]$

4. $\dfrac{8 + 2(3^2 - 4^2)}{15 - 6(3 + 1)}$

5. $\dfrac{3(20 + 4)}{2(9) - 4^2} \cdot \sqrt[3]{\dfrac{5(8) + 41}{12 - 3(16 - 13)}}$

Solutions

1. $\dfrac{4(3^2) - 6(2)}{6 + 2^2} = \dfrac{4(9) - 6(2)}{6 + 4} = \dfrac{36 - 12}{10} = \dfrac{24}{10} = \dfrac{12}{5}$ or $2\frac{2}{5}$

2. $\sqrt{\dfrac{8^2 - 3^2 - 5(6)}{2(5) + 6}} = \sqrt{\dfrac{64 - 9 - 5(6)}{2(5) + 6}} = \sqrt{\dfrac{64 - 9 - 30}{10 + 6}} = \sqrt{\dfrac{25}{16}} = \dfrac{5}{4}$

3. $5^2 - [2(3) + 1] = 5^2 - (6 + 1) = 5^2 - 7 = 25 - 7 = 18$

4. $\dfrac{8 + 2(3^2 - 4^2)}{15 - 6(3 + 1)} = \dfrac{8 + 2(9 - 16)}{15 - 6(4)} = \dfrac{8 + 2(-7)}{15 - 24} = \dfrac{8 - 14}{-9} = \dfrac{-6}{-9} = \dfrac{2}{3}$

5. $$\begin{aligned}\frac{3(20 + 4)}{2(9) - 4^2} \cdot \sqrt[3]{\frac{5(8) + 41}{12 - 3(16 - 13)}} &= \frac{3(20 + 4)}{2(9) - 16} \cdot \sqrt[3]{\frac{40 + 41}{12 - 3(3)}} \\ &= \frac{3(24)}{18 - 16}\sqrt[3]{\frac{40 + 41}{12 - 9}} = \frac{72}{2}\sqrt[3]{\frac{81}{3}} \\ &= 36\sqrt[3]{27} = 36(3) = 108\end{aligned}$$

To solve equations for the unknown, use inverse operations to isolate the variable. These inverse operations "undo" what has been done to the variable. That is, inverse operations are used to move quantities across the equal sign. For instance, in the equation $5x = 10$, x is multiplied by 5, so to move 5 across the equal sign, you need to "unmultiply" the 5. That is, divide both sides of the equation by 5 (equivalently, multiply each side of the equation by $\frac{1}{5}$). In the equation $5 + x = 10$, to move 5 across the equal sign, you must "unadd" 5. That is, subtract 5 from both sides of the equation (equivalently, add -5 to both sides of the equation).

In short, what is added must be subtracted; what is subtracted must be added; what is multiplied must be divided; and what is divided must be multiplied. There are other operation pairs (an operation and its inverse); some will be discussed later.

In much of this book, when the coefficient of x (the number multiplying x) is an integer, both sides of the equation will be divided by that integer. And when the coefficient is a fraction, both sides of the equation will be multiplied by the reciprocal of that fraction.

Examples

$$5x = 2$$ Divide both sides by 5 or multiply both sides by $\frac{1}{5}$

$$\frac{5x}{5} = \frac{2}{5}$$

$$x = \frac{2}{5}$$

$$\frac{2}{3}x = 6$$

$$\frac{3}{2} \cdot \frac{2}{3}x = \frac{3}{2} \cdot 6$$

$$x = 9$$

$$-3x = 18$$

$$\frac{-3}{-3}x = \frac{18}{-3}$$

$$x = -6$$

$$-x = 2$$

$$\frac{-x}{-1} = \frac{2}{-1}$$ (Remember that $-x$ can be written $-1x$; -1 is its own reciprocal.)

$$\frac{-1}{4}x = 7$$

$$-4 \cdot \frac{-1}{4}x = -4(7)$$ (Remember, reciprocals have the same sign.)

$$x = -28$$

Practice

1. $4x = 36$
2. $-2x = -26$
3. $\frac{3}{4}x = 24$
4. $\frac{-1}{3}x = 5$

Solutions

1. $4x = 36$

$$\frac{4}{4}x = \frac{36}{4}$$
$$x = 9$$

2. $-2x = -26$

$$\frac{-2}{-2}x = \frac{-26}{-2}$$
$$x = 13$$

3. $\frac{3}{4}x = 24$

$$\frac{4}{3} \cdot \frac{3}{4}x = \frac{4}{3} \cdot 24$$
$$x = 32$$

4. $\frac{-1}{3}x = 5$

$$(-3)\frac{-1}{3}x = (-3)5$$
$$x = -15$$

Some equations can be solved in a number of ways. However, the general method in this book will be the same:

1. Simplify both sides of the equation.
2. Collect all terms with variables in them on one side of the equation and all nonvariable terms on the other (this is done by adding/subtracting terms).
3. Factor out the variable.
4. Divide both sides of the equation by the variable's coefficient (this is what has been factored out in step 3).

Of course, you might need only one or two of these steps. In the previous examples and practice problems, only step 4 was used.

In the following examples, the number of the step used will be in parentheses. Although it will not normally be done here, it is a good idea to verify your solution in the original equation.

Examples

$$
\begin{aligned}
2(x-3)+7 &= 5x-8 \\
2x-6+7 &= 5x-8 \quad (1) \\
2x+1 &= 5x-8 \quad (1) \\
-2x \qquad & \quad -2x \\
1 &= 3x-8 \quad (2) \\
+8 \qquad & \quad +8 \\
9 &= 3x \qquad (2) \\
\frac{9}{3} &= \frac{3}{3}x \qquad (4) \\
3 &= x
\end{aligned}
$$

$$
\begin{aligned}
3(2x-1)-2 &= 10-(x+1) \\
6x-3-2 &= 10-x-1 \quad (1) \\
6x-5 &= 9-x \qquad (1) \\
+x \qquad & \quad +x \qquad (2) \\
7x-5 &= 9 \\
+5 \qquad & \quad +5 \quad (2) \\
7x &= 14 \\
\frac{7}{7}x &= \frac{14}{7} \quad (4) \\
x &= 2
\end{aligned}
$$

Practice

1. $2x + 16 = 10$
2. $\frac{3}{2}x - 1 = 5$
3. $6(8 - 2x) + 25 = 5(2 - 3x)$
4. $-4(8 - 3x) = 2x + 8$
5. $7(2x - 3) - 4(x + 5) = 8(x - 1) + 3$
6. $\frac{1}{2}(6x - 8) + 3(x + 2) = 4(2x - 1)$
7. $5x + 7 = 6(x - 2) - 4(2x - 3)$
8. $3(2x - 5) - 2(4x + 1) = -5(x + 3) - 2$
9. $-4(3x - 2) - (-x + 6) = -5x + 8$

Solutions

1. $$\begin{aligned} 2x + 16 &= 10 \\ -16 \quad & -16 \\ 2x &= -6 \\ \frac{2}{2}x &= \frac{-6}{2} \\ x &= -3 \end{aligned}$$

2. $$\begin{aligned} \frac{3}{2}x - 1 &= 5 \\ +1 \quad & +1 \\ \frac{3}{2}x &= 6 \\ \frac{2}{3} \cdot \frac{3}{2}x &= \frac{2}{3} \cdot 6 \\ x &= 4 \end{aligned}$$

3. $6(8-2x)+25=5(2-3x)$

$$48-12x+25=10-15x$$

$$73-12x=10-15x$$

$$-73 \qquad\qquad -73$$

$$-12x=-63-15x$$

$$+15x \qquad\qquad +15x$$

$$3x=-63$$

$$\frac{3}{3}x=\frac{-63}{3}$$

$$x=-21$$

4. $-4(8-3x)=2x+8$

$$-32+12x=2x+8$$

$$-2x \quad -2x$$

$$-32+10x=8$$

$$+32 \qquad\qquad +32$$

$$10x=40$$

$$\frac{10}{10}x=\frac{40}{10}$$

$$x=4$$

5. $7(2x-3)-4(x+5)=8(x-1)+3$

$$14x-21-4x-20=8x-8+3$$

$$10x-41=8x-5$$

$$-8x \qquad\qquad -8x$$

$$2x-41=-5$$

$$+41 \quad +41$$

$$2x=36$$

$$\frac{2}{2}x=\frac{36}{2}$$

$$x=18$$

6. $$\begin{aligned} \frac{1}{2}(6x-8)+3(x+2) &= 4(2x-1) \\ 3x-4+3x+6 &= 8x-4 \\ 6x+2 &= 8x-4 \\ -6x \qquad &\quad -6x \\ 2 &= 2x-4 \\ +4 \qquad &\quad +4 \\ 6 &= 2x \\ \frac{6}{2} &= \frac{2}{2}x \\ 3 &= x \end{aligned}$$

7. $$\begin{aligned} 5x+7 &= 6(x-2)-4(2x-3) \\ 5x+7 &= 6x-12-8x+12 \\ 5x+7 &= -2x \\ +2x \qquad &\quad +2x \\ 7x+7 &= 0 \\ -7 \qquad &\quad -7 \\ 7x &= -7 \\ \frac{7}{7}x &= \frac{-7}{7} \\ x &= -1 \end{aligned}$$

8. $$\begin{aligned} 3(2x-5)-2(4x+1) &= -5(x+3)-2 \\ 6x-15-8x-2 &= -5x-15-2 \\ -2x-17 &= -5x-17 \\ +5x \qquad &\quad +5x \\ 3x-17 &= -17 \\ +17 \qquad &\quad +17 \\ 3x &= 0 \\ \frac{3}{3}x &= \frac{0}{3} \\ x &= 0 \end{aligned}$$

9. $$-4(3x-2)-(-x+6)=-5x+8$$
$$-12x+8+x-6=-5x+8$$
$$-11x+2=-5x+8$$
$$+5x \qquad +5x$$
$$-6x+2=8$$
$$-2 \quad -2$$
$$-6x=6$$
$$\frac{-6}{-6}x=\frac{6}{-6}$$
$$x=-1$$

When the equation you are given has fractions and you prefer not to work with fractions, you can clear the fractions in the first step. Of course, the solution might be a fraction, but that fraction will not occur until the last step. Find the LCD of all fractions and multiply *both* sides of the equation by this number. Then, distribute this quantity on each side of the equation.

Examples

$$\frac{4}{5}x+1=-4$$
$$5\left[\frac{4}{5}x+1\right]=5(-4)$$
$$5\cdot\frac{4}{5}x+5(1)=-20$$
$$4x+5=-20$$
$$-5 \quad -5$$
$$4x=-25$$
$$\frac{4}{4}x=\frac{-25}{4}$$
$$x=\frac{-25}{4}$$

$$\frac{3}{2}x-\frac{1}{6}x+\frac{2}{9}=\frac{1}{3}$$ The LCD is 18.
$$18\left[\frac{3}{2}x-\frac{1}{6}x+\frac{2}{9}\right]=18\cdot\frac{1}{3}$$

$$18 \cdot \frac{3}{2}x - 18 \cdot \frac{1}{6}x + 18 \cdot \frac{2}{9} = 6$$
$$27x - 3x + 4 = 6$$
$$24x + 4 = 6$$
$$-4 \quad -4$$
$$24x = 2$$
$$\frac{24}{24}x = \frac{2}{24}$$
$$x = \frac{2}{24}$$
$$x = \frac{1}{12}$$

A common mistake is to fail to distribute the LCD. Another is to multiply only one side of the equation by the LCD.

In the first example, $\frac{4}{5}x + 1 = -4$, one common mistake is to multiply both sides by 5 but not to distribute 5 on the left-hand side.

$$5\left[\frac{4}{5}x + 1\right] = 5(-4)$$
$$4x + 1 = -20 \quad \text{(incorrect)}$$

Another common mistake is not to multiply both sides of the equation by the LCD.

$$5\left[\frac{4}{5}x + 1\right] = -4$$
$$4x + 5 = -4 \quad \text{(incorrect)}$$

In each case, the last line is *not* equivalent to the first line—that is, the solution to the last equation is not the solution to the first equation.

In some cases, you will need to use the associative property of multiplication with the LCD instead of the distributive property.

Example

$$\frac{1}{3}(x + 4) = \frac{1}{2}(x - 1)$$

The LCD is 6.

$$6\left[\frac{1}{3}(x+4)\right] = 6\left[\frac{1}{2}(x-1)\right]$$

On each side, there are three quantities being multiplied together. On the left, the quantities are 6, $\frac{1}{3}$ and $x+4$. By the associative law of multiplication, the 6 and $\frac{1}{3}$ can be multiplied, then that product is multiplied by $x+4$. Similarly, on the right, first multiply 6 and $\frac{1}{2}$, then multiply that product by $x-1$.

$$\left[6\left(\frac{1}{3}\right)\right](x+4) = \left[6\left(\frac{1}{2}\right)\right](x-1)$$

$$2(x+4) = 3(x-1)$$

$$2x+8 = 3x-3$$

$$-2x \qquad -2x$$

$$8 = x-3$$

$$+3 \qquad +3$$

$$11 = x$$

Practice

Solve for x after clearing the fraction.

1. $\frac{1}{2}x+3 = \frac{3}{5}x-1$

2. $\frac{1}{6}x-\frac{1}{3} = \frac{2}{3}x-\frac{5}{12}$

3. $\frac{1}{5}(2x-4) = \frac{1}{3}(x+2)$

4. $\frac{2}{3}(x-1)-\frac{1}{6}(2x+3) = \frac{1}{8}$

5. $\frac{3}{4}x-\frac{1}{3}x+1 = \frac{4}{5}x-\frac{3}{20}$

Solutions

1. $\frac{1}{2}x+3 = \frac{3}{5}x-1$

   The LCD is 10.

$$10\left(\frac{1}{2}x+3\right)=10\left(\frac{3}{5}x-1\right)$$
$$10\left(\frac{1}{2}x\right)+10(3)=10\left(\frac{3}{5}x\right)-10(1)$$
$$5x+30=6x-10$$
$$-5x \qquad -5x$$
$$30=x-10$$
$$+10 \qquad +10$$
$$40=x$$

2. $\frac{1}{6}x-\frac{1}{3}=\frac{2}{3}x-\frac{5}{12}$

The LCD is 12.

$$12\left(\frac{1}{6}x-\frac{1}{3}\right)=12\left(\frac{2}{3}x-\frac{5}{12}\right)$$
$$12\left(\frac{1}{6}x\right)-12\left(\frac{1}{3}\right)=12\left(\frac{2}{3}x\right)-12\left(\frac{5}{12}\right)$$
$$2x-4=8x-5$$
$$-2x \qquad -2x$$
$$-4=6x-5$$
$$+5 \qquad +5$$
$$1=6x$$
$$\frac{1}{6}=\frac{6}{6}x$$
$$\frac{1}{6}=x$$

3. $\frac{1}{5}(2x-4)=\frac{1}{3}(x+2)$

The LCD is 15.

$$15\left[\frac{1}{5}(2x-4)\right]=15\left[\frac{1}{3}(x+2)\right]$$
$$\left[15\left(\frac{1}{5}\right)\right](2x-4)=\left[15\left(\frac{1}{3}\right)\right](x+2)$$
$$3(2x-4)=5(x+2)$$

Solution 3 (*continued*)

$$\begin{aligned} 6x - 12 &= 5x + 10 \\ -5x \quad & \quad -5x \\ x - 12 &= 10 \\ +12 \quad & \quad +12 \\ x &= 22 \end{aligned}$$

4. $\frac{2}{3}(x-1) - \frac{1}{6}(2x+3) = \frac{1}{8}$

The LCD is 24.

$$24\left[\frac{2}{3}(x-1) - \frac{1}{6}(2x+3)\right] = 24\left(\frac{1}{8}\right)$$

$$\left[24\left(\frac{2}{3}\right)\right](x-1) - \left[24\left(\frac{1}{6}\right)\right](2x+3) = 3$$

$$16(x-1) - 4(2x+3) = 3$$

$$16x - 16 - 8x - 12 = 3$$

$$8x - 28 = 3$$

$$+28 \quad +28$$

$$8x = 31$$

$$x = \frac{31}{8} \quad \text{or} \quad 3\tfrac{7}{8}$$

5. $\frac{3}{4}x - \frac{1}{3}x + 1 = \frac{4}{5}x - \frac{3}{20}$

The LCD is 60.

$$60\left(\frac{3}{4}x - \frac{1}{3}x + 1\right) = 60\left(\frac{4}{5}x - \frac{3}{20}\right)$$

$$60\left(\frac{3}{4}x\right) - 60\left(\frac{1}{3}x\right) + 60(1) = 60\left(\frac{4}{5}x\right) - 60\left(\frac{3}{20}\right)$$

$$45x - 20x + 60 = 48x - 9$$

$$25x + 60 = 48x - 9$$

$$-25x \qquad -25x$$

Solution 5 (*continued*)

$$\begin{aligned} 60 &= 23x - 9 \\ +9 &\qquad\quad +9 \\ 69 &= 23x \\ \frac{69}{23} &= x \\ 3 &= x \end{aligned}$$

Decimals

Because decimal numbers are fractions in disguise, the same trick can be used to "clear the decimal" in equations with decimal numbers. Count the largest number of digits behind each decimal point and multiply both sides of the equation by 10 raised to the power of that number.

Examples

$0.25x + 0.6 = 0.1$

Because there are two digits behind the decimal in 0.25, we need to multiply both sides of the equation by $10^2 = 100$. Remember to distribute the 100 inside the parentheses.

$$\begin{aligned} 100(0.25x + 0.6) &= 100(0.1) \\ 100(0.25x) + 100(0.6) &= 100(0.1) \\ 25x + 60 &= 10 \\ -60 &\quad -60 \\ 25x &= -50 \\ x &= \frac{-50}{25} \\ x &= -2 \end{aligned}$$

$$x - 0.11 = 0.2x + 0.09$$

$$100(x - 0.11) = 100(0.2x + 0.09)$$

$$100x - 100(0.11) = 100(0.2x) + 100(0.09)$$

$$100x - 11 = 20x + 9$$

$$-20x \qquad -20x$$

$$80x - 11 = 9$$

$$+11 \quad +11$$

$$80x = 20$$

$$x = \frac{20}{80}$$

$$x = \frac{1}{4} \text{ or } 0.25$$

(Normally, decimal solutions are given in equations that have decimals in them.)

Practice

Solve for x after clearing the decimal. If your solution is a fraction, convert the fraction to a decimal.

1. $0.3(x - 2) + 0.1 = 0.4$
2. $0.12 - 0.4(x + 1) + x = 0.5x + 2$
3. $0.015x - 0.01 = 0.025x + 0.2$
4. $0.24(2x - 3) + 0.08 = 0.6(x + 8) - 1$
5. $0.01(2x + 3) - 0.003 = 0.11x$

Solutions

1. $0.3(x - 2) + 0.1 = 0.4$

   Multiply both sides by $10^1 = 10$

$$10[0.3(x-2)+0.1] = 10(0.4)$$
$$10(0.3)(x-2)+10(0.1) = 4$$
$$[10(0.3)](x-2)+1 = 4$$
$$3(x-2)+1 = 4$$
$$3x-6+1 = 4$$
$$3x-5 = 4$$
$$+5 \quad +5$$
$$3x = 9$$
$$x = \frac{9}{3}$$
$$x = 3$$

2. $0.12 - 0.4(x+1) + x = 0.5x + 2$

Multiply both sides by $10^2 = 100$.

$$100[0.12 - 0.4(x+1) + x] = 100(0.5x+2)$$
$$100(0.12) - 100[0.4(x+1)] + 100x = 100(0.5x) + 2(100)$$
$$12 - [100(0.4)](x+1) + 100x = 50x + 200$$
$$12 - 40(x+1) + 100x = 50x + 200$$
$$12 - 40x - 40 + 100x = 50x + 200$$
$$60x - 28 = 50x + 200$$
$$-50x \qquad -50x$$
$$10x - 28 = 200$$
$$+28 \quad +28$$
$$10x = 228$$
$$x = \frac{228}{10} = 22.8$$

3. $0.015x - 0.01 = 0.025x + 0.2$

Multiply both sides by $10^3 = 1000$.

$$1000(0.015x - 0.01) = 1000(0.025x + 0.2)$$

$$1000(0.015x) - 1000(0.01) = 1000(0.025x) + 1000(0.2)$$

$$15x - 10 = 25x + 200$$
$$-15x \qquad\qquad -15x$$

$$-10 = 10x + 200$$
$$-200 \qquad\qquad -200$$

$$-210 = 10x$$

$$-\frac{210}{10} = x$$

$$-21 = x$$

4. $0.24(2x - 3) + 0.08 = 0.6(x + 8) - 1$

   Multiply both sides by $10^2 = 100$.

$$100[0.24(2x - 3) + 0.08] = 100[0.6(x + 8) - 1]$$

$$100[0.24(2x - 3)] + 100(0.08) = 100[0.6(x + 8)] - 100(1)$$

$$[100(0.24)](2x - 3) + 8 = [100(0.6)](x + 8) - 100$$

$$24(2x - 3) + 8 = 60(x + 8) - 100$$

$$48x - 72 + 8 = 60x + 480 - 100$$

$$48x - 64 = 60x + 380$$
$$-48x \qquad\qquad -48x$$

$$-64 = 12x + 380$$
$$-380 \qquad\qquad -380$$

$$-444 = 12x$$

$$-\frac{444}{12} = x$$

$$-37 = x$$

5. $0.01(2x + 3) - 0.003 = 0.11x$

   Multiply both sides by $10^3 = 1000$.

$$1000[0.01(2x+3)-0.003] = 1000(0.11x)$$
$$1000[0.01(2x+3)] - 1000(0.003) = 110x$$
$$[1000(0.01)](2x+3) - 3 = 110x$$
$$10(2x+3) - 3 = 110x$$
$$20x + 30 - 3 = 110x$$
$$20x + 27 = 110x$$
$$-20x \qquad -20x$$
$$27 = 90x$$
$$\frac{27}{90} = x$$
$$\frac{3}{10} = x$$
$$0.3 = x$$

Formulas

At times math students are given a formula like $I = Prt$ and asked to solve for one of the variables; that is, to isolate that particular variable on one side of the equation. In $I = Prt$, the equation is solved for "I." The method used above for solving for x works on these, too. Many people are confused by the presence of multiple variables. The trick is to think of the variable for which you are trying to solve as x and all of the other variables as fixed numbers. For instance, if you were asked to solve for r in $I = Prt$, think of how you would solve something of the same form with numbers, say

$100 = (500)(x)(2)$:

$$100 = (500)(x)(2)$$
$$100 = [(500)(2)]x$$
$$\frac{100}{(500)(2)} = x.$$

The steps for solving for r in $I = Prt$ are identical:

$$I = Prt$$

$$I = (Pt)r$$

$$\frac{I}{Pt} = r.$$

All of the formulas used in the following examples and practice problems are formulas used in business, science, and mathematics.

Examples

Solve for q.

$$P = pq - c \quad (P \text{ and } p \text{ are different variables.})$$

$$\begin{array}{rl} +c & +c \end{array}$$

$$P + c = pq$$

$$\frac{P+c}{p} = \frac{pq}{p}$$

$$\frac{P+c}{p} = q$$

Solve for m.

$$y - y_1 = mx - mx_1$$

$$y - y_1 = m(x - x_1)$$

$$\frac{y - y_1}{x - x_1} = m$$

Solve for C.

$$\mathrm{F} = \frac{9}{5}\mathrm{C} + 32$$

$$\begin{array}{rl} -32 & -32 \end{array}$$

$$\mathrm{F} - 32 = \frac{9}{5}\mathrm{C}$$

$$\frac{5}{9}(\mathrm{F} - 32) = \frac{5}{9} \cdot \frac{9}{5}\mathrm{C}$$

$$\frac{5}{9}(\mathrm{F} - 32) = \mathrm{C}$$

Solve for b.

$$A = \frac{1}{2}(a+b)h \qquad (A \text{ and } a \text{ are different variables.})$$

$$A = \left(\frac{1}{2}a + \frac{1}{2}b\right)h \qquad \left(\frac{1}{2} \text{ is distributed}\right)$$

$$A = \frac{1}{2}ah + \frac{1}{2}bh \qquad (h \text{ is distributed})$$

$$A = \frac{1}{2}ah + \left(\frac{1}{2}h\right)b$$

$$-\frac{1}{2}ah \quad -\frac{1}{2}ah$$

$$A - \frac{1}{2}ah = \frac{1}{2}hb$$

$$A - \frac{ah}{2} = \frac{h}{2}b \quad \left(\frac{1}{2}h = \frac{h}{2} \text{ and } \frac{1}{2}ah = \frac{ah}{2}\right)$$

$$\frac{2}{h}\left(A - \frac{ah}{2}\right) = \frac{2}{h}\cdot\frac{h}{2}b$$

$$\frac{2}{h}\left(A - \frac{ah}{2}\right) = b$$

or

$$\frac{2A}{h} - \frac{2}{h}\cdot\frac{ah}{2} = b$$

$$\frac{2A}{h} - a = b$$

or

$$\frac{2A - ah}{h} = b$$

Practice

Solve for indicated variable.

1. $A = \frac{1}{2}bh; h$

2. $C = 2\pi r; r$

3. $V = \frac{\pi r^2 h}{3}; h$

4. $P = 2L + 2W; L$

5. $L = L_0[1 + a(dt)]; a$

6. $S = C + RC; C$

7. $A = P + PRT; R$

8. $H = \frac{kA(t_1 - t_2)}{L}; t_1$

9. $L = a + (n - 1)d; n$

10. $S = \frac{rL - a}{r - 1}; r$

Solutions

1.
$$A = \frac{1}{2}bh; h$$
$$A = \left(\frac{1}{2}b\right)h$$
$$A = \frac{b}{2}h$$
$$\frac{2}{b}A = \frac{2}{b} \cdot \frac{b}{2}h$$
$$\frac{2A}{b} = h$$

2.
$$C = 2\pi r; r$$
$$\frac{C}{2\pi} = \frac{2\pi}{2\pi}r$$
$$\frac{C}{2\pi} = r$$

3.
$$V = \frac{\pi r^2 h}{3}; h$$
$$\frac{3}{\pi r^2}V = \frac{3}{\pi r^2} \cdot \frac{\pi r^2 h}{3}$$
$$\frac{3V}{\pi r^2} = h$$

4.
$$P = 2L + 2W;\ L$$
$$P = 2L + 2W$$
$$-2W \qquad -2W$$
$$P - 2W = 2L$$
$$\frac{P - 2W}{2} = L$$

5.
$$L = L_0[1 + a(dt)];\ a$$
$$L = L_0[1 + a(dt)]$$
$$L = L_0(1) + L_0[a(dt)]$$
$$L = L_0 + (L_0 dt)a$$
$$-L_0 \qquad -L_0$$
$$L - L_0 = (L_0 dt)a$$
$$\frac{L - L_0}{L_0 dt} = \frac{L_0 dt}{L_0 dt} a$$
$$\frac{L - L_0}{L_0 dt} = a$$

6.
$$S = C + RC;\ C$$
$$S = C + RC$$
$$S = C(1 + R) \quad \text{(Factor out } C \text{ since we are solving for } C.)$$
$$\frac{S}{1 + R} = C\frac{1 + R}{1 + R}$$
$$\frac{S}{1 + R} = C$$

7.
$$A = P + PRT;\ R$$
$$A = P + PRT \quad \text{(Do not factor out } P \text{ since we are not solving for } P.)$$
$$-P \qquad -P$$
$$A - P = PRT$$
$$\frac{A - P}{PT} = \frac{PTR}{PT}$$
$$\frac{A - P}{PT} = R$$

8. $$H = \frac{kA(t_1 - t_2)}{L}; t_1$$

$$H = \frac{kA(t_1 - t_2)}{L}$$
$$HL = kA(t_1 - t_2)$$
$$HL = kAt_1 - kAt_2$$
$$+kAt_2 \qquad +kAt_2$$
$$HL + kAt_2 = kAt_1$$
$$\frac{HL + kAt_2}{kA} = \frac{kAt_1}{kA}$$
$$\frac{HL + kAt_2}{kA} = t_1$$

9. $$L = a + (n - 1)d; n$$

$$L = a + (n - 1)d$$
$$L = a + nd - d$$
$$-a \quad -a$$
$$L - a = nd - d$$
$$+d \qquad +d$$
$$L - a + d = nd$$
$$\frac{L - a + d}{d} = \frac{nd}{d}$$
$$\frac{L - a + d}{d} = n$$

10. $$S = \frac{rL - a}{r - 1}; r$$

$$S = \frac{rL - a}{r - 1}$$
$$(r - 1)S = rL - a$$
$$rS - S = rL - a$$
$$-rL \quad -rL$$
$$rS - rL - S = -a$$
$$+S \quad +S$$
$$rS - rL = S - a$$
$$r(S - L) = S - a$$
$$\frac{r(S - L)}{(S - L)} = \frac{S - a}{S - L}$$
$$r = \frac{S - a}{S - L}$$

Equations Leading to Linear Equations

Some equations are almost linear equations; after one or more steps these equations become linear equations. In this section, we will be converting rational expressions (one quantity divided by another quantity) into linear expressions and square root equations into linear equations. The solution(s) to these converted equations might not be the same as the solution(s) to the original equation. After certain operations, you *must* check the solution(s) to the converted equation in the original equation.

To solve a rational equation, clear the fraction. In this book, two approaches will be used First, if the equation is in the form of "fraction = fraction," cross multiply to eliminate the fraction. Second, if there is more than one fraction on one side of the equal sign, the LCD will be determined and each side of the equation will be multiplied by the LCD. These are not the only methods for solving rational equations.

The following is a rational equation in the form of one fraction equals another. We will use the fact that for b and d nonzero, $\frac{a}{b} = \frac{c}{d}$ if and only if $ad = bc$.

This method is called *cross multiplication.*

$$\frac{1}{x-1} = \frac{1}{2}$$

$$2(1) = 1(x-1) \qquad \text{(This is the cross multiplication step.)}$$

$$\begin{aligned} 2 &= x - 1 \\ +1 &\quad\;\; +1 \\ 3 &= x \end{aligned}$$

Check: $\frac{1}{3-1} = \frac{1}{2}$ is a true statement, so $x = 3$ is the solution.

Anytime you multiply (or divide) both sides of the equation by an expression with a variable in it, you *must* check your solution(s) in the original equation. When you cross multiply, you are implicitly multiplying both sides of the equations by the denominators of each fraction, so you must check your solution in this case as well. The reason is that sometimes a solution to the converted equation will cause a zero to be in a denominator of the original equation. Such solutions are called *extraneous solutions.* See what happens in the next example.

$$\frac{1}{x-2} = \frac{3}{x+2} - \frac{6x}{(x-2)(x+2)} \quad \text{The LCD is } (x-2)(x+2).$$

$$(x-2)(x+2)\frac{1}{x-2} = (x-2)(x+2)\left[\frac{3}{x+2} - \frac{6x}{(x-2)(x+2)}\right]$$

Multiply each side by the LCD.

$$x+2 = (x-2)(x+2)\frac{3}{x+2} - (x-2)(x+2)\frac{6x}{(x-2)(x+2)}.$$

Distribute the LCD.

$$\begin{aligned}
x+2 &= 3(x-2) - 6x \\
x+2 &= 3x - 6 - 6x \\
x+2 &= -3x - 6 \\
+3x & \quad\; +3x \\
4x+2 &= -6 \\
-2 & \quad\; -2 \\
4x &= -8 \\
x &= -2
\end{aligned}$$

But $x = -2$ leads to a zero in a denominator of the original equation, so $x = -2$ is not a solution to the original equation. The original equation has no solution.

Have you ever wondered why expressions like $\frac{2}{0}$ are not numbers? Let us see what complications arise when we try to see what "$\frac{2}{0}$" might mean. Say $\frac{2}{0} = x$.

$$\frac{2}{0} = \frac{x}{1}$$

Now cross multiply.

$$2(1) = 0(x)$$

Multiplication by zero *always* yields zero, so the right hand side is zero.

$$2 = 0 \quad \text{No value for } x \text{ can make this equation true.}$$

Or, if you try to "clear the fraction" by multiplying both sides of the equation by a common denominator, you will see that an absurd situation arises here, too.

$$0 \cdot \frac{2}{0} = 0x$$

So, $0 = 0x$, which is true for *any* x. Actually, the expression $\frac{0}{0}$ is not defined.

On some equations, you will want to raise both sides of the equation to a power in order to solve for x. Be careful to raise both sides of the equation to the same power, not simply the side with the root. Raising both sides of an equation to an even power is another operation which can introduce extraneous solutions. To see how this can happen, let us look at the equation $x = 4$. If we square both sides of the equation, we get the equation $x^2 = 16$. This equation has *two* solutions: $x = 4$ and $x = -4$.

Example

$\sqrt{x-1} = 6$

Remember that $(\sqrt{a})^2 = a$ if a is not negative. We will use this fact to eliminate the square root sign. So, to "undo" a square root, first isolate the square root on one side of the equation (in this example, it already is) then square both sides.

$$\begin{aligned} \left(\sqrt{x-1}\right)^2 &= 6^2 \\ x - 1 &= 36 \\ +1 \quad & +1 \\ x &= 37 \end{aligned}$$

Because we squared both sides, we need to make sure $x = 37$ is a solution to the original equation.

$\sqrt{37-1} = 6$

is a true statement, so $x = 37$ is the solution.

Quadratic equations, to be studied in the last chapter, have their variables squared—that is, the only powers on variables are one and two. Some quadratic equations are equivalent to linear equations.

Example

$$\begin{aligned} (6x-5)^2 &= (4x+3)(9x-2) \\ (6x-5)(6x-5) &= (4x+3)(9x-2) \\ 36x^2 - 30x - 30x + 25 &= 36x^2 - 8x + 27x - 6 \\ 36x^2 - 60x + 25 &= 36x^2 + 19x - 6 \end{aligned}$$

Because $36x^2$ is on each side of the equation, they cancel each other, and we are left with

$$-60x + 25 = 19x - 6,$$

an ordinary linear equation.

$$\begin{aligned} -60x + 25 &= 19x - 6 \\ +60x \quad\quad &\quad\quad +60x \\ 25 &= 79x - 6 \\ +6 \quad &\quad\quad +6 \\ 31 &= 79x \\ \frac{31}{79} &= x \end{aligned}$$

Because we neither multiplied (nor divided) both sides by an expression involving a variable nor raised both sides to a power, it is not necessary to check your solution. For accuracy, however, checking solutions is a good habit.

Practice

1. $\dfrac{18 - 5x}{3x + 2} = \dfrac{7}{3}$

2. $\dfrac{6}{5x - 2} = \dfrac{9}{7x + 6}$

3. $(x - 7)^2 - 4 = (x + 1)^2$

4. $\dfrac{6x + 7}{4x - 1} = \dfrac{3x + 8}{2x - 4}$

5. $\dfrac{9x}{3x - 1} = 4 + \dfrac{3}{3x - 1}$

6. $\dfrac{1}{x - 1} - \dfrac{2}{x + 1} = \dfrac{3}{x^2 - 1}$

7. $(2x - 1)^2 - 4x^2 = -4x + 1$

8. $\sqrt{7x + 1} = 13$

9. $\sqrt{x} - 6 = 10$

10. $\sqrt{2x-3}+1=6$

11. $\sqrt{7-2x}=3$

12. $\sqrt{3x+4}=\sqrt{2x+5}$

Solutions

Unless a solution is extraneous, the check step is not printed.

1. $\frac{18-5x}{3x+2}=\frac{7}{3}$ This equation is in the form "Fraction = Fraction," so cross multiply.

$$3(18-5x)=7(3x+2)$$

$$54-15x=21x+14$$

$$+15x \quad +15x$$

$$54=36x+14$$

$$-14 \qquad -14$$

$$40=36x$$

$$\frac{40}{36}=x$$

$$\frac{10}{9}=x$$

2. $\frac{6}{5x-2}=\frac{9}{7x+6}$ This equation is in the form "Fraction = Fraction," so cross multiply.

$$6(7x+6)=9(5x-2)$$

$$42x+36=45x-18$$

$$-42x \qquad -42x$$

$$36=3x-18$$

$$+18 \qquad +18$$

$$54=3x$$

$$\frac{54}{3}=x$$

$$18=x$$

3. $(x-7)^2 - 4 = (x+1)^2$

$$(x-7)(x-7) - 4 = (x+1)(x+1)$$

$$x^2 - 7x - 7x + 49 - 4 = x^2 + x + x + 1$$

$$x^2 - 14x + 45 = x^2 + 2x + 1 \quad x^2\text{'s cancel}$$

$$-14x + 45 = 2x + 1$$

$$-2x \qquad -2x$$

$$-16x + 45 = 1$$

$$-45 \qquad -45$$

$$-16x = -44$$

$$x = \frac{-44}{-16}$$

$$x = \frac{11}{4} \quad \text{or} \quad 2\tfrac{3}{4}$$

4. $\dfrac{6x+7}{4x-1} = \dfrac{3x+8}{2x-4}$ This is in the form "Fraction = Fraction," so cross multiply.

$$(6x+7)(2x-4) = (4x-1)(3x+8)$$

$$12x^2 - 24x + 14x - 28 = 12x^2 + 32x - 3x - 8$$

$$12x^2 - 10x - 28 = 12x^2 + 29x - 8 \qquad 12x^2\text{'s cancel}$$

$$-10x - 28 = 29x - 8$$

$$+10x \qquad +10x$$

$$-28 = 39x - 8$$

$$+8 \qquad +8$$

$$-20 = 39x$$

$$\frac{-20}{39} = x$$

5. $\dfrac{9x}{3x-1} = 4 + \dfrac{3}{3x-1}$ The LCD is $3x-1$.

$$(3x-1)\left[\frac{9x}{3x-1}\right] = (3x-1)\left[4 + \frac{3}{3x-1}\right]$$

Solution 5 (*continued*)

$$9x = (3x - 1)(4) + (3x - 1)\left[\frac{3}{3x - 1}\right]$$

$$9x = 12x - 4 + 3$$

$$9x = 12x - 1$$

$$-9x \quad -9x$$

$$0 = 3x - 1$$

$$+1 \qquad +1$$

$$1 = 3x$$

$$\frac{1}{3} = x$$

If we let $x = \frac{1}{3}$, then both denominators would be 0, so $x = \frac{1}{3}$ is *not* a solution to the original equation. The original equation has no solution.

6. $$\frac{1}{x - 1} - \frac{2}{x + 1} = \frac{3}{x^2 - 1}$$

$$\frac{1}{x - 1} - \frac{2}{x + 1} = \frac{3}{(x - 1)(x + 1)}$$ The LCD is $(x - 1)(x + 1)$.

$$(x - 1)(x + 1)\left[\frac{1}{x - 1} - \frac{2}{x + 1}\right] = (x - 1)(x + 1) \cdot \frac{3}{(x - 1)(x + 1)}$$

$$(x - 1)(x + 1) \cdot \frac{1}{x - 1} - (x - 1)(x + 1) \cdot \frac{2}{x + 1} = 3$$

$$1(x + 1) - 2(x - 1) = 3$$

$$x + 1 - 2x + 2 = 3$$

$$-x + 3 = 3$$

$$-3 \quad -3$$

$$-x = 0$$

$$x = 0$$

(0 and -0 are the same number)

7. $(2x - 1)^2 - 4x^2 = -4x + 1$

$$(2x-1)(2x-1)-4x^2=-4x+1$$
$$4x^2-2x-2x+1-4x^2=-4x+1$$
$$-4x+1=-4x+1$$

The last equation is true for any real number x. An equation true for any real number x is called an *identity*. So, $(2x-1)^2-4x^2=-4x+1$ is an identity.

8. $\sqrt{7x+1}=13$

$$\left[\sqrt{7x+1}\right]^2=13^2$$
$$7x+1=169$$
$$-1 \quad -1$$
$$7x=168$$
$$x=\frac{168}{7}$$
$$x=24$$

9. $\sqrt{x}-6=10$

$$\sqrt{x}-6=10$$ Isolate the square root *before* squaring both sides.
$$+6 \quad +6$$
$$\sqrt{x}=16$$
$$[\sqrt{x}]^2=16^2$$
$$x=256$$

10. $\sqrt{2x-3}+1=6$

$$\sqrt{2x-3}+1=6$$ Isolate the square root *before* squaring both sides.
$$-1 \quad -1$$
$$\sqrt{2x-3}=5$$
$$\left[\sqrt{2x-3}\right]^2=5^2$$
$$2x-3=25$$
$$+3 \quad +3$$
$$2x=28$$
$$x=\frac{28}{2}$$
$$x=14$$

11. $\sqrt{7-2x}=3$

$$\left[\sqrt{7-2x}\right]^2 = 3^2$$
$$7-2x=9$$
$$-7 \qquad -7$$
$$-2x=2$$
$$x=\frac{2}{-2}$$
$$x=-1$$

12. $\sqrt{3x+4}=\sqrt{2x+5}$

$$\left[\sqrt{3x+4}\right]^2 = \left[\sqrt{2x+5}\right]^2$$
$$3x+4=2x+5$$
$$-2x \qquad -2x$$
$$x+4=5$$
$$-4 \quad -4$$
$$x=1$$

Chapter Review

1. If $\frac{3}{x}=\frac{4}{x-1}$, then

(a) $x=-1$ (b) $x=1$ (c) $x=-3$

(d) There is no solution.

2. If $3(x-2)+5=2x$, then

(a) $x=1$ (b) $x=3$ (c) $x=-1$ (d) $x=-3$

3. $\frac{3(2^2+11)}{18-3^2}\cdot\sqrt{36}=$

(a) $\frac{74}{3}$ (b) 30 (c) 56 (d) 92

4. If $\frac{2}{3}x - 1 = \frac{5}{6}x + \frac{3}{2}$, then

(a) $x = -10$ (b) $x = -15$ (c) $x = -4$ (d) $x = -\frac{5}{3}$

5. If $\sqrt{2x-4} + 1 = 7$, then

(a) $x = 26$ (b) $x = 32$ (c) $x = 20$

(d) There is no solution.

6. If $\frac{9}{x-3} - \frac{2x}{x-3} = \frac{x}{x-3}$, then

(a) $x = 7$ (b) $x = 3$ (c) $x = -3$

(d) There is no solution.

7. If $0.16x + 1.1 = 0.2x + 0.95$, then

(a) $x = -\frac{15}{14}$ (b) $x = 21$ (c) $x = 6$ (d) $x = 3\frac{3}{4}$

8. If $A = \frac{1}{2}(2P + C)$, then $C =$

(a) $2A - 2P$ (b) $\frac{1}{2}A - 2P$ (c) $2A - P$ (d) $P - \frac{1}{2}A$

9. If $4(x-5) - 3(6-2x) = 2$, then

(a) $x = -12\frac{1}{2}$ (b) $x = 4$ (c) $x = -20$ (d) $x = 2\frac{1}{2}$

10. If $(x-3)(x+2) = (x+4)(x+1)$, then

(a) $x = \frac{5}{2}$ (b) $x = -\frac{3}{2}$ (c) $x = -\frac{5}{3}$ (d) $x = -\frac{5}{2}$

Solutions

1. (c) **2.** (a) **3.** (b) **4.** (b)
5. (c) **6.** (d) **7.** (d) **8.** (a)
9. (b) **10.** (c)

Linear Applications

To many algebra students, applications (word problems) seem impossible to solve. You might be surprised how easy solving many of them really is. If you follow the program in this chapter, you will find yourself becoming a pro at solving word problems. Mastering the problems in this chapter will also train you to solve applied problems in science courses and in more advanced mathematics courses.

Percents

A percent is a decimal number in disguise. In fact, the word "percent" literally means "per hundred." Remember that "per" means to divide, so 16% means $16 \div 100$ or $16/100 = 0.16$. Then 16% of 25 will be translated into (0.16)(25). (Remember that "of" means "multiply.") So, 16% of 25 is $(0.16)(25) = 4$.

Examples

82% of 44 is $(0.82)(44) = 36.08$

150% of 6 is $(1.50)(6) = 9$

$8\frac{3}{4}\%$ of 24 is $(0.0875)(24) = 2.1$

0.65% of 112 is $(0.0065)(112) = 0.728$

Practice

1. 64% of 50 is ________
2. 126% of 38 is ________
3. 0.42% of 16 is ________
4. 18.5% of 48 is ________
5. 213.6% of 90 is ________

Solutions

1. 64% of 50 is $(0.64)(50) = 32$
2. 126% of 38 is $(1.26)(38) = 47.88$
3. 0.42% of 16 is $(0.0042)(16) = 0.0672$
4. 18.5% of 48 is $(0.185)(48) = 8.88$
5. 213.6% of 90 is $(2.136)(90) = 192.24$

Increasing/Decreasing by a Percent

As consumers, we often see quantities being increased or decreased by some percentage. For instance, a cereal box boasts "25% More." An item might be on sale, saying "Reduced by 40%." When increasing a quantity by a percent, first compute what the percent is, then add it to the original quantity. When decreasing a quantity by a percent, again compute the percent then subtract it from the original quantity.

Examples

80 increased by 20%
(0.20)(80) = 16
So, 80 increased by 20% is 80 + 16 = 96.

24 increased by 35%
(0.35)(24) = 8.4
24 increased by 35% is 24 + 8.4 = 32.4

36 increased by 250%
(2.50)(36) = 90
36 increased by 250% is 36 + 90 = 126

64 decreased by 27%
(0.27)(64) = 17.28
64 decreased by 27% is 64 − 17.28 = 46.72

Practice

1. 46 increased by 60% is ________
2. 78 increased by 125% is ________
3. 16 decreased by 30% is ________
4. 54 increased by 21.3% is ________
5. 128 decreased by 8.16% is ________
6. 15 increased by 0.03% is ________
7. 24 decreased by 108.4% is ________

Solutions

1. 46 increased by 60% is 46 + (0.60)(46) = 46 + 27.6 = 73.6
2. 78 increased by 125% is 78 + (1.25)(78) = 78 + 97.5 = 175.5

3. 16 decreased by 30% is $16 - (0.30)(16) = 16 - 4.8 = 11.2$

4. 54 increased by 21.3% is $54 + (0.213)(54) = 54 + 11.502 = 65.502$

5. 128 decreased by 8.16% is $128 - (0.0816)(128) = 128 - 10.4448 = 117.5552$

6. 15 increased by 0.03% is $15 + (0.0003)(15) = 15 + 0.0045 = 15.0045$

7. 24 decreased by 108.4% is $24 - (1.084)(24) = 24 - 26.016 = -2.016$

Many word problems involving percents fit the above model—that is, a quantity being increased or decreased. Often you can solve these problems using one of the following formats:

$x + .\_\_\_x$ (for a quantity being increased by a percent)

or

$x - .\_\_x$ (for a quantity being decreased by a percent).

Examples

A $100 jacket will be reduced by 15% for a sale. What will the sale price be?

Let $x =$ sale price.

Then, $100 - (0.15)(100) = x$.

$$100 - 15 = x$$
$$85 = x$$

The sale price is $85.

More often, the sale price will be known and the original price is not known.

The sale price for a computer is $1200, which represents a 20% markdown. What is the original price?

Let x represent original price. Then the sale price is $x - 0.20x = 0.80x$. The sale price is also 1200. This gives us the equation $0.80x = 1200$.

$$0.80x = 1200$$
$$x = \frac{1200}{0.80}$$
$$x = 1500$$

The original price is \$1500.

In the first example, the percent was multiplied by the number given; and in the second, the percent was multiplied by the unknown. You must be very careful in deciding of which quantity you take the percent. Suppose the first problem was worded, "An item is marked down 20% for a sale. The sale price is \$80, what is the original price?" The equation to solve would be $x - 0.20x = 80$, where x represents the original price. A common mistake is to take 20% of \$80 and not 20% of the *original price*.

The data used in the following examples and practice problems are taken from the 117th edition of *Statistical Abstract of the United States*. Many quantities and percentages are approximate.

Examples

The average number of hours of television watched by U.S. adults during 1980 was 1470. By 1995, the number of hours of television viewing increased by about 7.1%. What was the average number of hours of television viewing in 1995?

$$1980 \text{ hours} + 7.1\% \text{ of } 1980 \text{ hours} = 1995 \text{ hours}$$

$$1470 + (0.071)(1470) = 1470 + 104.37 = 1574.37$$

The average number of hours U.S. adults spent watching television in 1995 was about 1574.

On January 1, 1995, the price of a first class postage stamp was \$0.32, which is a 60% increase from the cost on November 1, 1981. What was the cost of a first class stamp on November 1, 1981?

Let $x =$ 1st class price on November 1, 1981

\$0.32 is 60% more than this quantity.

$$0.32 = x + 0.60x$$

$$0.32 = 1x + 0.60x$$

$$0.32 = x(1 + 0.60)$$

$$0.32 = 1.60x$$

$$\frac{0.32}{1.60} = x$$

$$0.20 = x$$

The price of a first-class stamp on November 1, 1981 was \$0.20.

Practice

1. A local cable television company currently charges \$36 per month. It plans an increase in its monthly charge by 15%. What will the new rate be?

2. In 1980, the median age of U.S. residents was 30 years. By 1996, the median age had increased by about 15.3%. What was the median age in 1996?

3. In 1995, the death rate per 100,000 U.S. residents from major cardiovascular disease was 174.4, which is about a $31\frac{3}{4}\%$ decrease from 1980. What was the 1980 death rate per 100,000 from major cardiovascular disease?

4. A worker's take-home pay was \$480 after deductions totaling 40%. What is the worker's gross pay?

5. A cereal company advertises that its 16-ounce cereal represents 25% more than before. What was the original amount?

6. A couple does not wish to spend more than \$45 for dinner at their favorite restaurant. If a sales tax of $7\frac{1}{2}\%$ is added to the bill and they plan to tip 15% after the tax is added, what is the most they can spend for the meal?

7. A discount store prices its blank videotapes by raising the wholesale price by 40% and adding \$0.20. What must the tape's wholesale price be if the tape sells for \$3.00?

Solutions

1. \$36 will be increased by 15%:

$$36 + (0.15)(36) = 36 + 5.4 = 41.4$$

The new rate will be \$41.40.

2. 30 is increased by 15.3%:

$$30 + (0.153)(30) = 30 + 4.59 = 34.59$$

The median age of U.S. residents in 1996 was about 34.6 years.

3. The 1980 rate is decreased by $31\frac{3}{4}\%$. Let $x =$ 1980 rate.

$$\begin{aligned} x - 0.3175x &= 174.4 \\ 1x - 0.3175x &= 174.4 \\ x(1 - 0.3175) &= 174.4 \\ 0.6825x &= 174.4 \\ x &= \frac{174.4}{0.6825} \\ x &\approx 255.53 \end{aligned}$$

The 1980 death rate per 100,000 from major cardiovascular disease was about 255.5.

4. The gross pay is reduced by 40%. Let $x =$ gross pay.

$$\begin{aligned} x - 0.40x &= 480 \\ 1x - 0.40x &= 480 \\ x(1 - 0.40) &= 480 \\ 0.60x &= 480 \\ x &= \frac{480}{0.60} \\ x &= 800 \end{aligned}$$

The worker's gross pay is $800.

5. The original amount is increased by 25%. Let $x =$ original amount.

$$\begin{aligned} x + 0.25x &= 16 \\ 1x + 0.25x &= 16 \\ x(1 + 0.25) &= 16 \\ 1.25x &= 16 \\ x &= \frac{16}{1.25} \\ x &= 12.8 \end{aligned}$$

The original box of cereal contained 12.8 ounces.

6. The total bill is the cost of the meal plus the tax on the meal plus the tip.

   Let $x =$ cost of the meal. The tax, then, is $0.075x$.

   The tip is 15% of the meal plus tax: ($x + 0.075x = 1.075x$ is the price of the meal), so the tip is $0.15(1.075x) = 0.16125x$.

   The total bill is $x + 0.075x + 0.16125x$. We want this to equal 45:

   meal tax tip total

$$x + 0.075x + 0.16125x = 45$$
$$1x + 0.075x + 0.16125x = 45$$
$$x(1 + 0.075 + 0.16125) = 45$$
$$1.23625x = 45$$
$$x = \frac{45}{1.23625}$$
$$x \approx 36.40$$

   The couple can spend \$36.40 on their meal.

7. Let $x =$ wholesale price. 40% of the wholesale price is $0.40x$. The retail price is the wholesale price plus 40% of the wholesale price plus \$0.20:

$$x + 0.40x + 0.20 = 3.00$$
$$1x + 0.40x + 0.20 = 3.00$$
$$x(1 + 0.40) + 0.20 = 3.00$$
$$1.40x + 0.20 = 3.00$$
$$-0.020 \quad -0.20$$
$$1.40x = 2.80$$
$$x = \frac{2.80}{1.40}$$
$$x = 2$$

   The wholesale price is \$2.

At times the percent is the unknown. You are given two quantities and are asked what percent of one is of the other. Let x represent the percent as a decimal number.

Examples

5 is what percent of 8?

This sentence translates into $5 = x \cdot 8$
5 is what percent of 8

The equation to solve is $8x = 5$.

$$8x = 5$$
$$x = \frac{5}{8} = 0.625 = 62.5\%$$

5 is 62.5% of 8.

8 is what percent of 5?

$$8 = x \cdot 5$$
$$5x = 8$$
$$x = \frac{8}{5} = 1.6 = 160\%$$

8 is 160% of 5.

Practice

1. 2 is what percent of 5?
2. 5 is what percent of 2?
3. 3 is what percent of 15?
4. 15 is what percent of 3?
5. 1.8 is what percent of 18?

6. 18 is what percent of 1.8?

7. $\frac{1}{4}$ is what percent of 2?

8. 2 is what percent of $\frac{1}{4}$?

Solutions

1. $5x = 2$

$$x = \frac{2}{5} = 0.40 = 40\%$$

2 is 40% of 5.

2. $2x = 5$

$$x = \frac{5}{2} = 2.5 = 250\%$$

5 is 250% of 2.

3. $15x = 3$

$$x = \frac{3}{15} = \frac{1}{5} = 0.20 = 20\%$$

3 is 20% of 15.

4. $3x = 15$

$$x = \frac{15}{3} = 5 = 500\%$$

15 is 500% of 3.

5. $18x = 1.8$

$$x = \frac{1.8}{18}$$

$$x = \frac{1.8(10)}{18(10)}$$

$$x = \frac{18}{180} = 0.10 = 10\%$$

1.8 is 10% of 18.

6. $1.8x = 18$

$$x = \frac{18}{1.8}$$

$$x = \frac{18(10)}{1.8(10)}$$

$$x = \frac{180}{18} = 10 = 1000\%$$

18 is 1000% of 1.8.

7. $2x = \frac{1}{4}$

$$x = \frac{1}{2} \cdot \frac{1}{4}$$

$$x = \frac{1}{8} = 0.125 = 12.5\%$$

$\frac{1}{4}$ is 12.5% of 2.

8. $\frac{1}{4}x = 2$

$$x = 4(2)$$

$$x = 8 = 800\%$$

2 is 800% of $\frac{1}{4}$.

For some word problems, nothing more will be required of you than to substitute a given value into a formula, which is either given to you or is readily available. The most difficult part of these problems will be to decide which variable the given quantity will be. For example, the formula might look like $R = 8q$ and the value given to you is 440. Is $R = 440$ or is $q = 440$? The answer lies in the way the variables are described. In $R = 8q$, it might be that R represents revenue (in dollars) and q represents quantity (in units) sold of some item. "If 440 units were sold, what is the revenue?" Here 440 is q. You would then solve $R = 8(440)$. "If the revenue is \$440, how many units were sold?" Here 440 is R, and you would solve $440 = 8q$.

Examples

The cost formula for a manufacturer's product is $C = 5000 + 2x$, where C is the cost (in dollars) and x is the number of units manufactured.

(a) If no units are produced, what is the cost?
(b) If the manufacturer produces 3000 units, what is the cost?
(c) If the manufacturer has spent \$16,000 on production, how many units were manufactured?

Answer these questions by substituting the numbers into the formula.

(a) If no units are produced, then $x = 0$, and $C = 5000 + 2x$ becomes $C = 5000 + 2(0) = 5000$. The cost is \$5,000.
(b) If the manufacturer produces 3000 units, then $x = 3000$, and $C = 5000 + 2x$ becomes $C = 5000 + 2(3000) = 5000 + 6000 = 11{,}000$. The manufacturer's cost would be \$11,000.
(c) The manufacturer's cost is \$16,000, so $C = 16{,}000$. Substitute $C = 16{,}000$ into $C = 5000 + 2x$ to get $16{,}000 = 5000 + 2x$.

$$16{,}000 = 5000 + 2x$$
$$-5000 \quad -5000$$
$$11{,}000 = 2x$$
$$\frac{11{,}000}{2} = x$$
$$5500 = x$$

There were 5500 units produced.

The profit formula for a manufacturer's product is $P = 2x - 4000$ where x is the number of units sold and P is the profit (in dollars).

(a) What is the profit when 12,000 units were sold?
(b) What is the loss when 1500 units were sold?
(c) How many units must be sold for the manufacturer to have a profit of \$3000?
(d) How many units must be sold for the manufacturer to break even?

(This question could have been phrased, "How many units must be sold in order for the manufacturer to cover its costs?")

(a) If 12,000 units are sold, then $x = 12{,}000$. The profit equation then becomes $P = 2(12{,}000) - 4000 = 24{,}000 - 4000 = 20{,}000$. The profit is \$20,000.
(b) Think of a loss as a negative profit. When 1500 units are sold, $P = 2x - 4000$ becomes $P = 2(1500) - 4000 = 3000 - 4000 = -1000$. The manufacturer loses \$1000 when 1500 units are sold.
(c) If the profit is \$3000, then $P = 3000$; $P = 2x - 4000$ becomes $3000 = 2x - 4000$.

$$3000 = 2x - 4000$$
$$+4000 \qquad +4000$$
$$7000 = 2x$$
$$\frac{7000}{2} = x$$
$$3500 = x$$

A total of 3500 units were sold.

(d) The break-even point occurs when the profit is zero, that is when $P = 0$. Then $P = 2x - 4000$ becomes $0 = 2x - 4000$.

$$0 = 2x - 4000$$
$$+4000 \qquad +4000$$
$$4000 = 2x$$
$$\frac{4000}{2} = x$$
$$2000 = x$$

The manufacturer must sell 2000 units in order to break even.

A box has a square bottom. The height has not yet been determined, but the bottom is 10 inches by 10 inches. The volume formula is $V = lwh$, because each of the length and width is 10, lw becomes $10 \cdot 10 = 100$. The formula for the box's volume is $V = 100h$.

(a) If the height of the box is to be 6 inches, what is its volume?
(b) If the volume is to be 450 cubic inches, what should its height be?
(c) If the volume is to be 825 cubic inches, what should its height be?

(a) The height is 6 inches, so $h = 6$. Then $V = 100h$ becomes $V = 100(6) = 600$.
The box's volume is 600 cubic inches.

(b) The volume is 450 cubic inches, so $V = 450$, and $V = 100h$ becomes $450 = 100h$.

$$450 = 100h$$
$$\frac{450}{100} = h$$
$$4.5 = h$$

The box's height would need to be 4.5 inches.

(c) The volume is 825, so $V = 100h$ becomes $825 = 100h$.

$$825 = 100h$$

$$\frac{825}{100} = h$$

$$8.25 = h$$

The height should be 8.25 inches.

Suppose a square has a perimeter of 18 cm. What is the length of each of its sides? (Recall the formula for the perimeter of a square: $P = 4l$ where l is the length of each of its sides.)

$P = 18$, so $P = 4l$ becomes $18 = 4l$.

$$18 = 4l$$

$$\frac{18}{4} = l$$

$$4.5 = l$$

The length of each of its sides is 4.5 cm.

The relationship between degrees Fahrenheit and degrees Celsius is given by the formula $\mathrm{C} = \frac{5}{9}(\mathrm{F} - 32)$. At what temperature will degrees Fahrenheit and degrees Celsius be the same?

Both expressions "C" and "$\frac{5}{9}(\mathrm{F} - 32)$" represent degrees Celsius, so when Fahrenheit equals Celsius, we can say that $\mathrm{F} = \frac{5}{9}(\mathrm{F} - 32)$. Before, our equation had two variables. By substituting "$\frac{5}{9}(\mathrm{F} - 32)$" in place of "C," we have reduced the number of variables in the equation to one. Now we can solve it.

$$\mathrm{F} = \frac{5}{9}(\mathrm{F} - 32)$$

The LCD is 9.

$$9\mathrm{F} = 9\left[\frac{5}{9}(\mathrm{F} - 32)\right]$$

$$9\mathrm{F} = \left[9\left(\frac{5}{9}\right)\right](\mathrm{F} - 32)$$

$$9\mathrm{F} = 5(\mathrm{F} - 32)$$

$$9\mathrm{F} = 5\mathrm{F} - 5(32)$$

$$
\begin{aligned}
9F &= 5F - 160 \\
-5F &\quad -5F \\
4F &= -160 \\
F &= \frac{-160}{4} \\
F &= -40
\end{aligned}
$$

At −40 degrees Fahrenheit and degrees Celsius are the same.

Practice

1. The daily charge for a small rental car is $C = 18 + 0.35x$ where x is the number of miles driven.
   (a) If the car was driven 80 miles, what was the charge?
   (b) Suppose that a day's bill was \$39. How many miles were driven?

2. The profit obtained for a company's product is given by $P = 25x - 8150$, where P is the profit in dollars and x is the number of units sold. How many units must be sold in order for the company to have a profit of \$5000 from this product?

3. A salesman's weekly salary is based on the formula $S = 200 + 0.10s$, where S is the week's salary in dollars and s is the week's sales level in dollars. One week, his salary was \$410. What was the sales level for that week?

4. The volume of a box with a rectangular bottom is given by $V = 120h$, where V is the volume in cubic inches and h is the height in inches. If the volume of the box is to be 1140 cubic inches, what should its height be?

5. The volume of a certain cylinder with radius 2.8 cm is given by $V = 7.84\pi h$, where h is the height of the cylinder in centimeters. If the volume needs to be 25.088π cubic centimeters, what does the height need to be?

6. At what temperature will the Celsius reading be twice as high as the Fahrenheit reading?

7. At what temperature will degrees Fahrenheit be twice degrees Celsius?

Solutions

1. (a) Here $x = 80$, so $C = 18 + 0.35x$ becomes $C = 18 + 0.35(80)$.

$$C = 18 + 0.35(80)$$
$$C = 18 + 28$$
$$C = 46$$

The charge is \$46.

(b) The cost is \$39, so $C = 18 + 0.35x$ becomes $39 = 18 + 0.35x$.

$$\begin{aligned} 39 &= 18 + 0.35x \\ -18 & \quad -18 \\ 21 &= 0.35x \\ \frac{21}{0.35} &= x \\ 60 &= x \end{aligned}$$

Sixty miles were driven.

2. The profit is \$5000, so $P = 25x - 8150$ becomes $5000 = 25x - 8150$.

$$\begin{aligned} 5000 &= 25x - 8150 \\ +8150 & \qquad\quad +8150 \\ 13{,}150 &= 25x \\ \frac{13{,}150}{25} &= x \\ 526 &= x \end{aligned}$$

The company must sell 526 units.

3. The salary is \$410, so $S = 200 + 0.10s$ becomes $410 = 200 + 0.10s$.

$$\begin{aligned} 410 &= 200 + 0.10s \\ -200 & \quad -200 \\ 210 &= 0.10s \\ \frac{210}{0.10} &= s \\ 2100 &= s \end{aligned}$$

The week's sales level was \$2100.

4. The volume is 1140 cubic inches, so $V = 120h$ becomes $1140 = 120\text{h}$.

$$1140 = 120h$$

$$\frac{1140}{120} = h$$

$$9.5 = h$$

The box needs to be 9.5 inches tall.

5. The volume is 25.088π cm^3, so $V = 7.84\pi h$ becomes $25.088\pi = 7.84\pi h$.

$$25.088\pi = 7.84\pi h$$

$$\frac{25.088\pi}{7.84\pi} = h$$

$$3.2 = h$$

The height of the cylinder needs to be 3.2 cm.

6. $\frac{5}{9}(\text{F} - 32)$ represents degrees Celsius, and 2F represents twice degrees Fahrenheit. We want these two quantities to be equal.

$$\frac{5}{9}(\text{F} - 32) = 2\text{F}$$

The LCD is 9.

$$9\left[\left(\frac{5}{9}\right)(\text{F} - 32)\right] = 9(2\text{F})$$

$$\left[9\left(\frac{5}{9}\right)\right](\text{F} - 32) = 18\text{F}$$

$$5(\text{F} - 32) = 18\text{F}$$

$$\begin{aligned} 5\text{F} - 160 &= 18\text{F} \\ -5\text{F} \qquad &\quad -5\text{F} \end{aligned}$$

$$-160 = 13\text{F}$$

$$\frac{-160}{13} = \text{F}$$

When $\text{F} = -\frac{160}{13}$, $\text{C} = \frac{5}{9}\left(-\frac{160}{13} - 32\right) = -\frac{320}{13}$ or $-24\frac{8}{13}$.

7. Degrees Celsius is represented by $\frac{5}{9}(\mathrm{F}-32)$ so twice degrees Celsius is represented by $2[\frac{5}{9}(\mathrm{F}-32)]$. We want for this to equal degrees Fahrenheit.

$$\mathrm{F}=2\left[\frac{5}{9}(\mathrm{F}-32)\right]$$

$$\mathrm{F}=\left[2\left(\frac{5}{9}\right)\right](\mathrm{F}-32)$$

$$\mathrm{F}=\frac{10}{9}(\mathrm{F}-32)$$

$$9\mathrm{F}=9\left[\frac{10}{9}(\mathrm{F}-32)\right]$$

$$9\mathrm{F}=\left[9\left(\frac{10}{9}\right)\right](\mathrm{F}-32)$$

$$9\mathrm{F}=10(\mathrm{F}-32)$$

$$\begin{aligned} 9\mathrm{F} &= 10\mathrm{F}-320 \\ -10\mathrm{F} &\quad -10\mathrm{F} \end{aligned}$$

$$-\mathrm{F}=-320$$

$$(-1)(-\mathrm{F})=(-1)(-320)$$

$$\mathrm{F}=320$$

When $\mathrm{F}=320$, $\mathrm{C}=\frac{5}{9}(320-32)=160$.

Many problems require the student to use common sense to solve them—that is, mathematical reasoning. For instance, when a problem refers to consecutive integers, the student is expected to realize that any two consecutive integers differ by one. If two numbers are consecutive, normally x is set equal to the first and $x+1$, the second.

Examples

The sum of two consecutive integers is 25. What are the numbers?

Let $x=$ first number.

$x+1=$ second number

Their sum is 25, so $x+(x+1)=25$.

$$x + (x + 1) = 25$$
$$2x + 1 = 25$$
$$-1 \quad -1$$
$$2x = 24$$
$$x = \frac{24}{2}$$
$$x = 12$$

The first number is 12 and the second number is $x + 1 = 12 + 1 = 13$.

The sum of three consecutive integers is 27. What are the numbers?
Let $x =$ first number.
$x + 1 =$ second number
$x + 2 =$ third number
Their sum is 27, so $x + (x + 1) + (x + 2) = 27$.

$$x + (x + 1) + (x + 2) = 27$$
$$3x + 3 = 27$$
$$-3 \quad -3$$
$$3x = 24$$
$$x = \frac{24}{3}$$
$$x = 8$$

The first number is 8; the second is $x + 1 = 8 + 1 = 9$; the third is $x + 2 = 8 + 2 = 10$.

Practice

1. Find two consecutive numbers whose sum is 57.

2. Find three consecutive numbers whose sum is 48.

3. Find four consecutive numbers whose sum is 90.

Solutions

1. Let $x =$ first number.
   $x + 1 =$ second number
   Their sum is 57, so $x + (x + 1) = 57$.

$$x + (x + 1) = 57$$
$$2x + 1 = 57$$
$$-1 \quad -1$$
$$2x = 56$$
$$x = \frac{56}{2}$$
$$x = 28$$

The first number is 28 and the second is $x + 1 = 28 + 1 = 29$.

2. Let x = first number.
   $x + 1$ = second number
   $x + 2$ = third number
   Their sum is 48, so $x + (x + 1) + (x + 2) = 48$.

$$x + (x + 1) + (x + 2) = 48$$
$$3x + 3 = 48$$
$$-3 \quad -3$$
$$3x = 45$$
$$x = \frac{45}{3}$$
$$x = 15$$

   The first number is 15; the second, $x + 1 = 15 + 1 = 16$; and the third, $x + 2 = 15 + 2 = 17$.

3. Let x = first number.
   $x + 1$ = second number
   $x + 2$ = third number
   $x + 3$ = fourth number
   Their sum is 90, so $x + (x + 1) + (x + 2) + (x + 3) = 90$.

$$x + (x + 1) + (x + 2) + (x + 3) = 90$$
$$4x + 6 = 90$$
$$-6 \quad -6$$
$$4x = 84$$
$$x = \frac{84}{4}$$
$$x = 21$$

The first number is 21; the second, $x+1=21+1=22$; the third, $x+2=21+2=23$; and the fourth, $x+3=21+3=24$.

Examples

The sum of two numbers is 70. One number is eight more than the other. What are the two numbers?

Problems such as this are similar to the above in that we are looking for two or more numbers and we have a little information about how far apart the numbers are. In the problems above, the numbers differed by one. Here, two numbers differ by eight.

Let $x=$ first number. (The term "first" is used because it is the first number we are looking for; it is not necessarily the "first" in order.) The other number is eight more than this, so $x+8$ represents the other number. Their sum is 70, so $x+(x+8)=70$.

$$x+(x+8)=70$$

$$2x+8=70$$

$$-8 \quad -8$$

$$2x=62$$

$$x=\frac{62}{2}$$

$$x=31$$

The numbers are 31 and $x+8=39$.

The sum of two numbers is 63. One of the numbers is twice the other.

Let $x=$ first number.

$2x=$ other number

Their sum is 63, so $x+2x=63$.

$$x+2x=63$$

$$3x=63$$

$$x=\frac{63}{3}$$

$$x=21$$

The numbers are 21 and $2x=2(21)=42$.

Practice

1. The sum of two numbers is 85. One number is 15 more than the other. What are the two numbers?

2. The sum of two numbers is 48. One number is three times the other. What are the numbers?

Solutions

1. Let $x =$ first number.
$x + 15 =$ second number
Their sum is 85, so $x + (x + 15) = 85$.

$$\begin{aligned} x + (x + 15) &= 85 \\ 2x + 15 &= 85 \\ -15 \quad & \;\; -15 \\ 2x &= 70 \\ x &= \frac{70}{2} \\ x &= 35 \end{aligned}$$

The numbers are 35 and $x + 15 = 35 + 15 = 50$.

2. Let $x =$ first number.
$3x =$ second number
Their sum is 48, so $x + 3x = 48$.

$$\begin{aligned} x + 3x &= 48 \\ 4x &= 48 \\ x &= \frac{48}{4} \\ x &= 12 \end{aligned}$$

The numbers are 12 and $3x = 3(12) = 36$.

Examples

The difference between two numbers is 13. Twice the smaller plus three times the larger is 129.

If the difference between two numbers is 13, then one of the numbers is 13 more than the other. The statement "The difference between two numbers is 13," could have been given as, "One number is 13 more than the other." As before, let x represent the first number. Then, $x + 13$ represents the other. "Twice the smaller" means "$2x$" (x is the smaller quantity because the other quantity is 13 more than x). Three times the larger number is $3(x + 13)$. "Twice the smaller plus three times the larger is 129" becomes $2x + 3(x + 13) = 129$.

$$2x + 3(x + 13) = 129$$
$$2x + 3x + 39 = 129$$
$$5x + 39 = 129$$
$$-39 \quad -39$$
$$5x = 90$$
$$x = \frac{90}{5}$$
$$x = 18$$

The numbers are 18 and $x + 13 = 18 + 13 = 31$.

The sum of two numbers is 14. Three times the smaller plus twice the larger is 33. What are the two numbers?

Let x represent the smaller number. How can we represent the larger number? We know that the sum of the smaller number and larger number is 14. Let "?" represent the larger number and we'll get "?" in terms of x.

The smaller number plus the larger number is 14.

$$x \qquad + \qquad ? \qquad = 14$$

$$x + ? = 14$$
$$-x \qquad -x$$
$$? = 14 - x$$

So, $14 - x$ is the larger number. Three times the smaller is $3x$. Twice the larger is $2(14 - x)$. Their sum is 33, so $3x + 2(14 - x) = 33$.

$$3x + 2(14 - x) = 33$$
$$3x + 28 - 2x = 33$$
$$x + 28 = 33$$
$$-28 \quad -28$$
$$x = 5$$

The smaller number is 5 and the larger is $14 - x = 14 - 5 = 9$.

Practice

1. The sum of two numbers is 10. Three times the smaller plus 5 times the larger number is 42. What are the numbers?

2. The difference between two numbers is 12. Twice the smaller plus four times the larger is 108. What are the two numbers?

3. The difference between two numbers is 8. The sum of one and a half times the smaller and four times the larger is 54. What are the numbers?

4. The sum of two numbers is 11. When twice the larger is subtracted from 5 times the smaller, the difference is 6. What are the numbers?

Solutions

1. Let x represent the smaller number. The larger number is then $10 - x$.

$$3x + 5(10 - x) = 42$$
$$3x + 50 - 5x = 42$$
$$-2x + 50 = 42$$
$$-50 \quad -50$$
$$-2x = -8$$
$$x = \frac{-8}{-2}$$
$$x = 4$$

The numbers are 4 and $10 - x = 10 - 4 = 6$.

2. The difference between the numbers is 12, so one number is 12 more than the other. Let x represent the smaller number. Then $x + 12$ is the larger. Twice the smaller is $2x$, and four times the larger is $4(x + 12)$.

$$2x + 4(x + 12) = 108$$
$$2x + 4x + 48 = 108$$
$$6x + 48 = 108$$
$$-48 \quad -48$$

$$6x = 60$$
$$x = \frac{60}{6}$$
$$x = 10$$

The smaller number is 10 and the larger is $x + 12 = 10 + 12 = 22$.

3. The difference between the numbers is 8, so one of the numbers is 8 more than the other. Let x represent smaller number. The larger number is $x + 8$. One and a half of the smaller number is $1\frac{1}{2}x$; four times the larger is $4(x + 8)$.

$$1\frac{1}{2}x + 4(x + 8) = 54$$
$$\frac{3}{2}x + 4x + 32 = 54$$
$$2\left(\frac{3}{2}x + 4x + 32\right) = 2(54)$$
$$\left[2\left(\frac{3}{2}x\right)\right] + 2(4x) + 2(32) = 108$$
$$3x + 8x + 64 = 108$$
$$11x + 64 = 108$$
$$-64 \quad -64$$
$$11x = 44$$
$$x = \frac{44}{11}$$
$$x = 4$$

The smaller number is 4 and the larger, $x + 8 = 4 + 8 = 12$.

4. Let $x =$ smaller number. Then $11 - x$ is the larger. Five times the smaller is $5x$, and twice the larger is $2(11 - x)$. "Twice the larger subtracted from 5 times the smaller" becomes "$5x - 2(11 - x)$."

$$5x - 2(11 - x) = 6$$
$$5x - 22 + 2x = 6$$
$$7x - 22 = 6$$
$$+22 \quad +22$$

$$7x = 28$$

$$x = \frac{28}{7}$$

$$x = 4$$

The smaller number is 4 and the larger is $11 - x = 11 - 4 = 7$.

Algebra students are often asked to compute people's ages. The steps in solving such problems are usually the same as those used above.

Examples

Jill is twice as old as Jim and Jim is three years older than Ken. The sum of their ages is 61. What are their ages?

Three quantities are being compared, so find one age and relate the other two ages to it. Ken's age is being compared to Jim's and Jim's to Jill's. The easiest route to take is to let x represent Jim's age. We can write Jill's age in terms of Jim's age: $2x$. Jim is three years older than Ken, so Ken is three years younger than Jim. This makes Ken's age as $x - 3$.

$$x + 2x + (x - 3) = 61$$

$$4x - 3 = 61$$

$$+3 \quad +3$$

$$4x = 64$$

$$x = \frac{64}{4}$$

$$x = 16$$

Jim's age is 16. Jill's age is $2x = 2(16) = 32$. Ken's age is $x - 3 = 16 - 3 = 13$.

Karen is four years older than Robert, and Jerri is half as old as Robert. The sum of their ages is 44. Find Karen's, Robert's, and Jerri's ages.

Both Karen's and Jerri's ages are being compared to Robert's age, so let x represent Robert's age. Karen is four years older than Robert, so Karen's age is $x + 4$. Jerri is half as old as Robert, so Jerri's age is $\frac{1}{2}x$.

$$x + (x + 4) + \frac{x}{2} = 44$$
$$2x + 4 + \frac{x}{2} = 44$$
$$2\left(2x + 4 + \frac{x}{2}\right) = 2(44)$$
$$2(2x) + 2(4) + 2\left(\frac{x}{2}\right) = 88$$
$$4x + 8 + x = 88$$
$$5x + 8 = 88$$
$$-8 \quad -8$$
$$5x = 80$$
$$x = \frac{80}{5}$$
$$x = 16$$

Robert's age is 16; Karen's age is $x + 4 = 16 + 4 = 20$; and Jerri's, $\frac{1}{2}x = \frac{1}{2}(16) = 8$.

Practice

1. Andy is three years older than Bea and Bea is five years younger than Rose. If Rose is 28, how old are Andy and Bea?

2. Michele is four years younger than Steve and three times older than Sean. If the sum of their ages is 74, how old are they?

3. Monica earns three times per hour as John. John earns $2 more per hour than Alicia. Together they earn $43 per hour. How much is each one's hourly wage?

Solutions

1. Because Rose is 28 and Bea is five years younger than Rose, Bea is $28 - 5 = 23$ years old. Andy is three years older than Bea, so Andy is $23 + 3 = 26$ years old.

2. Let $x =$ Michele's age. Steve is four years older, so his age is $x + 4$. Sean is one-third Michele's age, so his age is $\frac{1}{3}x = \frac{x}{3}$.

$$x + (x + 4) + \frac{x}{3} = 74$$
$$2x + 4 + \frac{x}{3} = 74$$
$$3\left(2x + 4 + \frac{x}{3}\right) = 3(74)$$
$$3(2x) + 3(4) + 3\left(\frac{x}{3}\right) = 222$$
$$6x + 12 + x = 222$$
$$7x + 12 = 222$$
$$-12 \quad -12$$
$$7x = 210$$
$$x = \frac{210}{7}$$
$$x = 30$$

Michele is 30 years old; Steve is $x + 4 = 34$; and Sean is $\frac{x}{3} = \frac{30}{3} = 10$.

You can avoid the fraction $\left(\frac{x}{3}\right)$ in this problem if you let x represent Sean's age. Then Michele's age would be $3x$; and Steve's, $3x + 4$.

3. Monica's earnings are being compared to John's, and John's to Alicia's. The easiest thing to do is to let x represent Alicia's hourly wage. Then John's hourly wage would be $x + 2$. Monica earns three times as much as John, so her hourly wage is $3(x + 2)$.

$$x + (x + 2) + 3(x + 2) = 43$$
$$x + x + 2 + 3x + 6 = 43$$
$$5x + 8 = 43$$
$$-8 \quad -8$$
$$5x = 35$$
$$x = \frac{35}{5}$$
$$x = 7$$

Alicia earns \$7 per hour; John, $x + 2 = 7 + 2 = \$9$; and Monica $3(x + 2) = 3(7 + 2) = \$27$.

Grade computation problems are probably the most useful to students. In these problems, the formula for the course grade and all but one grade are given. The student is asked to compute the unknown grade in order to ensure a particular course average.

Examples

A student has grades of 72, 74, 82, and 90. What does the next grade have to be to obtain an average of 80?

We will be taking the average of five numbers: 72, 74, 82, 90 and the next grade. Call this next grade x. We want this average to be 80.

$$\frac{72 + 74 + 82 + 90 + x}{5} = 80$$

$$5\left(\frac{318 + x}{5}\right) = 5(80)$$

$$318 + x = 400$$

$$-318 \qquad -318$$

$$x = 82$$

The student needs an 82 to raise his/her average to 80.

A student has grades of 78, 83, 86, 82, and 88. If the next grade counts twice as much as each of the others, what does this grade need to be in order to yield an average of 85?

Even though there will be a total of six grades, the last one will count twice as much as the others, so it is like having a total of seven grades; that is, the divisor needs to be seven. Let x represent the next grade.

$$\frac{78 + 83 + 86 + 82 + 88 + 2x}{7} = 85$$

$$\frac{417 + 2x}{7} = 85$$

$$7\left(\frac{417 + 2x}{7}\right) = 7(85)$$

$$417 + 2x = 595$$

$$-417 \qquad -417$$

$$2x = 178$$

$$x = \frac{178}{2}$$

$$x = 89$$

The student needs a grade of 89 to raise the average to 85.

A major project accounts for one-third of the course grade. The rest of the course grade is determined by the quiz average. A student has quiz grades of 82, 80, 99, and 87, each counting equally. What does the project grade need to be to raise the student's average to 90?

The quiz average accounts for two-thirds of the grade and the project, one-third. The equation to use, then, is $\frac{2}{3}$ quiz average $+\frac{1}{3}$ project grade $= 90$. The quiz average is $\dfrac{82+80+99+87}{4} = 87$.

Let x represent the project grade.

$$\frac{2}{3}(87) + \frac{1}{3}x = 90$$

$$58 + \frac{x}{3} = 90$$

$$3\left(58 + \frac{x}{3}\right) = 3(90)$$

$$3(58) + 3\left(\frac{x}{3}\right) = 270$$

$$174 + x = 270$$

$$\underline{-174 \qquad -174}$$

$$x = 96$$

The student needs a grade of 96 for a course grade of 90.

Practice

1. A student's grades are 93, 89, 96, and 98. What does the next grade have to be to raise her average to 95?

2. A student's grades are 79, 82, 77, 81, and 78. What does the next grade have to be to raise the average to 80?

3. A presentation grade counts toward one-fourth of the course grade. The average of the four tests counts toward the remaining three-fourths of the course grade. If a student's test scores are 61, 63, 65, and 83, what does he need to make on the presentation grade to raise his average to 70?

4. The final exam accounts for one-third of the course grade. The average of the four tests accounts for another third, and a presenta-

tion accounts for the final third A student's test scores are 68, 73, 80, and 95. His presentation grade is 75. What does the final exam grade need to be to raise his average to 80?

5. A book report counts toward one-fifth of a student's course grade. The remaining four-fifths of the courses' average is determined by the average of six quizzes. One student's book report grade is 90 and has quiz grades of 72, 66, 69, 80, and 85. What does she need to earn on her sixth quiz to raise her average to 80?

Solutions

1. Let $x =$ the next grade.

$$\frac{93+89+96+98+x}{5} = 95$$

$$\frac{376+x}{5} = 95$$

$$5\left(\frac{376+x}{5}\right) = 5(95)$$

$$376 + x = 475$$

$$\begin{array}{rr} -376 & -376 \end{array}$$

$$x = 99$$

The last grade needs to be 99 in order to raise her average to 95.

2. Let $x =$ the next grade.

$$\frac{79+82+77+81+78+x}{6} = 80$$

$$\frac{397+x}{6} = 80$$

$$6\left(\frac{397+x}{6}\right) = 6(80)$$

$$397 + x = 480$$

$$\begin{array}{rr} -397 & -397 \end{array}$$

$$x = 83$$

The next grade needs to be 83 to raise the average to 80.

3. Let x represent the presentation grade. The test average is $(61 + 63 + 65 + 83)/4 = 68$. Then $\frac{3}{4}$ test average $+\frac{1}{4}$ presentation grade $= 70$ becomes $\frac{3}{4}(68) + \frac{1}{4}x = 70$.

$$\frac{3}{4}(68) + \frac{1}{4}x = 70$$
$$51 + \frac{x}{4} = 70$$
$$-51 \qquad -51$$
$$\frac{x}{4} = 19$$
$$x = 4(19)$$
$$x = 76$$

He needs a 76 on his presentation to have a course grade of 70.

4. Let x represent the final exam grade. The test average is $(68 + 73 + 80 + 95)/4 = 79$. Then $\frac{1}{3}$ test average $+\frac{1}{3}$ presentation grade $+\frac{1}{3}$ final exam grade is 80 becomes

$$\frac{1}{3}(79) + \frac{1}{3}(75) + \frac{1}{3}(x) = 80.$$
$$\frac{79}{3} + 25 + \frac{x}{3} = 80$$
$$3\left(\frac{79}{3} + 25 + \frac{x}{3}\right) = 3(80)$$
$$3\left(\frac{79}{3}\right) + 3(25) + 3 \cdot \frac{x}{3} = 240$$
$$79 + 75 + x = 240$$
$$154 + x = 240$$
$$-154 \qquad -154$$
$$x = 86$$

The final exam grade needs to be 86 to obtain an average of 80.

5. Let $x =$ sixth quiz grade. The course grade $\frac{4}{5}$ quiz grade $+ \frac{1}{5}$ book report becomes

$$\frac{4}{5}\left(\frac{72 + 66 + 69 + 80 + 85 + x}{6}\right) + \frac{1}{5}(90).$$

Simplified, the above is

$$\frac{4}{5}\left(\frac{372+x}{6}\right)+18=\frac{4(372)+4x}{30}+18=\frac{1488+4x}{30}+18.$$

We want this quantity to equal 80.

$$\frac{1488+4x}{30}+18=80$$

$$30\left(\frac{1488+4x}{30}+18\right)=30(80)$$

$$30\left(\frac{1488+4x}{30}\right)+30(18)=2400$$

$$1488+4x+540=2400$$

$$2028+4x=2400$$

$$-2028 \qquad -2028$$

$$4x=372$$

$$x=\frac{372}{4}$$

$$x=93$$

The student needs a 93 on her quiz to raise her average to 80.

Coin problems are also common algebra applications. Usually the total number of coins is given as well as the total dollar value. The question is normally "How many of each coin is there?"

Let x represent the number of one specific coin and put the number of other coins in terms of x. The steps involved are:

1. Let x represent the number of a specific coin;
2. Write the number of other coins in terms of x (this skill was developed in the "age" problems);
3. Multiply the value of the coin by its number; this gives the total amount of money represented by each coin;
4. Add all of the terms obtained in Step 3 and set equal to the total money value;
5. Solve for x;
6. Answer the question. Don't forget this step! It is easy to feel like you are done when you have solved for x, but sometimes the answer to the question requires one more step.

Examples

As in all word problems, units of measure must be consistent. In the following problems, this means that all money will need to be in terms of dollars or in terms of cents. In the examples that follow, dollars will be used.

Terri has \$13.45 in dimes and quarters. If there are 70 coins in all, how many of each coin does she have?

Let x represent the number of dimes. Because the number of dimes and quarters is 70, $70 - x$ represents the number of quarters. Terri has x dimes, so she has $\$0.10x$ in dimes. She has $70 - x$ quarters, so she has $\$0.25(70 - x)$ in quarters. These two amounts must sum to \$13.45.

$$\underset{\text{(amount in dimes)}}{0.10x} + \underset{\text{(amount in quarters)}}{0.25(70 - x)} = 13.45$$

$$\begin{aligned} 0.10x + 0.25(70 - x) &= 13.45 \\ 0.10x + 17.5 - 0.25x &= 13.45 \\ -0.15x + 17.5 &= 13.45 \\ -17.5 \quad & \; -17.50 \\ -0.15x &= -4.05 \\ x &= \frac{-4.05}{-0.15} \\ x &= 27 \end{aligned}$$

Terri has 27 dimes and $70 - x = 70 - 27 = 43$ quarters.

Bobbie has \$1.54 in quarters, dimes, nickels, and pennies. He has twice as many dimes as quarters and three times as many nickels as dimes. The number of pennies is the same as the number of dimes. How many of each coin does he have?

Nickels are being compared to dimes, and dimes are being compared to quarters, so we will let x represent the number of quarters. Bobbie has twice as many dimes as quarters, so $2x$ is the number of dimes he has. He has three times as many nickels as dimes, namely triple $2x$: $3(2x) = 6x$. He has the same number of pennies as dimes, so he has $2x$ pennies.

How much of the total \$1.54 does Bobbie have in each coin? He has x quarters, each worth \$0.25, so he has a total of $0.25x$ (dollars) in quarters. He has $2x$ dimes, each worth \$0.10; this gives him $0.10(2x) = 0.20x$ (dollars) in dimes. Bobbie has $6x$ nickels, each worth \$0.05. The total amount of money in nickels, then, is $0.05(6x) = 0.30x$ (dollars). Finally, he has $2x$ pennies, each worth \$0.01. The pennies count as $0.01(2x) = 0.02x$ (dollars).

The total amount of money is \$1.54, so

$0.25x$	$+$	$0.20x$	$+$	$0.30x$	$+$	$0.02x$	$=$	$1.54.$
(amount in quarters)		(amount in dimes)		(amount in nickels)		(amount in pennies)		

$$\begin{aligned}
0.25x + 0.20x + 0.30x + 0.02x &= 1.54\\
0.77x &= 1.54\\
x &= \frac{1.54}{0.77}\\
x &= 2
\end{aligned}$$

Bobbie has 2 quarters; $2x = 2(2) = 4$ dimes; $6x = 6(2) = 12$ nickels; and $2x = 2(2) = 4$ pennies.

Practice

1. A vending machine has \$19.75 in dimes and quarters. There are 100 coins in all. How many dimes and quarters are in the machine?
2. Ann has \$2.25 in coins. She has the same number of quarters as dimes. She has half as many nickels as quarters. How many of each coin does she have?
3. Sue has twice as many quarters as nickels and half as many dimes as nickels. If she has a total of \$4.80, how many of each coin does she have?

Solutions

1. Let x represent the number of dimes. Then $100 - x$ is the number of quarters. There is $0.10x$ dollars in dimes and $0.25(100 - x)$ dollars in quarters.

$$
\begin{aligned}
0.10x + 0.25(100 - x) &= 19.75 \\
0.10x + 25 - 0.25x &= 19.75 \\
-0.15x + 25 &= 19.75 \\
-25 \quad & \quad -25.00 \\
-0.15x &= -5.25 \\
x &= \frac{-5.25}{-0.15} \\
x &= 35
\end{aligned}
$$

There are 35 dimes and $100 - x = 100 - 35 = 65$ quarters.

2. Let x represent the number of quarters. There are as many dimes as quarters, so x also represents the number of dimes. There are half as many nickels as dimes, so $\frac{1}{2}x$ (or $0.50x$, as a decimal number) is the number of nickels.

$$
\begin{aligned}
0.25x &= \text{amount in quarters} \\
0.10x &= \text{amount in dimes} \\
0.05(0.50x) &= \text{amount in nickels}
\end{aligned}
$$

$$
\begin{aligned}
0.25x + 0.10x + 0.05(0.50x) &= 2.25 \\
0.25x + 0.10x + 0.025x &= 2.25 \\
0.375x &= 2.25 \\
x &= \frac{2.25}{0.375} \\
x &= 6
\end{aligned}
$$

There are 6 quarters, 6 dimes, $0.50x = 0.50(6) = 3$ nickels.

3. As both the number of quarters and dimes are being compared to the number of nickels, let x represent the number of nickels. Then $2x$ represents the number of quarters and $\frac{1}{2}x$ (or $0.50x$) is the number of dimes.

$$
\begin{aligned}
0.05x &= \text{amount of money in nickels} \\
0.10(0.50x) &= \text{amount of money in dimes} \\
0.25(2x) &= \text{amount of money in quarters}
\end{aligned}
$$

$$\begin{aligned} 0.05x + 0.10(0.50x) + 0.25(2x) &= 4.80 \\ 0.05x + 0.05x + 0.50x &= 4.80 \\ 0.60x &= 4.80 \\ x &= \frac{4.80}{0.60} \\ x &= 8 \end{aligned}$$

There are 8 nickels, $0.50x = 0.50(8) = 4$ dimes, and $2x = 2(8) = 16$ quarters.

Some money problems involve one quantity divided into two investments paying different interest rates. Such questions are phrased "How much was invested at ___%?" or "How much was invested at each rate?"

Examples

A woman had \$10,000 to invest. She deposited her money into two accounts—one paying 6% interest and the other $7\frac{1}{2}\%$ interest. If at the end of the year the total interest earned was \$682.50, how much was originally deposited in each account?

You could either let x represent the amount deposited at 6% or at $7\frac{1}{2}\%$. Here, we will let x represent the amount deposited into the 6% account. Because the two amounts must sum to 10,000, $10{,}000 - x$ is the amount deposited at $7\frac{1}{2}\%$. The amount of interest earned at 6% is $0.06x$, and the amount of interest earned at $7\frac{1}{2}\%$ is $0.075(10{,}000 - x)$. The total amount of interest is \$682.50, so $0.06x + 0.075(10{,}000 - x) = 682.50$.

$$\begin{aligned} 0.06x + 0.075(10{,}000 - x) &= 682.50 \\ 0.06x + 750 - 0.075x &= 682.50 \\ -0.015x + 750 &= 682.50 \\ -750 \quad & -750.00 \\ -0.015x &= -67.50 \\ x &= \frac{-67.50}{-0.015} \\ x &= 4500 \end{aligned}$$

The woman deposited \$4500 in the 6% account and $10{,}000 - x = 10{,}000 - 4500 = \$5500$ in the $7\frac{1}{2}\%$ account.

Practice

1. A businessman invested \$50,000 into two funds which yielded profits of $16\frac{1}{2}\%$ and 18%. If the total profit was \$8520, how much was invested in each fund?

2. A college student deposited \$3500 into two savings accounts, one with an annual yield of $4\frac{3}{4}\%$ and the other with an annual yield of $5\frac{1}{4}\%$. If he earned \$171.75 total interest the first year, how much was deposited in each account?

3. A banker plans to lend \$48,000 at a simple interest rate of 16% and the remainder at 19%. How should she allot the loans in order to obtain a return of $18\frac{1}{2}\%$?

Solutions

1. Let x represent the amount invested at $16\frac{1}{2}\%$. Then $50{,}000 - x$ represents the amount invested at 18%. The profit from the $16\frac{1}{2}\%$ account is $0.165x$, and the profit from the 18% investment is $0.18(50{,}000 - x)$. The sum of the profits is \$8520.

$$
\begin{aligned}
0.165x + 0.18(50{,}000 - x) &= 8520 \\
0.165x + 9000 - 0.180x &= 8520 \\
-0.015x + 9000 &= 8520 \\
-9000 \quad & \quad -9000 \\
-0.015x &= -480 \\
x &= \frac{-480}{-0.015} \\
x &= 32{,}000
\end{aligned}
$$

The amount invested at $16\frac{1}{2}\%$ is \$32,000, and the amount invested at 18% is $50{,}000 - x = 50{,}000 - 32{,}000 = \$18{,}000$.

2. Let x represent the amount deposited at $4\frac{3}{4}\%$. Then the amount deposited at $5\frac{1}{4}\%$ is $3500 - x$. The interest earned at $4\frac{3}{4}\%$ is $0.0475x$; the interest earned at $5\frac{1}{4}\%$ is $0.0525(3500 - x)$. The sum

of these two quantities is 171.75.

$$0.0475x + 0.0525(3500 - x) = 171.75$$
$$0.0475x + 183.75 - 0.0525x = 171.75$$
$$183.75 - 0.005x = 171.75$$
$$-183.75 \qquad -183.75$$
$$-0.005x = -12$$
$$x = \frac{-12}{-0.005}$$
$$x = 2400$$

\$2400 was deposited in the $4\frac{3}{4}\%$ account, and $3500 - x = 3500 - 2400 =$ \$1100 was deposited in the $5\frac{1}{4}\%$ account.

3. Let x represent the amount to be loaned at 16%, so $48{,}000 - x$ represents the amount to be loaned at 19%. The total amount of return should be $18\frac{1}{2}\%$ of 48,000 which is $0.185(48{,}000) = 8880$.

$$0.16x + 0.19(48{,}000 - x) = 8880$$
$$0.16x + 9120 - 0.19x = 8880$$
$$9120 - 0.03x = 8880$$
$$-9120 \qquad -9120$$
$$-0.03x = -240$$
$$x = \frac{-240}{-0.03}$$
$$x = 8000$$

\$8000 should be loaned at 16%, and $48{,}000 - x = 48{,}000 - 8000 =$ \$40,000 should be loaned at 19%.

Mixture problems involve mixing two different concentrations to obtain some concentration in between. Often these problems are stated as alcohol or acid solutions, but there are many more types. For example, you might want to know how many pure peanuts should be mixed with a 40% peanut mixture to obtain a 50% peanut mixture. You might have a two-cycle engine requiring a particular oil and gas mixture. Or, you might have a recipe calling for 1% fat milk and all you have on hand is 2% fat milk and $\frac{1}{2}\%$ fat milk. These problems can be solved using the method illustrated below.

There will be three quantities—the two concentrations being mixed together and the final concentration. One of the three quantities will be a fixed number. Let the variable represent one of the two concentrations being mixed The other unknown quantity will be written as some combination of the variable and the fixed quantity. If one of the quantities being mixed is known, then let x represent the other quantity being mixed and the final solution will be "$x +$ known quantity." If the final solution is known, again let x represent one of the quantities being mixed, the other quantity being mixed will be of the form "final solution quantity $- x$."

For example, in the following problem, the amount of one of the two concentrations being mixed will be known.

"How many liters of 10% acid solution should be mixed with 75 liters of 30% acid solution to yield a 25% acid solution?" Let x represent the number of liters of 10% acid solution. Then $x + 75$ will represent the number of liters of the final solution. If the problem were stated, "How many liters of 10% acid solution and 30% solution should be mixed together with to produce 100 liters of 25% solution?" We can let x represent either the number of liters of 10% solution or 30% solution. We will let x represent the number of liters of 10% solution. How do we represent the number of liters of 30% solution? For the moment, let "?" represent the number of liters of 30% solution. We know that the final solution must be 100 liters, so the two amounts have to sum to 100: $x + ? = 100$.

$$\begin{aligned} x + ? &= 100 \\ -x \quad & \quad -x \\ ? &= 100 - x \end{aligned}$$

Now we see that $100 - x$ represents the number of liters of 30% solution.

Draw three boxes. Write the percentages given above the boxes and the volume inside the boxes. Multiply the percentages (converted to decimal numbers) and the volumes. Write these quantities below the boxes; this will give you the equation to solve. Incidentally, when you multiply the percent by the volume, you are getting the volume of pure acid/alcohol/milk-fat/etc.

Examples

How much 10% acid solution should be added to 30 liters of 25% acid solution to achieve a 15% solution?

Let x represent the amount of 10% solution. Then the total amount of solution will be $30 + x$.

10%		25%		15%
x liters	+	30 liters	=	$30 + x$ liters
$0.10x$	+	$0.25(30)$	=	$0.15(30 + x)$

(There are $0.10x$ liters of pure acid in the 10% mixture, 0.25(30) liters of pure acid in the 25% mixture, and $0.15(x + 30)$ liters of pure acid in the 15% mixture.)

$$\begin{aligned}
0.10x + 0.25(30) &= 0.15(x + 30) \\
0.10x + 7.5 &= 0.15x + 4.5 \\
-0.10x \quad & \quad -0.10x \\
7.5 &= 0.05x + 4.5 \\
-4.5 \quad & \quad -4.5 \\
3 &= 0.05x \\
\frac{3}{0.05} &= x \\
60 &= x
\end{aligned}$$

Add 60 liters of 10% acid solution to 30 liters of 25% acid solution to achieve a 15% acid solution.

How much 10% acid solution and 30% acid solution should be mixed together to yield 100 liters of a 25% acid solution?

Let x represent the amount of 10% acid solution. Then $100 - x$ represents the amount of 30% acid solution.

10%		30%		25%
x liters	+	$100 - x$ liters	=	100 liters
$0.10x$	+	$0.30(100 - x)$	=	$0.25(100)$

$$\begin{aligned}
0.10x + 0.30(100 - x) &= 0.25(100) \\
0.10x + 30 - 0.30x &= 25 \\
30 - 0.20x &= 25 \\
-30 \quad & \quad -30 \\
-0.20x &= -5 \\
x &= \frac{-5}{-0.20} \\
x &= 25
\end{aligned}$$

Add 25 liters of 10% solution to $100 - x = 100 - 25 = 75$ liters of 30% solution to obtain 100 liters of 25% solution.

How much pure alcohol should be added to six liters of 30% alcohol solution to obtain a 40% alcohol solution?

100%		30%		40%
x liters	+	6 liters	=	$x+6$ liters
$1.0x$	+	$0.30(6)$	=	$0.40(x+6)$

$$1.00x + 0.30(6) = 0.40(x+6)$$
$$1.00x + 1.80 = 0.40x + 2.4$$
$$-0.40x \qquad -0.40x$$
$$0.60x + 1.8 = 2.4$$
$$-1.8 \quad -1.8$$
$$0.60x = 0.6$$
$$x = \frac{0.6}{0.6}$$
$$x = 1$$

Add one liter of pure alcohol to six liters of 30% alcohol solution to obtain a 40% alcohol solution.

How much water should be added to 9 liters of 45% solution to weaken it to a 30% solution?

Think of water as a "0% solution."

0%		45%		30%
x liters	+	9 liters	=	$x+9$ liters
$0x$	+	$0.45(9)$	=	$0.30(x+9)$

$$0x + 0.45(9) = 0.30(x+9)$$
$$0 + 4.05 = 0.30x + 2.70$$
$$-2.70 \qquad -2.70$$
$$1.35 = 0.30x$$
$$\frac{1.35}{0.30} = x$$
$$4.5 = x$$

Add 4.5 liters of water to weaken 9 liters of 45% solution to a 30% solution.

How much pure acid and 30% acid solution should be mixed together to obtain 28 quarts of 40% acid solution?

100%		30%		40%
x quarts	+	$28-x$ quarts	=	28 quarts
$1.00x$	+	$0.30(28-x)$	=	$0.40(28)$

$$
\begin{aligned}
1.00x + 0.30(28 - x) &= 0.40(28) \\
x + 8.40 - 0.30x &= 11.2 \\
0.70x + 8.40 &= 11.2 \\
-8.40 \quad & \quad -8.4 \\
0.70x &= 2.8 \\
x &= \frac{2.8}{0.70} \\
x &= 4
\end{aligned}
$$

Add 4 quarts of pure acid to $28 - x = 28 - 4 = 24$ quarts of 30% acid solution to yield 28 quarts of a 40% solution.

Practice

1. How much 60% acid solution should be added to 8 liters of 25% acid solution to produce a 40% acid solution?

2. How many quarts of $\frac{1}{2}$% fat milk should be added to 4 quarts of 2% fat milk to produce 1% fat milk?

3. How much 30% alcohol solution should be mixed with 70% alcohol solution to produce 12 liters of 60% alcohol solution?

4. How much 65% acid solution and 25% acid solution should be mixed together to produce 180 ml of 40% acid solution?

5. How much water should be added to 10 liters of 45% alcohol solution to produce a 30% solution?

6. How much decaffeinated coffee (assume this means 0% caffeine) and 50% caffeine coffee should be mixed to produce 25 cups of 40% caffeine coffee?

7. How much pure acid should be added to 18 ounces of 35% acid solution to produce 50% acid solution?

8. How many peanuts should be mixed with a nut mixture that is 40% peanuts to produce 36 ounces of a 60% peanut mixture?

Solutions

1.

60%		25%		40%
x liters	+	8 liters	=	$x+8$ liters
$0.60x$	+	$0.25(8)$	=	$0.40(x+8)$

$$0.60x + 0.25(8) = 0.40(x+8)$$

$$\begin{aligned} 0.60x + 2 &= 0.40x + 3.2 \\ -0.40x \quad & \quad -0.40x \\ 0.20x + 2.0 &= 3.2 \\ -2.0 \quad & \quad -2.0 \\ 0.20x &= 1.2 \\ x &= \frac{1.2}{0.20} \\ x &= 6 \end{aligned}$$

Add 6 liters of 60% solution to 8 liters of 25% solution to produce a 40% solution.

2.

0.5%		2%		1%
x quarts	+	4 quarts	=	$x+4$ quarts
$0.005x$	+	$0.02(4)$	=	$0.01(x+4)$

$$
\begin{aligned}
0.005x + 0.02(4) &= 0.01(x+4) \\
0.005x + 0.08 &= 0.010x + 0.04 \\
-0.005x \qquad & \quad -0.005x \\
0.08 &= 0.005x + 0.04 \\
-0.04 \qquad & \qquad\quad -0.04 \\
0.04 &= 0.005x \\
\frac{0.04}{0.005} &= x \\
8 &= x
\end{aligned}
$$

Add 8 quarts of $\frac{1}{2}$% fat milk to 4 quarts of 2% milk to produce 1% milk.

3.

30%		70%		60%
x liters	+	$12 - x$ liters	=	12 liters
$0.30x$	+	$0.70(12-x)$	=	$0.60(12)$

$$
\begin{aligned}
0.30x + 0.70(12 - x) &= 0.60(12) \\
0.30x + 8.4 - 0.70x &= 7.2 \\
-0.40x + 8.4 &= 7.2 \\
-8.4 \quad & \; -8.4 \\
-0.40x &= -1.2 \\
x &= \frac{-1.2}{-0.40} \\
x &= 3
\end{aligned}
$$

Add 3 liters of 30% alcohol solution to $12 - x = 12 - 3 = 9$ liters of 70% alcohol solution to produce 12 liters of 60% alcohol solution.

4.

65%		25%		40%
x ml	+	$180 - x$ ml	=	180 ml
$0.65x$	+	$0.25(180-x)$	=	$0.40(180)$

$$0.65x + 0.25(180 - x) = 0.40(180)$$
$$0.65x + 45 - 0.25x = 72$$
$$0.40x + 45 = 72$$
$$\quad -45 \quad -45$$
$$0.40x = 27$$
$$x = \frac{27}{0.40}$$
$$x = 67.5$$

Add 67.5 ml of 65% acid solution to $180 - x = 180 - 67.5 = 112.5$ ml of 25% solution to produce 180 ml of 40% acid solution.

5.

0%		45%		30%
x liters	+	10 liters	=	$x + 10$ liters
$0x$	+	$0.45(10)$	=	$0.30(x + 10)$

$$0x + 0.45(10) = 0.30(x + 10)$$
$$0 + 4.5 = 0.30x + 3$$
$$-3.0 \qquad -3$$
$$1.5 = 0.30x$$
$$\frac{1.5}{0.30} = x$$
$$5 = x$$

Add 5 liters of water to 10 liters of 45% alcohol solution to produce a 30% alcohol solution.

6.

0%		50%		40%
x cups	+	$25 - x$ cups	=	25 cups
$0x$	+	$0.50(25 - x)$	=	$0.40(25)$

$$0x + 0.50(25 - x) = 0.40(25)$$

$$0 + 12.5 - 0.50x = 10.0$$
$$-12.5 \qquad -12.5$$

$$-0.50x = -2.5$$

$$x = \frac{-2.5}{-0.50}$$

$$x = 5$$

Mix 5 cups of decaffeinated coffee with $25 - x = 25 - 5 = 20$ cups of 50% caffeine coffee to produce 25 cups of 40% caffeine coffee.

7.

100%		35%		50%
x ounces	+	18 ounces	=	$x + 18$ ounces
$1.00x$	+	$0.35(18)$	=	$0.50(x + 18)$

$$1.00x + 0.35(18) = 0.50(x + 18)$$

$$1.00x + 6.3 = 0.50x + 9$$
$$-0.50x \qquad -0.50x$$

$$0.50x + 6.3 = 9.0$$
$$-6.3 \quad -6.3$$

$$0.50x = 2.7$$

$$x = \frac{2.7}{0.50}$$

$$x = 5.4$$

Add 5.4 ounces of pure acid to 18 ounces of 35% acid solution to produce a 50% acid solution.

8.

100%		40%		60%
x ounces	+	$36 - x$ ounces	=	36 ounces
$1.00x$	+	$0.40(36 - x)$	=	$0.60(36)$

$$
\begin{aligned}
1.00x + 0.40(36 - x) &= 0.60(36) \\
1.00x + 14.4 - 0.40x &= 21.6 \\
0.60x + 14.4 &= 21.6 \\
-14.4 \quad & \quad -14.4 \\
0.60x &= 7.2 \\
x &= \frac{7.2}{0.6} \\
x &= 12
\end{aligned}
$$

Add 12 ounces of peanuts to $36 - x = 36 - 12 = 24$ ounces of a 40% peanut mixture to produce 36 ounces of a 60% peanut mixture.

Work Problems

Work problems are another staple of algebra courses. A work problem is normally stated as two workers (two people, machines, hoses, drains, etc.) working together and working separately to complete a task. Often one worker performs faster than the other. Sometimes the problem states how fast each can complete the task alone and you are asked to find how long it takes for them to complete the task together. At other times, you are told how long one worker takes to complete the task alone and how long it takes for both to work together to complete it; you are asked how long the second worker would take to complete the task alone. The formula is quantity (work done—usually "1") = rate times time: $Q = rt$. The method outlined below will help you solve most, if not all, work problems. The following chart is useful in solving these problems.

Worker	**Quantity**	**Rate**	**Time**
1st Worker			
2nd Worker			
Together			

There are four equations in this chart. One of them will be the one you will use to solve for the unknown. Each horizontal line in the chart represents the equation $Q = rt$ for that particular line. The fourth equation comes from the sum of each worker's rate set equal to the together rate. Often, the fourth equation is the one you will need to solve. Remember, as in all word problems, that all units of measure must be consistent.

Examples

Joe takes 45 minutes to mow a lawn. His older brother Jerry takes 30 minutes to mow the lawn. If they work together, how long will it take for them to mow the lawn?

The quantity in each of the three cases is 1—there is one yard to be mowed. Use the formula $Q = rt$ and the data given in the problem to fill in all nine boxes. Because we are looking for the time (in minutes) it takes for them to mow the lawn together, let t represent the number of minutes needed to mow the lawn together.

Worker	Quantity	Rate	Time
Joe	1		45
Jerry	1		30
Together	1		t

Because $Q = rt$, $r = Q/t$. But $Q = 1$, so $r = 1/t$. This makes Joe's rate 1/45 and Jerry's rate 1/30. The together rate is $1/t$.

Worker	Quantity	Rate	Time
Joe	1	1/45	45
Jerry	1	1/30	30
Together	1	$1/t$	t

Of the four equations on the chart, only "Joe's rate + Jerry's rate = Together rate" has enough information in it to solve for t.

The equation to solve is $1/45 + 1/30 = 1/t$. The LCD is $90t$.

$$\frac{1}{45} + \frac{1}{30} = \frac{1}{t}$$

$$90t\left(\frac{1}{45} + \frac{1}{30}\right) = 90t\left(\frac{1}{t}\right)$$

$$90t\left(\frac{1}{45}\right) + 90t\left(\frac{1}{30}\right) = 90$$

$$2t + 3t = 90$$

$$5t = 90$$

$$t = \frac{90}{5}$$

$$t = 18$$

They can mow the yard in 18 minutes.

Tammy can wash a car in 40 minutes. When working with Jim, they can wash the same car in 15 minutes. How long would Jim need to wash the car by himself?

Let t represent the number of minutes Jim needs to wash the car alone.

Worker	Quantity	Rate	Time
Tammy	1	1/40	40
Jim	1	$1/t$	t
Together	1	1/15	15

The equation to solve is $1/40 + 1/t = 1/15$. The LCD is $120t$.

$$\frac{1}{40} + \frac{1}{t} = \frac{1}{15}$$

$$120t\left(\frac{1}{40} + \frac{1}{t}\right) = 120t\left(\frac{1}{15}\right)$$

$$120t\left(\frac{1}{40}\right) + 120t\left(\frac{1}{t}\right) = 8t$$

$$3t + 120 = 8t$$
$$\underline{-3t \qquad\qquad -3t}$$
$$120 = 5t$$
$$\frac{120}{5} = t$$
$$24 = t$$

Jim needs 24 minutes to wash the car alone.

Kellie can mow the campus yard in $2\frac{1}{2}$ hours. When Bobby helps, they can mow the yard in $1\frac{1}{2}$ hours. How long would Bobby need to mow the yard by himself?

Let t represent the number of hours Bobby needs to mow the yard himself. Kellie's time is $2\frac{1}{2}$ or $\frac{5}{2}$. Then her rate is $\frac{1}{5/2}$.

$$\frac{1}{5/2} = 1 \div \frac{5}{2} = 1 \cdot \frac{2}{5} = \frac{2}{5}$$

The together time is $1\frac{1}{2}$ or $\frac{3}{2}$, so the together rate is $\frac{1}{3/2}$.

$$\frac{1}{3/2} = 1 \div \frac{3}{2} = 1 \cdot \frac{2}{3} = \frac{2}{3}$$

Worker	Quantity	Rate	Time
Kellie	1	2/5	$2\frac{1}{2}$
Bobby	1	$1/t$	t
Together	1	2/3	$1\frac{1}{2}$

The equation to solve is $2/5 + 1/t = 2/3$. The LCD is $15t$.

$$\frac{2}{5} + \frac{1}{t} = \frac{2}{3}$$
$$15t\left(\frac{2}{5} + \frac{1}{t}\right) = 15t\left(\frac{2}{3}\right)$$
$$15t\left(\frac{2}{5}\right) + 15t\left(\frac{1}{t}\right) = 10t$$

$$\begin{aligned} 6t + 15 &= 10t \\ -6t \quad & \quad -6t \\ 15 &= 4t \\ \frac{15}{4} &= t \end{aligned}$$

Bobby needs $15/4 = 3\frac{3}{4}$ hours or 3 hours 45 minutes to mow the yard by himself.

Practice

1. Sherry and Denise together can mow a yard in 20 minutes. Alone, Denise can mow the yard in 30 minutes. How long would Sherry need to mow the yard by herself?

2. Together, Ben and Brandon can split a pile of wood in 2 hours. If Ben could split the same pile of wood in 3 hours, how long would it take Brandon to split the pile alone?

3. A boy can weed the family garden in 90 minutes. His sister can weed it in 60 minutes. How long will they need to weed the garden if they work together?

4. Robert needs 40 minutes to assemble a bookcase. Paul needs 20 minutes to assemble the same bookcase. How long will it take them to assemble the bookcase if they work together?

5. Together, two pipes can fill a reservoir in $\frac{3}{4}$ of an hour. Pipe I needs one hour ten minutes ($1\frac{1}{6}$ hours) to fill the reservoir by itself. How long would Pipe II need to fill the reservoir by itself?

6. A pipe can drain a reservoir in 6 hours 30 minutes ($6\frac{1}{2}$ hours). A larger pipe can drain the same reservoir in 4 hours 20 minute ($4\frac{1}{3}$ hours). How long will it take to drain the reservoir if both pipes are used?

Solutions

In the following, t will represent the unknown time.

1.

Worker	Quantity	Rate	Time
Sherry	1	$1/t$	t
Denise	1	1/30	30
Together	1	1/20	20

The equation to solve is $1/t + 1/30 = 1/20$. The LCD is $60t$.

$$\frac{1}{t} + \frac{1}{30} = \frac{1}{20}$$

$$60t\left(\frac{1}{t} + \frac{1}{30}\right) = 60t\left(\frac{1}{20}\right)$$

$$60t\left(\frac{1}{t}\right) + 60t\left(\frac{1}{30}\right) = 3t$$

$$60 + 2t = 3t$$

$$-2t \quad -2t$$

$$60 = t$$

Alone, Sherry can mow the yard in 60 minutes.

2.

Worker	Quantity	Rate	Time
Ben	1	1/3	3
Brandon	1	$1/t$	t
Together	1	1/2	2

The equation to solve is $1/3 + 1/t = 1/2$. The LCD is $6t$.

$$\frac{1}{3} + \frac{1}{t} = \frac{1}{2}$$

$$6t\left(\frac{1}{3} + \frac{1}{t}\right) = 6t\left(\frac{1}{2}\right)$$

$$6t\left(\frac{1}{3}\right) + 6t\left(\frac{1}{t}\right) = 3t$$

$$2t + 6 = 3t$$

$$-2t \qquad -2t$$

$$6 = t$$

Brandon can split the wood-pile by himself in 6 hours.

3.

Worker	Quantity	Rate	Time
Boy	1	1/90	90
Girl	1	1/60	60
Together	1	$1/t$	t

The equation to solve is $1/90 + 1/60 = 1/t$. The LCD is $180t$.

$$\frac{1}{90} + \frac{1}{60} = \frac{1}{t}$$

$$180t\left(\frac{1}{90} + \frac{1}{60}\right) = 180t\left(\frac{1}{t}\right)$$

$$180t\left(\frac{1}{90}\right) + 180t\left(\frac{1}{60}\right) = 180$$

$$2t + 3t = 180$$

$$5t = 180$$

$$t = \frac{180}{5}$$

$$t = 36$$

Working together, the boy and girl need 36 minutes to weed the garden.

4.

Worker	Quantity	Rate	Time
Robert	1	1/40	40
Paul	1	1/20	20
Together	1	$1/t$	t

The equation to solve is $1/40 + 1/20 = 1/t$. The LCD is $40t$.

$$\frac{1}{40}+\frac{1}{20}=\frac{1}{t}$$

$$40t\left(\frac{1}{40}+\frac{1}{20}\right)=40t\left(\frac{1}{t}\right)$$

$$40t\left(\frac{1}{40}\right)+40t\left(\frac{1}{20}\right)=40$$

$$t+2t=40$$

$$3t=40$$

$$t=\frac{40}{3}=13\tfrac{1}{3}$$

Together Robert and Paul can assemble the bookcase in $13\frac{1}{3}$ minutes or 13 minutes 20 seconds.

5.

Worker	Quantity	Rate	Time
Pipe I	1	6/7 $\frac{1}{7/6}=6/7$	7/6
Pipe II	1	$1/t$	t
Together	1	4/3 $\frac{1}{3/4}=4/3$	3/4

The equation to solve is $6/7+1/t=4/3$. The LCD is $21t$.

$$\frac{6}{7}+\frac{1}{t}=\frac{4}{3}$$

$$21t\left(\frac{6}{7}+\frac{1}{t}\right)=21t\left(\frac{4}{3}\right)$$

$$21t\left(\frac{6}{7}\right)+21t\left(\frac{1}{t}\right)=28t$$

$$18t+21=28t$$

$$-18t \qquad -18t$$

$$21=10t$$

$$\frac{21}{10}=t$$

Alone, Pipe II can fill the reservoir in $2\frac{1}{10}$ hours or 2 hours, 6 minutes. ($\frac{1}{10}$ of an hour is $\frac{1}{10}$ of 60 minutes and $\frac{1}{10} \cdot 60 = 6$.)

6.

Worker	Quantity	Rate	Time
Pipe I	1	$2/13$ $\frac{1}{13/2} = 2/13$	$6\frac{1}{2} = 13/2$
Pipe II	1	$3/13$ $\frac{1}{13/3} = 3/13$	$4\frac{1}{3} = 13/3$
Together	1	$1/t$	t

The equation to solve is $2/13 + 3/13 = 1/t$. The LCD is $13t$.

$$\frac{2}{13} + \frac{3}{13} = \frac{1}{t}$$

$$13t\left(\frac{2}{13} + \frac{3}{13}\right) = 13t\left(\frac{1}{t}\right)$$

$$13t\left(\frac{2}{13}\right) + 13t\left(\frac{3}{13}\right) = 13$$

$$2t + 3t = 13$$

$$5t = 13$$

$$t = \frac{13}{5}$$

Together the pipes can drain the reservoir in $2\frac{3}{5}$ hours or 2 hours 36 minutes. ($\frac{3}{5}$ of hour is $\frac{3}{5}$ of 60 minutes and $\frac{3}{5} \cdot 60 = 36$.)

Some work problems require part of the work being performed by one worker before the other worker joins in, or both start the job and one finishes the job. In these cases, the together quantity and one of the individual quantities will not be "1." Take the time the one worker works alone divided by the time that worker requires to do the job alone, then subtract from 1. This is the proportion left over for both to work together.

Examples

Jerry needs 40 minutes to mow the lawn. Lou can mow the same lawn in 30 minutes. If Jerry works alone for 10 minutes then Lou joins in, how long will it take for them to finish the job?

Because Jerry worked for 10 minutes, he did $10/40 = \frac{1}{4}$ of the job alone. So, there is $1 - \frac{1}{4} = \frac{3}{4}$ of the job remaining when Lou started working. Let t represent the number of minutes they worked together—after Lou joins in. Even though Lou does not work the entire job, his rate is still $1/30$.

Worker	Quantity	Rate	Time
Jerry	1	1/40	40
Lou	1	1/30	30
Together	3/4	$\frac{3/4}{t} = \frac{3}{4t}$	t

The equation to solve is $1/40 + 1/30 = 3/4t$. The LCD is $120t$.

$$\frac{1}{40} + \frac{1}{30} = \frac{3}{4t}$$

$$120t\left(\frac{1}{40} + \frac{1}{30}\right) = 120t\left(\frac{3}{4t}\right)$$

$$120t\left(\frac{1}{40}\right) + 120t\left(\frac{1}{30}\right) = 90$$

$$3t + 4t = 90$$

$$7t = 90$$

$$t = \frac{90}{7}$$

Together, they will work $90/7 = 12\frac{6}{7}$ minutes.

A pipe can fill a reservoir in 6 hours. Another pipe can fill the same reservoir in 4 hours. If the second pipe is used alone for $2\frac{1}{2}$ hours, then the first pipe joins the second to finish the job, how long will the first pipe be used?

The amount of time Pipe I is used is the same as the amount of time both pipes work together. Let t represent the number of hours both pipes are used. Alone, the second pipe performed $2\frac{1}{2}$ parts of a 4-part job:

$$\frac{2\frac{1}{2}}{4} = \frac{5}{2} \div 4 = \frac{5}{2} \cdot \frac{1}{4} = \frac{5}{8}. \quad 1 - \frac{5}{8} = \frac{3}{8} \text{ of the job remains.}$$

Worker	Quantity	Rate	Time
Pipe I	1	1/6	6
Pipe II	1	1/4	4
Together	3/8	$\frac{3}{8t}$	t

The equation to solve is $1/6 + 1/4 = 3/8t$. The LCD is $24t$.

$$\frac{1}{6} + \frac{1}{4} = \frac{3}{8t}$$

$$24t\left(\frac{1}{6} + \frac{1}{4}\right) = 24t\left(\frac{3}{8t}\right)$$

$$24t\left(\frac{1}{6}\right) + 24t\left(\frac{1}{4}\right) = 9$$

$$4t + 6t = 9$$

$$10t = 9$$

$$t = \frac{9}{10}$$

Both pipes together will be used for $\frac{9}{10}$ hours or $\frac{9}{10} \cdot 60 = 54$ minutes. Hence, Pipe I will be used for 54 minutes.

Press A can print 100 fliers per minute. Press B can print 150 fliers per minute. The presses will be used to print 150,000 fliers.

(a) How long will it take for both presses to complete the run if they work together?

(b) If Press A works alone for 24 minutes then Press B joins in, how long will it take both presses to complete the job?

These problems are different from the previous work problems because the *rates* are given, not the times.

Before, we used $Q = rt$ implies $r = Q/t$. Here, we will use $Q = rt$ to fill in the Quantity boxes.

(a) Press A's rate is 100, and Press B's rate is 150. The together quantity is 150,000. Let t represent the number of minutes both presses work together; this is also how much time each individual press will run.

Press A's quantity is $100t$, and Press B's quantity is $150t$. The together rate is $r = Q/t = 150{,}000/t$.

Worker	Quantity	Rate	Time
Press A	$100t$	100	t
Press B	$150t$	150	t
Together	150,000	$150{,}000/t$	t

In this problem, the quantity produced by Press A plus the quantity produced by Press B will equal the quantity produced together. This gives the equation $100t + 150t = 150{,}000$. (Another equation that works is $100 + 150 = 150{,}000/t$.)

$$100t + 150t = 150{,}000$$

$$250t = 150{,}000$$

$$t = \frac{150{,}000}{250}$$

$$t = 600$$

The presses will run for 600 minutes or 10 hours.

(b) Because Press A works alone for 24 minutes, it has run $24 \times 100 = 2400$ fliers. When Press B begins its run, there are $150{,}000 - 2400 = 147{,}600$ fliers left to run. Let t represent the number of minutes both presses are running. This is also how much time Press B spends on the run. The boxes will represent work done together.

Worker	Quantity	Rate	Time
Press A	$100t$	100	t
Press B	$150t$	150	t
Together	147,600	$147{,}600/t$	t

The equation to solve is $100t + 150t = 147{,}600$. (Another equation that works is $100 + 150 = 147{,}600/t$.)

$$100t + 150t = 147{,}600$$
$$250t = 147{,}600$$
$$t = \frac{147{,}600}{250} = 590\tfrac{2}{5} = 590.4 \text{ minutes}$$

The presses will work together for 590.4 minutes or 9 hours 50 minutes 24 seconds. (This is 590 minutes and $0.4(60) = 24$ seconds.)

Practice

1. Neil can paint a wall in 45 minutes; Scott, in 30 minutes. If Neil begins painting the wall and Scott joins in after 15 minutes, how long will it take both to finish the job?

2. Two hoses are used to fill a water trough. The first hose can fill it in 20 minutes while the second hose needs only 16 minutes. If the second hose is used for the first 4 minutes and then the first hose is also used, how long will the first hose be used?

3. Jeremy can mow a lawn in one hour. Sarah can mow the same lawn in one and a half hours. If Jeremy works alone for 20 minutes then Sarah starts to help, how long will it take for them to finish the lawn?

4. A mold press can produce 1200 buttons an hour. Another mold press can produce 1500 buttons an hour. They need to produce 45,000 buttons.

   (a) How long will be needed if both presses are used to run the job?

(b) If the first press runs for 3 hours then the second press joins in, how long will it take for them to finish the run?

Solutions

1. Neil worked alone for $15/45 = 1/3$ of the job, so $1 - 1/3 = 2/3$ of the job remains. Let t represent the number of minutes both will work together.

Worker	Quantity	Rate	Time
Neil	1	1/45	45
Scott	1	1/30	30
Together	2/3	$\frac{2}{3t}$ $\frac{2/3}{t} = \frac{2}{3t}$	t

The equation to solve is $1/45 + 1/30 = 2/3t$. The LCD is $90t$.

$$\frac{1}{45} + \frac{1}{30} = \frac{2}{3t}$$

$$90t\left(\frac{1}{45} + \frac{1}{30}\right) = 90t\left(\frac{2}{3t}\right)$$

$$90t\left(\frac{1}{45}\right) + 90t\left(\frac{1}{30}\right) = 60$$

$$2t + 3t = 60$$

$$5t = 60$$

$$t = \frac{60}{5} = 12$$

It will take Scott and Neil 12 minutes to finish painting the wall.

2. Hose 2 is used alone for $\frac{4}{16} = \frac{1}{4}$ of the job, so $1 - \frac{1}{4} = \frac{3}{4}$ of the job remains. Let t represent the number of minutes both hoses will be used.

Worker	Quantity	Rate	Time
Hose 1	1	1/20	20
Hose 2	1	1/16	16
Together	3/4	$\frac{3}{4t}$ $\frac{3/4}{t} = \frac{3}{4t}$	t

The equation to solve is $1/20 + 1/16 = 3/4t$. The LCD is $80t$.

$$\frac{1}{20} + \frac{1}{16} = \frac{3}{4t}$$

$$80t\left(\frac{1}{20} + \frac{1}{16}\right) = 80t\left(\frac{3}{4t}\right)$$

$$80t\left(\frac{1}{20}\right) + 80t\left(\frac{1}{16}\right) = 60$$

$$4t + 5t = 60$$

$$9t = 60$$

$$t = \frac{60}{9} = 6\tfrac{2}{3}$$

Both hoses will be used for $6\frac{2}{3}$ minutes or 6 minutes 40 seconds. Therefore, Hose 1 will be used for $6\frac{2}{3}$ minutes.

3. Some of the information given in this problem is given in hours and other information in minutes. We must use only one unit of measure. Using minutes as the unit of measure will make the computations a little less messy. Let t represent the number of minutes both Sarah and Jeremy work together. Alone, Jeremy completed $\frac{20}{60} = \frac{1}{3}$ of the job, so $1 - \frac{1}{3} = \frac{2}{3}$ of the job remains to be done.

Worker	Quantity	Rate	Time
Jeremy	1	1/60	60
Sarah	1	1/90	90
Together	2/3	$\frac{2}{3t}$ $\frac{2/3}{t} = \frac{2}{3t}$	t

The equation to solve is $1/60 + 1/90 = 2/3t$. The LCD is $180t$.

$$\frac{1}{60} + \frac{1}{90} = \frac{2}{3t}$$

$$180t\left(\frac{1}{60} + \frac{1}{90}\right) = 180t \cdot \frac{2}{3t}$$

$$180t\left(\frac{1}{60}\right) + 180t\left(\frac{1}{90}\right) = 120$$

$$3t + 2t = 120$$

$$5t = 120$$

$$t = \frac{120}{5}$$

$$t = 24$$

They will need 24 minutes to finish the lawn.

4. (a) Let t represent the number of hours the presses need, working together, to complete the job.

Worker	Quantity	Rate	Time
Press 1	$1200t$	1200	t
Press 2	$1500t$	1500	t
Together	45,000	$45{,}000/t$	t

The equation to solve is $1200t + 1500t = 45{,}000$. (Another equation that works is $1200 + 1500 = 45{,}000/t$.)

$$1200t + 1500t = 45{,}000$$

$$2700t = 45{,}000$$

$$t = \frac{45{,}000}{2700}$$

$$t = 16\tfrac{2}{3}$$

They will need $16\frac{2}{3}$ hours or 16 hours 40 minutes ($\frac{2}{3}$ of an hour is $\frac{2}{3}$ of 60 minutes—$\frac{2}{3} \cdot 60 = 40$) to complete the run.

(b) Press 1 has produced $3(1200) = 3600$ buttons alone, so there remains $45{,}000 - 3600 = 41{,}400$ buttons to be produced. Let t represent the number of hours the presses, running together, need to complete the job.

Worker	Quantity	Rate	Time
Press A	$1200t$	1200	t
Press B	$1500t$	1500	t
Together	41,400	$41{,}400/t$	t

The equation to solve is $1200t + 1500t = 41{,}400$. (Another equation that works is $1200 + 1500 = 41{,}400/t$.)

$$1200t + 1500t = 41{,}400$$

$$2700t = 41{,}400$$

$$t = \frac{41{,}400}{2700}$$

$$t = 15\tfrac{1}{3}$$

The presses will need $15\frac{1}{3}$ hours or 15 hours 20 minutes to complete the run.

Distance Problems

Another common word problem type is the distance problem, sometimes called the uniform rate problem. The underlying formula is $d = rt$ (distance equals rate times time). From $d = rt$, we get two other relationships: $r = d/t$ and $t = d/r$. These problems come in many forms: two bodies traveling in opposite directions, two bodies traveling in the same direction, two bodies traveling away from each other or toward each other at right angles. Sometimes the bodies leave at the same time, sometimes one gets a head start. Usually they are traveling at different rates, or speeds. As in all applied problems, the units of measure must be consistent throughout the problem. For instance, if your rates are given to you in miles per hour and your time is given in minutes, you should convert minutes to hours. You could convert miles per hour into miles per minute, but this would be awkward.

Examples

When the bodies move in the same direction, the rate at which the distance between them is changing is the difference between their rates.

A cyclist starts at a certain point and rides at a rate of 10 mph. Twelve minutes later, another cyclist starts from the same point in the same direction and rides at 16 mph. How long will it take for the second cyclist to catch up with the first?

When the second cyclist begins, the first has traveled $10(\frac{12}{60}) = 2$ miles. The "10" is the rate and the "$\frac{12}{60}$" is the twelve minutes converted to hours. Because the cyclists are moving in the same direction, the rate at which the distance between them is shrinking is $16 - 10 = 6$ mph. Then, the question boils down to "How long will it take for something traveling 6 mph to cover 2 miles?"

Let t represent the number of hours the second cyclist is traveling.

$$d = rt$$

$$2 = 6t$$

$$\frac{2}{6} = t$$

$$\frac{1}{3} = t$$

It will take the second cyclist $\frac{1}{3}$ of an hour or 20 minutes to catch up the first cyclist.

A car passes an intersection heading north at 40 mph. Another car passes the same intersection 15 minutes later heading north traveling at 45 mph. How long will it take for the second car to overtake the first?

In 15 minutes, the first car has traveled $40(\frac{15}{60}) = 10$ miles. The second car is gaining on the first at a rate of $45 - 40 = 5$ mph. So the question becomes "How long will it take a body traveling 5 mph to cover 10 miles?"

Let t represent the number of hours the second car has traveled after passing the intersection.

$$d = rt$$

$$10 = 5t$$

$$\frac{10}{5} = t$$

$$2 = t$$

It will take the second car two hours to overtake the first.

Practice

1. Lori starts jogging from a certain point and runs 5 mph. Jeffrey jogs from the same point 15 minutes later at a rate of 8 mph. How long will it take Jeffrey to catch up to Lori?

2. A truck driving east at 50 mph passes a certain mile marker. A motorcyclist also driving east passes that same mile marker 45 minutes later. If the motorcyclist is driving 65 mph, how long will it take for the motorcyclist to pass the truck?

Solutions

1. Lori has jogged

   $$r \quad t \ = d$$

   $$5\,(\tfrac{15}{60}) = \tfrac{5}{4}$$

   miles before Jeffrey began. Jeffrey is catching up to Lori at the rate of $8 - 5 = 3$ mph. How long will it take a body traveling 3 mph to cover $\frac{5}{4}$ miles?

Let t represent the number of hours Jeffrey jogs.

$$3t = \frac{5}{4}$$

$$t = \frac{1}{3} \cdot \frac{5}{4}$$

$$t = \frac{5}{12}$$

Jeffrey will catch up to Lori in $\frac{5}{12}$ hours or $(\frac{5}{12})(60) = 25$ minutes.

2. The truck traveled $50(\frac{45}{60}) = \frac{75}{2}$ miles. The motorcyclist is catching up to the truck at a rate of $65 - 50 = 15$ mph. How long will it take a body moving at a rate of 15 mph to cover $\frac{75}{2}$ miles?

   Let t represent the number of hours the motorcyclist has been driving since passing the mile marker.

$$\frac{75}{2} = 15t$$

$$\frac{1}{15} \cdot \frac{75}{2} = t$$

$$\frac{5}{2} = t$$

$$2\tfrac{1}{2} = t$$

   The motorcyclist will overtake the truck in $2\frac{1}{2}$ hours or 2 hours 30 minutes.

When two bodies are moving in opposite directions, whether towards each other or away from each other, the rate at which the distance between them is changing, whether growing larger or smaller, is the sum of their individual rates.

Examples

Two cars meet at an intersection, one heading north; the other, south. If the northbound driver drives at 30 mph and the southbound driver at 40 mph, when will they be 35 miles apart?

The distance between them is growing at the rate of $30 + 40 = 70$ mph. The question then becomes, "how long will it take a body moving 70 mph to travel 35 miles?"

Let t represent the number of hours the cars travel after leaving the intersection.

$$70t = 35$$
$$t = \frac{35}{70}$$
$$t = \frac{1}{2}$$

In half an hour, the cars will be 35 miles apart.

Katy left her house on a bicycle heading north at 8 mph. At the same time, her sister Molly headed south at 12 mph. How long will it take for them to be 24 miles apart?

The distance between them is increasing at the rate of $8 + 12 = 20$ mph. The question then becomes "How long will it take a body moving 20 mph to travel 24 miles?"

Let t represent the number of hours each girl is traveling.

$$20t = 24$$
$$t = \frac{24}{20}$$
$$t = \frac{6}{5} = 1\tfrac{1}{5}$$

The girls will be 24 miles apart after $1\frac{1}{5}$ hours or 1 hour 12 minutes.

Practice

1. Two airplanes leave an airport simultaneously, one heading east; the other, west. The eastbound plane travels at 140 mph and the westbound plane travels at 160 mph. How long will it take for the planes to be 750 miles apart?

2. Mary began walking home from school, heading south at a rate of 4 mph. Sharon left school at the same time heading north at 6 mph. How long will it take for them to be 3 miles apart?

3. Two freight trains pass each other on parallel tracks. One train is traveling west, going 40 mph. The other is traveling east, going 60 mph. When will the trains be 325 miles apart?

Solutions

1. The planes are moving apart at a rate of $140 + 160 = 300$ mph. Let t represent the number of hours the planes are flying.

$$300t = 750$$
$$t = \frac{750}{300}$$
$$t = 2\tfrac{1}{2}$$

In $2\frac{1}{2}$ hours, or 2 hours 30 minutes, the planes will be 750 miles apart.

2. The distance between the girls is increasing at the rate of $4 + 6 = 10$ mph. Let t represent the number of hours the girls are walking.

$$10t = 3$$
$$t = \frac{3}{10}$$

Mary and Sharon will be 3 miles apart in $\frac{3}{10}$ of an hour or $60(\frac{3}{10}) = 18$ minutes.

3. The distance between the trains is increasing at the rate of $40 + 60 = 100$ mph. Let t represent the number of hours the trains travel after leaving the station.

$$100t = 325$$
$$t = \frac{325}{100}$$
$$t = 3\tfrac{1}{4}$$

The trains will be 325 miles apart after $3\frac{1}{4}$ hours or 3 hours 15 minutes.

When two bodies travel towards each other (from opposite directions) the rate at which the distance between them is shrinking is also the sum of their individual rates.

Examples

Dale left his high school at 3:45 and walked towards his brother's school at 5 mph. His brother, Jason, left his elementary school at the same time and walked toward Dale's high school at 3 mph. If their schools are 2 miles apart, when will they meet?

The rate at which the brothers are moving towards each other is $3 + 5 = 8$ mph. Let t represent the number of hours the boys walk.

$$8t = 2$$
$$t = \frac{2}{8}$$
$$t = \frac{1}{4}$$

The boys will meet after $\frac{1}{4}$ an hour or 15 minutes; that is, at 4:00.

A jet airliner leaves Dallas going to Houston, flying at 400 mph. At the same time, another jet airliner leaves Houston, flying to Dallas, at the same rate. How long will it take for the two airliners to meet? (Dallas and Houston are 250 miles apart.) The distance between the jets is decreasing at the rate of $400 + 400 = 800$ mph. Let t represent the number of hours they are flying.

$$800t = 250$$
$$t = \frac{250}{800}$$
$$t = \frac{5}{16}$$

The planes will meet after $\frac{5}{16}$ hours or $60(\frac{5}{16}) = 18\frac{3}{4}$ minutes or 18 minutes 45 seconds.

Practice

1. Jessie leaves her house on a bicycle, traveling at 8 mph. She is going to her friend Kerrie's house. Coincidentally, Kerrie leaves her house at the same time and rides her bicycle at 7 mph to Jessie's house. If they live 5 miles apart, how long will it take for the girls to meet?

2. Two cars 270 miles apart enter an interstate highway traveling towards one another. One car travels at 65 mph and the other at 55 mph. When will they meet?

3. At one end of town, a jogger jogs southward at the rate of 6 mph. At the opposite end of town, at the same time, another jogger heads

northward at the rate of 9 mph. If the joggers are 9 miles apart, how long will it take for them to meet?

Solutions

1. The distance between the girls is decreasing at the rate of $8 + 7 = 15$ mph. Let t represent the number of hours they are on their bicycles.

$$15t = 5$$
$$t = \frac{5}{15}$$
$$t = \frac{1}{3}$$

The girls will meet in $\frac{1}{3}$ of an hour or 20 minutes.

2. The distance between the cars is decreasing at the rate of $65 + 55 = 120$ mph. Let t represent the number of hours the cars have traveled since entering the highway.

$$120t = 270$$
$$t = \frac{270}{120}$$
$$t = 2\tfrac{1}{4}$$

The cars will meet after $2\frac{1}{4}$ hours or 2 hours 15 minutes.

3. The distance between the joggers is decreasing at the rate of $6 + 9 = 15$ mph. Let t represent the number of the hours they are jogging.

$$15t = 9$$
$$t = \frac{9}{15}$$
$$t = \frac{3}{5}$$

The joggers will meet after $\frac{3}{5}$ of an hour or $60(\frac{3}{5}) = 36$ minutes.

For distance problems in which the bodies are moving away from each other or toward each other at right angles (for example, one heading east, the other north), the Pythagorean Theorem is used. This topic will be covered in the last chapter.

Some distance problems involve the complication of the two bodies starting at different times. For these, you need to compute the head start of the first one and let t represent the time they are both moving (which is the same as the amount of time the second is moving). Subtract the head start from the distance in question then proceed as if they started at the same time.

Examples

A car driving eastbound passes through an intersection at 6:00 at the rate of 30 mph. Another car driving westbound passes through the same intersection ten minutes later at the rate of 35 mph. When will the cars be 18 miles apart?

The eastbound driver has a 10-minute head start. In 10 minutes ($\frac{10}{60}$ hours), that driver has traveled $30(\frac{10}{60}) = 5$ miles. So when the westbound driver passes the intersection, there is already 5 miles between them, so the question is now "How long will it take for there to be $18 - 5 = 13$ miles between two bodies moving away from each other at the rate of $30 + 35 = 65$ mph?"

Let t represent the number of hours after the second car has passed the intersection.

$$65t = 13$$

$$t = \frac{13}{65}$$

$$t = \frac{1}{5}$$

In $\frac{1}{5}$ of an hour or $60(\frac{1}{5}) = 12$ minutes, an additional 13 miles is between them. So 12 minutes after the second car passes the intersection, there will be a total of 18 miles between the cars. That is, at 6:22 the cars will be 18 miles apart.

Two employees ride their bikes to work. At 10:00 one leaves work and rides southward home at 9 mph. At 10:05 the other leaves work and rides home northward at 8 mph. When will they be 5 miles apart?

The first employee has ridden $9(\frac{5}{60}) = \frac{3}{4}$ miles by the time the second employee has left. So we now need to see how long, after 10:05, it takes for an additional $5 - \frac{3}{4} = 4\frac{1}{4} = \frac{17}{4}$ miles to be between them. Let t represent the number of hours after 10:05. When both employees are riding, the distance between them is increasing at the rate of $9 + 8 = 17$ mph.

$$17t = \frac{17}{4}$$

$$t = \frac{1}{17} \cdot \frac{17}{4}$$

$$t = \frac{1}{4}$$

After $\frac{1}{4}$ hour, or 15 minutes, they will be an additional $4\frac{1}{4}$ miles apart. That is, at 10:20, the employees will be 5 miles apart.

Two boys are 1250 meters apart when one begins walking toward the other. If one walks at a rate of 2 meters per second and the other, who starts walking toward the first boy four minutes later, walks at the rate of 1.5 meters per second, how long will it take for them to meet?

The boy with the head start has walked for $4(60) = 240$ seconds. (Because the rate is given in meters per second, all times will be converted to seconds.) So, he has traveled $240(2) = 480$ meters. At the time the other boy begins walking, there remains $1250 - 480 = 770$ meters to cover. When the second boy begins to walk, they are moving toward one another at the rate of $2 + 1.5 = 3.5$ meters per second. The question becomes "How long will it take a body moving 3.5 meters per second to travel 770 meters?"

Let t represent the number of seconds the second boy walks.

$$3.5t = 770$$

$$t = \frac{770}{3.5}$$

$$t = 220$$

The boys will meet 220 seconds, or 3 minutes 40 seconds, after the second boy starts walking.

A plane leaves City A towards City B at 9:10, flying at 200 mph. Another plane leaves City B towards City A at 9:19, flying at 180 mph. If the cities are 790 miles apart, when will the planes pass each other?

In 9 minutes the first plane has flown $200(\frac{9}{60}) = 30$ miles, so when the second plane takes off, there are $790 - 30 = 760$ miles between them. The planes are traveling towards each other at $200 + 180 = 380$ mph. Let t represent the number of hours the second plane flies.

$$380t = 760$$

$$t = \frac{760}{380}$$

$$t = 2$$

Two hours after the second plane has left the planes will pass each other; that is, at 11:19 the planes will pass each other.

Practice

1. Two joggers start jogging on a trail. One jogger heads north at the rate of 7 mph. Eight minutes later, the other jogger begins at the same point and heads south at the rate of 9 mph. When will they be two miles apart?

2. Two boats head toward each other from opposite ends of a lake, which is six miles wide. One boat left at 2:05 going 12 mph. The other boat left at 2:09 at a rate of 14 mph. What time will they meet?

3. The Smiths leave the Tulsa city limits, heading toward Dallas, at 6:05 driving 55 mph. The Hewitts leave Dallas and drive to Tulsa at 6:17, driving 65 mph. If Dallas and Tulsa are 257 miles apart, when will they pass each other?

Solutions

1. The first jogger had jogged $7(\frac{8}{60}) = \frac{56}{60} = \frac{14}{15}$ miles when the other jogger began. So there is $2 - \frac{14}{15} = 1\frac{1}{5} = \frac{16}{15}$ miles left to cover. The distance between them is growing at a rate of $7 + 9 = 16$ mph. Let t represent the number of hours the second jogger jogs.

$$16t = \frac{16}{15}$$

$$t = \frac{1}{16} \cdot \frac{16}{15}$$

$$t = \frac{1}{15}$$

In $\frac{1}{15}$ of an hour, or $60(\frac{1}{15}) = 4$ minutes after the second jogger began, the joggers will be two miles apart.

2. The first boat got a $12(\frac{4}{60}) = \frac{4}{5}$ mile head start. When the second boat leaves, there remains $6 - \frac{4}{5} = 5\frac{1}{5} = \frac{26}{5}$ miles between them. When the second boat leaves the distance between them is decreasing at a rate of $12 + 14 = 26$ mph. Let t represent the number of hours the second boat travels.

$$26t = \frac{26}{5}$$

$$t = \frac{1}{26} \cdot \frac{26}{5}$$

$$t = \frac{1}{5}$$

In $\frac{1}{5}$ hour, or $\frac{1}{5}(60) = 20$ minutes, after the second boat leaves, the boats will meet. That is, at 2:29 both boats will meet.

3. The Smiths have driven $55(\frac{12}{60}) = 11$ miles outside of Tulsa by the time the Hewitts have left Dallas. So, when the Hewitts leave Dallas, there are $257 - 11 = 246$ miles between the Smiths and Hewitts. When the Hewitts leave Dallas, the distance between them is decreasing at the rate of $55 + 65 = 120$ mph. Let t represent the number of hours after the Hewitts have left Dallas.

$$120t = 246$$

$$t = \frac{246}{120}$$

$$t = 2\frac{1}{20}$$

$2\frac{1}{20}$ hours, or 2 hours 3 minutes $(60 \cdot \frac{1}{20} = 3)$, after the Hewitts leave Dallas, the Smiths and Hewitts will pass each other. In other words, at 8:20, the Smiths and Hewitts will pass each other.

There are some distance/rate problems for which there are three unknowns. You must reduce the number of unknowns to one. The clues on how to do so are given in the problem.

Examples

A semi-truck traveled from City A to City B at 50 mph. On the return trip, it averaged only 45 mph and took 15 minutes longer. How far is it from City A to City B?

There are three unknowns—the distance between City A and City B, the time spent traveling from City A to City B, and the time spent traveling from City B to City A. We must eliminate two of these unknowns. Let t represent the number of hours spent on the trip from City A to City B. We know that it took 15 minutes longer traveling from City B to City A (the return trip), so $t + \frac{15}{60}$ represents the number of hours traveling from City B to City A. We also know that the distance from City A to City B is the same as from City B to City A. Let d represent the distance between the two cities. We now have the following two equations.

From City A to City B:	From City B to City A:
$d = 50t$	$d = 45(t + \frac{15}{60})$

But if the distance between them is the same, then $50t$ = Distance from City A to City B is equal to the distance from City B to City A $= 45(t + \frac{15}{60})$. Therefore,

$$50t = 45\left(t + \frac{15}{60}\right)$$

$$50t = 45\left(t + \frac{1}{4}\right)$$

$$\begin{aligned} 50t &= 45t + \frac{45}{4} \\ -45t &\quad -45t \end{aligned}$$

$$5t = \frac{45}{4}$$

$$t = \frac{1}{5} \cdot \frac{45}{4}$$

$$t = \frac{9}{4}$$

We now know the time, but the problem asked for the distance. The distance from City A to City B is given by $d = 50t$, so $d = 50(\frac{9}{4}) = \frac{225}{2} = 112\frac{1}{2}$. The cities are $112\frac{1}{2}$ miles apart.

Another approach to this problem would be to let t represent the number of hours the semi spent traveling from City B to City A. Then $t - \frac{15}{60}$ would represent the number of hours the semi spent traveling from City A to City B. The equation to solve would be $50(t - \frac{15}{60}) = 45t$.

Kaye rode her bike to the library. The return trip took 5 minutes less. If she rode to the library at the rate of 10 mph and home from the library at the rate of 12 mph, how far is her house from the library?

Again there are three unknowns—the distance between Kaye's house and the library, the time spent riding to the library and the time spent riding home. Let t represent the number of hours spent riding to the library. She spent 5 minutes less riding home, so $t - \frac{5}{60}$ represents the number of hours spent riding home. Let d represent the distance between Kaye's house and the library.

The trip to the library is given by $d = 10t$, and the trip home is given by $d = 12(t - \frac{5}{60})$. As these distances are equal, we have that $10t = d = 12(t - \frac{5}{60})$.

$$10t = 12\left(t - \frac{5}{60}\right)$$

$$10t = 12\left(t - \frac{1}{12}\right)$$

$$10t = 12t - 1$$

$$-12t \quad -12t$$

$$-2t = -1$$

$$t = \frac{-1}{-2}$$

$$t = \frac{1}{2}$$

The distance from home to the library is $d = 10t = 10(\frac{1}{2}) = 5$ miles.

Practice

1. Terry, a marathon runner, ran from her house to the high school then back. The return trip took 5 minutes longer. If her speed was 10 mph to the high school and 9 mph to her house, how far is Terry's house from the high school?

2. Because of heavy morning traffic, Toni spent 18 minutes more driving to work than driving home. If she averaged 30 mph on her drive to work and 45 mph on her drive home, how far is her home from her work?

3. Leo walked his grandson to school. If he averaged 3 mph on the way to school and 5 mph on his way home, and if it took 16 minutes longer to get to school, how far is it between his home and his grandson's school?

Solutions

1. Let t represent the number of hours spent running from home to school. Then $t + \frac{5}{60}$ represents the number of hours spent running from school to home. The distance to school is given by $d = 10t$, and the distance home is given by $d = 9(t + \frac{5}{60})$.

$$10t = 9\left(t + \frac{5}{60}\right)$$

$$10t = 9\left(t + \frac{1}{12}\right)$$

$$10t = 9t + \frac{9}{12}$$

$$10t = 9t + \frac{3}{4}$$

$$-9t \quad -9t$$

$$t = \frac{3}{4}$$

The distance between home and school is $10t = 10(\frac{3}{4}) = \frac{15}{2} = 7\frac{1}{2}$ miles.

2. Let t represent the number of hours Toni spent driving to work. Then $t - \frac{18}{60}$ represents the number of hours driving home. The distance from home to work is given by $d = 30t$, and the distance from work to home is given by $d = 45(t - \frac{18}{60})$.

$$30t = 45\left(t - \frac{18}{60}\right)$$

$$30t = 45\left(t - \frac{3}{10}\right)$$

$$30t = 45t - \frac{135}{10}$$

$$30t = 45t - \frac{27}{2}$$

$$-45t \quad -45t$$

$$-15t = \frac{-27}{2}$$

$$t = \frac{1}{-15} \cdot \frac{-27}{2}$$

$$t = \frac{9}{10}$$

The distance from Toni's home and work is $30t = 30(\frac{9}{10}) = 27$ miles.

3. Let t represent the number of hours Leo spent walking his grandson to school. Then $t - \frac{16}{60}$ represents the number of hours Leo spent walking home. The distance from home to school is given by $d = 3t$, and the distance from school to home is given by $d = 5(t - \frac{16}{60})$.

$$3t = 5\left(t - \frac{16}{60}\right)$$

$$3t = 5\left(t - \frac{4}{15}\right)$$

$$3t = 5t - \frac{20}{15}$$

$$3t = 5t - \frac{4}{3}$$

$$-5t \quad -5t$$

$$-2t = \frac{-4}{3}$$

$$t = \frac{1}{-2} \cdot \frac{-4}{3}$$

$$t = \frac{2}{3}$$

The distance from home to school is $3t = 3(\frac{2}{3}) = 2$ miles.

In the above examples and practice problems, the number for t was substituted in the first distance equation to get d. It does not matter which equation you used to find d, you should get the same value. If you do not, then you have made an error somewhere.

Geometric Figures

Algebra problems involving geometric figures are very common. In algebra, you normally deal with rectangles, triangles, and circles. On occasion, you will be asked to solve problems involving other shapes like right circular cylinders and right circular cones. If you master solving the more common types of geometric problems, you will find that the more exotic shapes are just as easy.

In many of these problems, you will have several unknowns which you must reduce to one unknown. In the problems above, you reduced a problem of three unknowns to one unknown by relating one quantity to another (the time on one direction related to the time on the return trip) and by setting the equal distances equal to each other. We will use similar techniques here.

Example

A rectangle is $1\frac{1}{2}$ times as long as it is wide. The perimeter is 100 cm^2. Find the dimensions of the rectangle.

The formula for the perimeter of a rectangle is given by $P = 2l + 2w$. We are told the perimeter is 100, so the equation now becomes $100 = 2l + 2w$. We are also told that the length is $1\frac{1}{2}$ times the width, so $l = 1.5w$. We can substitute this l into the equation: $100 = 2l + 2w = 2(1.5w) + 2w$. We have reduced an equation with three unknowns to one with a single unknown.

$$100 = 2(1.5w) + 2w$$
$$100 = 3w + 2w$$
$$100 = 5w$$
$$\frac{100}{5} = w$$
$$20 = w$$

The width is 20 cm and the length is $1.5w = 1.5(20) = 30$ cm.

Practice

1. A box's width is $\frac{2}{3}$ its length. The perimeter of the box is 40 inches. What are the box's length and width?

2. A rectangular yard is twice as long as it is wide. The perimeter is 120 feet. What are the yard's dimensions?

Solutions

1. The perimeter of the box is 40 inches, so $P = 2l + 2w$ becomes $40 = 2l + 2w$. The width is $\frac{2}{3}$ its length, and $w = (2l/3)$, so $40 = 2l + 2w$ becomes $40 = 2l + 2(2l/3) = 2l + (4l/3)$.

$$40 = 2l + \frac{4l}{3}$$
$$40 = \frac{10l}{3} \left(2 + \frac{4}{3} = \frac{6}{3} + \frac{4}{3} = \frac{10}{3}\right)$$
$$\frac{3}{10} \cdot 40 = \frac{3}{10} \cdot \frac{10l}{3}$$
$$12 = l$$

The length of the box is 12 inches and its width is $\frac{2}{3}l = \frac{2}{3}(12) = 8$ inches.

2. The perimeter of the yard is 120 feet, so $P = 2l + 2w$ becomes $120 = 2l + 2w$. The length is twice the width, so $l = 2w$, and $120 = 2l + 2w$ becomes $120 = 2(2w) + 2w$.

$$120 = 2(2w) + 2w$$
$$120 = 4w + 2w$$
$$120 = 6w$$
$$\frac{120}{6} = w$$
$$20 = w$$

The yard's width is 20 feet and its length is $2l = 2(20) = 40$ feet.

Some geometric problems involve changing one or more dimensions. In the following problems, one or more dimensions are changed and you are given information about how this change has affected the figure's area. Next you will decide how the two areas are related Then you will be able to reduce your problem from several unknowns to just one.

Example

A rectangle is twice as long as it is wide. If the length is decreased by 4 inches and its width is decreased by 3 inches, the area is decreased by 88 square inches. Find the original dimensions.

The area formula for a rectangle is $A = lw$. Let A represent the original area; l, the original length; and w, the original width. We know that the original length is twice the original width, so $l = 2w$ and $A = lw$ becomes $A = 2ww = 2w^2$. The new length is $l - 4 = 2w - 4$ and the new width is $w - 3$, so the new area is $(2w - 4)(w - 3)$. But the new area is also 88 square inches less than the old area, so $A - 88$ represents the new area, also. We then have for the new area, $A - 88 = (2w - 4)(w - 3)$. But the A can be replaced with $2w^2$. We now have the equation $2w^2 - 88 = (2w - 4)(w - 3)$, an equation with one unknown.

$$
\begin{aligned}
2w^2 - 88 &= (2w - 4)(w - 3) \\
2w^2 - 88 &= 2w^2 - 6w - 4w + 12 \quad \text{(Use the FOIL method.)} \\
2w^2 - 88 &= 2w^2 - 10w + 12 \quad (2w^2 \text{ on each side cancels}) \\
-12 & \qquad\qquad\quad -12 \\
-100 &= -10w \\
\frac{-100}{-10} &= w \\
10 &= w
\end{aligned}
$$

The width of the original rectangle is 10 inches and its length is $2w = 2(10) = 20$ inches.

A square's length is increased by 3 cm, which causes its area to increase by 33 cm^2. What is the length of the original square?

A square's length and width are the same, so the area formula for the square is $A = ll = l^2$. Let l represent the original length. The new length is $l + 3$. The original area is $A = l^2$ and its new area is $(l + 3)^2$. The new area is also the original area plus 33, so $(l + 3)^2 = \text{new area} = A + 33 = l^2 + 33$. We now have the equation, with one unknown: $(l + 3)^2 = l^2 + 33$.

$$
\begin{aligned}
(l + 3)^2 &= l^2 + 33 \\
(l + 3)(l + 3) &= l^2 + 33 \\
l^2 + 6l + 9 &= l^2 + 33 \quad (l^2 \text{ on each side cancels}) \\
-9 & \qquad\quad -9 \\
6l &= 24 \\
l &= \frac{24}{6} \\
l &= 4
\end{aligned}
$$

The original length is 4 cm.

Practice

1. A rectangular piece of cardboard starts out with its width being three-fourths its length. Four inches are cut off its length and two inches from its width. The area of the cardboard is 72 square inches smaller than before it was trimmed What was its original length and width?

2. A rectangle's length is one-and-a-half times its width. The length is increased by 4 inches and its width by 3 inches. The resulting area is 97 square inches more than the original rectangle. What were the original dimensions?

Solutions

1. Let l represent the original length and w, the original width. The original area is $A = lw$. The new length is $l - 4$ and the new width is $w - 2$. The new area is then $(l - 4)(w - 2)$. But the new area is 72 square inches smaller than the original area, so $(l - 4)(w - 2) = A - 72 = lw - 72$. So far, we have $(l - 4)(w - 2) = lw - 72$.

   The original width is three-fourths its length, so $w = (\frac{3}{4})l$. We will now replace w with $(\frac{3}{4})l = \frac{3l}{4}$.

$$(l - 4)\left(\frac{3l}{4} - 2\right) = l\left(\frac{3l}{4}\right) - 72$$

$$\frac{3l^2}{4} - 2l - 4\left(\frac{3l}{4}\right) + 8 = \frac{3l^2}{4} - 72 \quad \left(\frac{3l^2}{4} \text{ on each side cancels}\right)$$

$$-2l - 3l + 8 = -72$$

$$-5l + 8 = -72$$

$$-8 \quad -8$$

$$-5l = -80$$

$$l = \frac{-80}{-5}$$

$$l = 16$$

The original length was 16 inches and the original width was $\frac{3}{4}l = \frac{3}{4}(16) = 12$ inches.

2. Let l represent the original length and w, the original width. The original area is then given by $A = lw$. The new length is $l + 4$ and the new width is $w + 3$. The new area is now $(l + 4)(w + 3)$. But the new area is also the old area plus 97 square inches, so $A + 97 = (l + 4)(w + 3)$. But $A = lw$, so $A + 97$ becomes $lw + 97$. We now have

$$lw + 97 = (l + 4)(w + 3).$$

Since the original length is $1\frac{1}{2} = \frac{3}{2}$ of the original width, $l = \frac{3}{2}w$. Replace each l by $\frac{3}{2}w$.

$$\frac{3}{2}ww + 97 = \left(\frac{3}{2}w + 4\right)(w + 3)$$

$$\frac{3}{2}w^2 + 97 = \left(\frac{3}{2}w + 4\right)(w + 3)$$

$$\frac{3}{2}w^2 + 97 = \frac{3}{2}w^2 + \frac{9}{2}w + 4w + 12 \quad \left(\frac{3}{2}w^2 \text{ on each side cancels}\right)$$

$$97 = \frac{9}{2}w + 4w + 12$$

$$97 = \frac{17}{2}w + 12 \quad \left(\frac{9}{2} + 4 = \frac{9}{2} + \frac{8}{2} = \frac{17}{2}\right)$$

$$-12 \qquad -12$$

$$85 = \frac{17}{2}w$$

$$\frac{2}{17} \cdot 85 = \frac{2}{17} \cdot \frac{17}{2}w$$

$$10 = w$$

The original width is 10 inches and the original length is $\frac{3}{2}w = \frac{3}{2}(10) = 15$ inches.

Example

The radius of a circle is increased by 3 cm. As a result, the area is increased by 45π cm^2. What was the original radius?

Remember that the area of a circle is $A = \pi r^2$, where r represents the radius. So, let r represent the original radius. The new radius is then represented by $r + 3$. The new area is represented by $\pi(r + 3)^2$. But the new area is also the original area plus 45π cm^2. This gives us

$A + 45\pi = \pi(r+3)^2$. Because $A = \pi r^2$, $A + 45\pi$ becomes $\pi r^2 + 45\pi$. Our equation, then, is $\pi r^2 + 45\pi = \pi(r+3)^2$.

$$\pi r^2 + 45\pi = \pi(r+3)^2$$
$$\pi r^2 + 45\pi = \pi(r+3)(r+3)$$
$$\pi r^2 + 45\pi = \pi(r^2 + 6r + 9)$$
$$\pi r^2 + 45\pi = \pi r^2 + 6r\pi + 9\pi \quad (\pi r^2 \text{ on each side cancels})$$
$$45\pi = 6r\pi + 9\pi$$
$$-9\pi \qquad -9\pi$$
$$36\pi = 6r\pi$$
$$\frac{36\pi}{6\pi} = r$$
$$6 = r$$

The original radius was 6 cm.

Practice

A circle's radius is increased by 5 inches and as a result, its area is increased by 155π square inches. What is the original radius?

Solution

Let r represent the original radius. Then $r + 5$ represents the new radius, and $A = \pi r^2$ represents the original area. The new area is 155π square inches more than the original area, so $155\pi + A = \pi(r+5)^2 = 155\pi + \pi r^2$.

$$\pi(r+5)^2 = 155\pi + \pi r^2$$
$$\pi(r+5)(r+5) = 155\pi + \pi r^2$$
$$\pi(r^2 + 10r + 25) = 155\pi + \pi r^2$$
$$\pi r^2 + 10r\pi + 25\pi = 155\pi + \pi r^2 \quad (\pi r^2 \text{ on each side cancels})$$
$$10r\pi + 25\pi = 155\pi$$
$$-25\pi \quad -25\pi$$
$$10r\pi = 130\pi$$
$$r = \frac{130\pi}{10\pi}$$
$$r = 13$$

The original radius is 13 inches.

Chapter Review

1. How much 10% alcohol solution should be mixed with 14 ounces of 18% solution to get a 12% solution?

   (a) $\frac{14}{15}$ of an ounce (b) 42 ounces (c) 4.2 ounces
   (d) 14 ounces

2. The perimeter of a rectangle is 56 inches and the width is three-fourths of the length. What is the length?

   (a) 12 inches (b) 16 inches (c) 8.64 inches
   (d) $36\frac{3}{4}$ inches

3. What is 15% of 30?

   (a) 4.5 (b) 2 (c) 200 (d) 450

4. If the daily profit formula for a certain product is $P = 7q - 5600$, where P is the profit in dollars and q is the number sold per day, how many units must be sold per day to break even?

   (a) 600 (b) 700 (c) 800
   (d) No quantity can be sold to break even

5. At 3:00 one car heading north on a freeway passes an exit ramp averaging 60 mph. Another northbound car passes the same exit ramp 10 minutes later averaging 65 mph. When will the second car pass the first?

   (a) 4:00 (b) 4:15 (c) 5:00 (d) 5:10

6. The difference between two numbers is 12 and twice the larger is three times the smaller. What is the smaller number?

   (a) 12 (b) 24 (c) 36 (d) 48

7. A jacket is on sale for $84, which is 20% off the original price. What is the original price?

   (a) $110 (b) $105 (c) $70 (d) $100.80

8. When the radius of a circle is increased by 2 cm, the area is increased by $16\pi\,\text{cm}^2$. What is the radius of the original circle?

(a) 4 cm (b) 5 cm (c) 2 cm (d) 3 cm

9. Theresa is 4 years older than Linda but 7 years younger than Charles. The sum of their ages is 51. How old is Charles?

(a) 16 (b) 22 (c) 23 (d) 25

10. Nicole can mow a lawn in 20 minutes. Christopher can mow the same lawn in 30 minutes. How long would it take them to mow the lawn if they work together?

(a) 12 minutes (b) 50 minutes (c) 25 minutes
(d) 15 minutes

11. A piggybank contains $3.20. There are twice as many nickels as quarters and half as many dimes as quarters. How many dimes are in the piggybank?

(a) 6 (b) 16 (c) 8 (d) 4

12. A student's course grade is based on four exams and one paper. Each exam is worth 15% and the paper is worth 40% of the course grade. If the grade on the paper is 80, the grade on the first exam is 70, the grade on the second exam is 85 and the grade on the third exam is 75, what grade does the student need to make on the fourth exam to earn a course grade of 80?

(a) 88 (b) 90 (c) 92 (d) 94

13. A car and passenger train pass each other at noon. The train is eastbound and its average speed is 45 mph. The car is westbound and its average speed is 60 mph. When will the car and train be 14 miles apart?

(a) 1:00 (b) 12:56 (c) 12:14 (d) 12:08

14. 16 is what percent of 80?

(a) 25 (b) 5 (c) 20 (d) 15

15. The relationship between degrees Fahrenheit and degrees Celsius is given by the formula $C = \frac{5}{9}(F - 32)$. For what temperature will degrees Celsius be 20 more than degrees Fahrenheit?

(a) F = $-85°$ (b) F = $-15°$ (c) F = $-65°$
(d) F = $-20°$

Solutions

1. (b)	**2.** (b)	**3.** (a)	**4.** (c)
5. (d)	**6.** (b)	**7.** (b)	**8.** (d)
9. (c)	**10.** (a)	**11.** (d)	**12.** (b)
13. (d)	**14.** (c)	**15.** (a)	

CHAPTER 9

Linear Inequalities

The solution to algebraic inequalities consists of a range (or ranges) of numbers. The solution to linear inequalities will be of the form $x < a$, $x \leq a$, $x > a$, or $x \geq a$, where a is a number. The inequality $x < a$ means all numbers smaller than a but not including a; $x \leq a$ means all numbers smaller than a including a itself. Similarly the inequality $x > a$ means all numbers larger than a but not a itself, and $x \geq a$ means all numbers larger than a including a itself.

The solutions to some algebra and calculus problems are inequalities. Sometimes you will be asked to shade these inequalities on the real number line and sometimes you will be asked to give your solution in interval notation. Every interval on the number line can be represented by an inequality and every inequality is represented by an interval on the number line. First we will represent inequalities by shaded regions on the number line. Later we will represent inequalities by intervals.

The inequality $x < a$ is represented on the number line by shading to the left of the number a with an open dot at a.

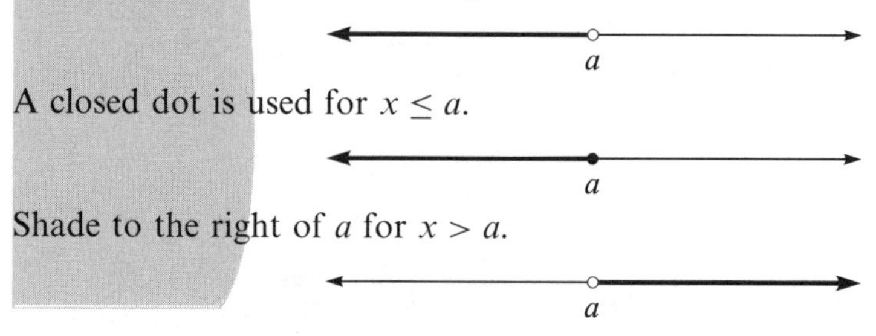

A closed dot is used for $x \leq a$.

Shade to the right of a for $x > a$.

Use a closed dot for $x \geq a$.

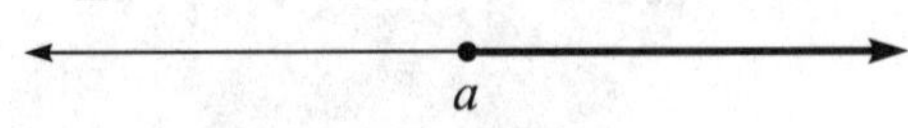

Examples

$x < 0$

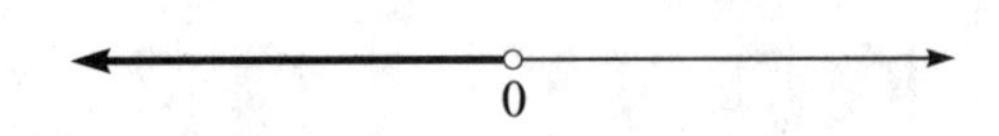

$x \geq 0$

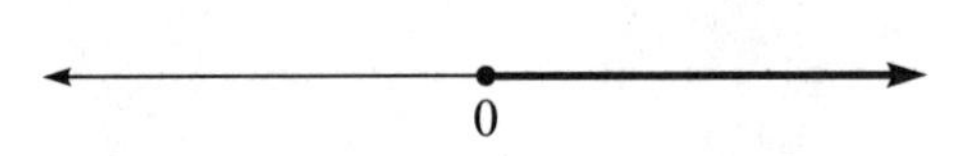

$x > 3$

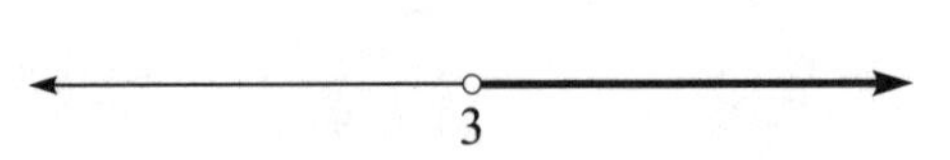

$x \leq -2$

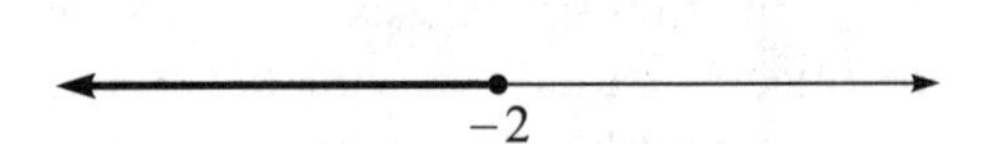

Practice

Shade the region on the number line.

1. $x > 4$
2. $x > -5$
3. $x \leq 1$
4. $x < -3$
5. $x \geq 10$

Solutions

1. $x > 4$

4

2. $x > -5$

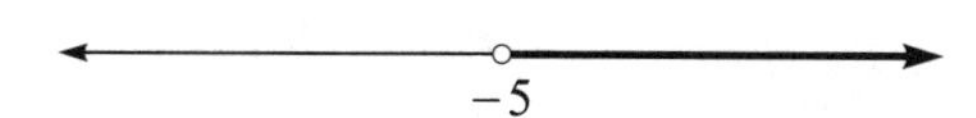

3. $x \leq 1$

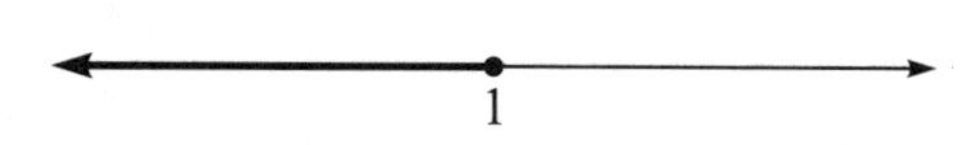

4. $x < -3$

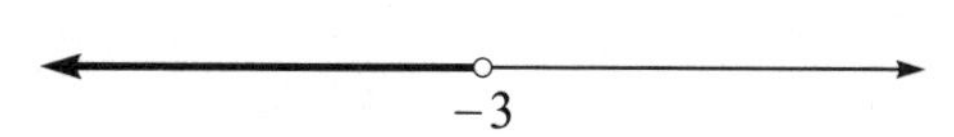

5. $x \geq 10$

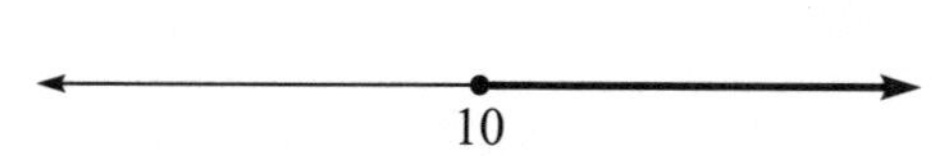

Solving Linear Inequalities

Linear inequalities are solved much the same way as linear equations with one exception: when multiplying or dividing both sides of an inequality by a negative number the inequality sign must be reversed For example $2 < 3$ but $-2 > -3$. Adding and subtracting the same quantity to both sides of an inequality never changes the direction of the inequality sign.

Examples

$$2x - 7 > 5x + 2$$
$$\quad +7 \qquad +7$$

$$2x > 5x + 9$$
$$-5x \quad -5x$$

$$-3x > 9$$

$$\frac{-3}{-3}x < \frac{9}{-3}$$ The sign changed at this step.

$$x < -3$$

$$\frac{-1}{2}(4x-6)+2>x-7$$

$$-2x+3+2>x-7$$

$$-2x+5>x-7$$
$$-x \qquad -x$$

$$-3x+5>-7$$
$$-5 \quad -5$$

$$-3x>-12$$

$$\frac{-3}{-3}x<\frac{-12}{-3}$$ The sign changed at this step.

$$x<4$$

$$2x+1\le 5$$
$$-1 \quad -1$$

$$2x\le 4$$

$$x\le\frac{4}{2}$$

$$x\le 2$$

Practice

Solve the inequality and graph the solution on the number line.

1. $7x-4\le 2x+8$
2. $\frac{2}{3}(6x-9)+4>5x+1$
3. $0.2(x-5)+1\ge 0.12$
4. $10x-3(2x+1)\ge 8x+1$
5. $3(x-1)+2(x-1)\le 7x+7$

Solutions

1.
$$\begin{aligned} 7x - 4 &\le 2x + 8 \\ -2x \quad & \quad -2x \\ 5x - 4 &\le 8 \\ +4 \quad & \quad +4 \\ 5x &\le 12 \\ x &\le \frac{12}{5} \end{aligned}$$

$\frac{12}{5}$

2.
$$\begin{aligned} \frac{2}{3}(6x - 9) + 4 &> 5x + 1 \\ 4x - 6 + 4 &> 5x + 1 \\ 4x - 2 &> 5x + 1 \\ +2 \quad & \quad +2 \\ 4x &> 5x + 3 \\ -5x \quad & \quad -5x \\ -x &> 3 \\ x &< -3 \end{aligned}$$

-3

3.
$$\begin{aligned} 0.2(x - 5) + 1 &\ge 0.12 \\ 0.2x - 1 + 1 &\ge 0.12 \\ 0.2x &\ge 0.12 \\ x &\ge \frac{0.12}{0.2} \\ x &\ge \frac{3}{5} \\ x &\ge 0.60 \end{aligned}$$

0.60

4. $10x - 3(2x + 1) \geq 8x + 1$

$$10x - 6x - 3 \geq 8x + 1$$

$$4x - 3 \geq 8x + 1$$

$$+3 \qquad +3$$

$$4x \geq 8x + 4$$

$$-8x \qquad -8x$$

$$-4x \geq 4$$

$$\frac{-4}{-4}x \leq \frac{4}{-4}$$

$$x \leq -1$$

−1

5. $3(x - 1) + 2(x - 1) \leq 7x + 7$

$$3x - 3 + 2x - 2 \leq 7x + 7$$

$$5x - 5 \leq 7x + 7$$

$$+5 \qquad +5$$

$$5x \leq 7x + 12$$

$$-7x \qquad -7x$$

$$-2x \leq 12$$

$$\frac{-2}{-2}x \geq \frac{12}{-2}$$

$$x \geq -6$$

−6

The symbol for infinity is "∞," and "$-\infty$" is the symbol for negative infinity. These symbols mean that the numbers in the interval are getting larger in the positive or negative direction. The intervals for the previous examples and practice problems are called *infinite* intervals.

An interval consists of, in order, an open parenthesis "(" or open bracket "[," a number or "$-\infty$," a comma, a number or "∞," and a closing parenthesis ")" or closing bracket "]." A parenthesis is used for strict inequalities ($x < a$ and $x > a$) and a bracket is used for an "or equal to" inequality ($x \leq a$ and $x \geq a$). A parenthesis is always used next to an infinity symbol.

Inequality	Interval
$x <$ number	$(-\infty,$ number$)$
$x >$ number	(number, ∞)
$x \leq$ number	$(-\infty,$ number$]$
$x \geq$ number	[number, ∞)

Examples

$x < 3 \quad (-\infty, 3)$ $\qquad$ $x > -6 \quad (-6, \infty)$

$x \leq 100 \quad (-\infty, 100]$ $\qquad$ $x \geq 4 \quad [4, \infty)$

Practice

Give the interval notation for the inequality.

1. $x \geq 5$
2. $x < 1$
3. $x \leq 4$
4. $x \geq -10$
5. $x \leq -2$
6. $x > 9$
7. $x < -8$
8. $x > \frac{1}{2}$

Solutions

1. $x \geq 5 \qquad [5, \infty)$

2. $x < 1$ $(-\infty, 1)$

3. $x \leq 4$ $(-\infty, 4]$

4. $x \geq -10$ $[-10, \infty)$

5. $x \leq -2$ $(\infty, -2]$

6. $x > 9$ $(9, \infty)$

7. $x < -8$ $(-\infty, -8)$

8. $x > \frac{1}{2}$ $(\frac{1}{2}, \infty)$

The table below gives the relationship between an inequality, its region on the number line, and its interval notation.

Inequality	Number Line Region	Interval Notation
$x < a$	a	$(-\infty, a)$
$x \leq a$	a	$(-\infty, a]$
$x > a$	a	(a, ∞)
$x \geq a$	a	$[a, \infty)$

Ordinarily the variable is written on the left in an inequality but not always. For instance to say that x is less than 3 ($x < 3$) is the same as saying 3 is greater than x ($3 > x$).

Inequality	Equivalent Inequality
$x < a$	$a > x$
$x \leq a$	$a \geq x$
$x > a$	$a < x$
$x \geq a$	$a \leq x$

Applications

Linear inequality word problems are solved much the same way as linear equality word problems. There are two important differences. Multiplying and dividing both sides of an inequality by a negative quantity requires that the sign reverse. You must also decide which inequality sign to use: $<$, $>$, $\leq$, and $\geq$. The following tables should help.

A < B	A > B
A is less than B	A is greater than B
A is smaller than B	A is larger than B
B is greater than A	B is less than A
B is larger than A	A is more than B

A ≤ B	B ≤ A
A is less than or equal to B	B is less than or equal to A
A is not more than B	B is not more than A
B is at least A	A is at least B
B is A or more	A is B or more
A is no greater than B	B is no greater than A
B is no less than A	A is no less than B

Some word problems give two alternatives and ask for what interval of the variable is one alternative more attractive than the other. If the alternative is between two costs, for example, in order for the cost of A to be more attractive than the cost of B, solve "Cost of A $<$ Cost of B." If the cost of A to be no more than the cost of B (also the cost of A to be at least as attractive as the cost of B), solve "Cost of A $\leq$ Cost of B." If the alternative is between two incomes of some kind, for the income of A to be more attractive than the income of B, solve "Income of A > Income of B." If the income of A is to be at least as attractive as the income of B (also the income of A to be no less attractive than the income of B), solve "Income of A ≥ Income of B."

Some of the following examples and practice problems are business problems. Let us review a few business formulas. Revenue is normally the price per unit times the number of units sold For instance if an item sells for \$3.25 each and x represents the number of units sold, the revenue is represented by $3.25x$ (dollars). Cost tends to consist of overhead costs (sometimes called *fixed costs*) and production costs (sometimes called *variable costs*). The overhead costs will be a fixed number (no variable). The production costs is usually computed as the cost per unit times the number of units sold. The total cost is usually the overhead costs plus the production costs. Profit is revenue minus cost. If a problem asks how many units must be sold to make a profit, solve "Revenue > Total Cost."

Examples

A manufacturing plant, which produces compact disks, has monthly overhead costs of \$6000. Each disk costs 18 cents to produce and sells for 30 cents. How many disks must be sold in order for the plant to make a profit?

Let $x =$ number of CDs produced and sold monthly
Cost $= 6000 + 0.18x$ and Revenue $= 0.30x$
Revenue > Cost

$$\begin{aligned} 0.30x &> 6000 + 0.18x \\ -0.18x & \qquad\quad -0.18x \\ 0.12x &> 6000 \\ x &> \frac{6000}{0.12} \\ x &> 50{,}000 \end{aligned}$$

The plant should produce and sell more than 50,000 CDs per month in order to make a profit.

Mary inherited \$16,000 and will deposit it into two accounts, one paying $5\frac{1}{2}\%$ interest and the other paying $6\frac{3}{4}\%$ interest. What is the most she can deposit into the $5\frac{1}{2}\%$ account so that her interest at the end of a year will be at least \$960?

Let $x =$ amount deposited in the $5\frac{1}{2}\%$ account
$0.055x =$ interest earned at $5\frac{1}{2}\%$
$16{,}000 - x =$ amount deposited in the $6\frac{3}{4}\%$ account
$0.0675(16{,}000 - x) =$ interest earned at $6\frac{3}{4}\%$
Interest earned at $5\frac{1}{2}\%$+ Interest earned at $6\frac{3}{4}\% \geq 960$

$$0.055x + 0.0675(16{,}000 - x) \geq 960$$
$$0.055x + 1080 - 0.0675x \geq 960$$
$$-0.0125x + 1080 \geq 960$$
$$-1080 \quad -1080$$
$$-0.0125x \geq -120$$
$$\frac{-0.0125}{-0.0125}x \leq \frac{-120}{-0.0125}$$
$$x \leq 9600$$

Mary can invest no more than \$9600 in the $5\frac{1}{2}\%$ account in order to receive at least \$960 interest at the end of the year.

An excavating company can rent a piece of equipment for \$45,000 per year. The company could purchase the equipment for monthly costs of \$2500 plus \$20 for each hour it is used How many hours per year must the equipment be used to justify purchasing it rather than renting it?

Let $x =$ number of hours per year the equipment is used

The monthly payments amount to $12(2500) = 30{,}000$ dollars annually. The annual purchase cost is $30{,}000 + 20x$.

$$\text{Purchase cost} < \text{Rent cost}$$
$$30{,}000 + 20x < 45{,}000$$
$$-30{,}000 \qquad -30{,}000$$
$$20x < 15{,}000$$
$$x < \frac{15{,}000}{20}$$
$$x < 750$$

The equipment should be used less than 750 hours annually to justify purchasing it rather than renting it.

An amusement park sells an unlimited season pass for \$240. A daily ticket sells for \$36. How many times would a customer need to use the ticket in order for the season ticket to cost less than purchasing daily tickets?

Let $x =$ number of daily tickets purchased per season
$36x =$ daily ticket cost

Season ticket cost < daily ticket cost

$$240 < 36x$$

$$\frac{240}{36} < x$$

$$6\tfrac{2}{3} < x$$

A customer would need to use the ticket more than $6\frac{2}{3}$ times (or 7 or more times) in order for the season ticket to cost less than purchasing daily tickets.

Bank A offers a $6\frac{1}{2}\%$ certificate of deposit and Bank B offers a $5\frac{3}{4}\%$ certificate of deposit but will give a \$25 bonus at the end of the year. What is the least amount a customer would need to deposit at Bank A to make Bank A's offer no less attractive than Bank B's offer?

Let $x =$ amount to deposit

If x dollars is deposited at Bank A, the interest at the end of the year would be $0.065x$. If x dollars is deposited at Bank B, the interest at the end of the year would be $0.0575x$. The total income from Bank B would be $25 + 0.0575x$.

Income from Bank A $\geq$ Income from Bank B

$$0.0650x \geq 25 + 0.0575x$$

$$-0.0575x \geq \quad -0.0575x$$

$$0.0075x \geq 25$$

$$x \geq \frac{25}{0.0075}$$

$$x \geq 3333.33$$

A customer would need to deposit at least \$3333.33 in Bank A to earn no less than would be earned at Bank B.

Practice

1. A scholarship administrator is using a \$500,000 endowment to purchase two bonds. A corporate bond pays 8% interest per year and a safer treasury bond pays $5\frac{1}{4}\%$ interest per year. If he needs at least \$30,000 annual interest payments, what is the least he can spend on the corporate bond?

2. Kelly sells corn dogs at a state fair. Booth rental and equipment rental total \$200 per day. Each corn dog costs 35 cents to make and sells for \$2. How many corn dogs should she sell in order to have a daily profit of at least \$460?

3. The owner of a snow cone stand pays \$200 per month to rent his equipment and \$400 per month for a stall in a flea market. Each snow cone costs 25 cents to make and sells for \$1.50. How many snow cones does he need to sell in order to make a profit?

4. A tee-shirt stand can sell a certain sports tee shirt for \$18. Each shirt costs \$8 in materials and labor. Monthly fixed costs are \$1500. How many tee shirts must be sold to guarantee a monthly profit of at least \$3500?

5. A car rental company rents a certain car for \$40 per day with unlimited mileage or \$24 per day plus 80 cents per mile. What is the most a customer can drive the car per day for the \$24 option to cost no more than the unlimited mileage option?

6. An internet service provider offers two plans. One plan costs \$25 per month and allows unlimited internet access. The other plan costs \$12 per month and allows 50 free hours plus 65 cents for each additional hour. How many hours per month would a customer need to use in order for the unlimited access plan be less expensive than the other plan?

7. Sharon can purchase a pair of ice skates for \$60. It costs her \$3 to rent a pair each time she goes to the rink. How many times would she need to use the skates to make purchasing them more attractive than renting them?

8. The James family has \$210 budgeted each month for electricity. They have a monthly base charge of \$28 plus 7 cents per kilowatt-hour. How many kilowatt-hours can they use each month to stay within their budget?

9. A warehouse store charges an annual fee of \$40 to shop there. A shopper without paying this fee can still shop there if he pays a 5% buyer's premium on his purchases. How much would a shopper need to spend at the store to make paying the annual \$40 fee at least as attractive as paying the 5% buyer's premium?

10. A sales clerk at an electronics store is given the option for her salary to be changed from a straight annual salary of \$25,000 to an annual base salary of \$15,000 plus an 8% commission on sales. What would her annual sales level need to be in order for this option to be at least as attractive as the straight salary option?

Solutions

1. Let $x =$ amount invested in the corporate bond
$500{,}000 - x =$ amount invested in the treasury bond
$0.08x =$ annual interest from the corporate bond
$0.0525(500{,}000 - x) =$ annual interest from the treasury bond
Corporate bond interest + Treasury bond interest $\geq 30{,}000$

$$\begin{aligned} 0.08x + 0.0525(500{,}000 - x) &\geq 30{,}000 \\ 0.08x + 26{,}250 - 0.0525x &\geq 30{,}000 \\ -26{,}250 \qquad\qquad & \quad -26{,}250 \\ 0.0275x &\geq 3750 \\ x &\geq \frac{3750}{0.0275} \\ x &\geq 136{,}363.64 \end{aligned}$$

The administrator should invest at least \$136,363.64 in the corporate bond in order to receive at least \$30,000 per year in interest payments.

2. Let $x =$ number of corn dogs sold per day
$2x =$ revenue
$200 + 0.35x =$ overhead costs + production costs = total cost
$2x - (200 + 0.35x) =$ profit

$$\begin{aligned} \text{Profit} &\geq 460 \\ 2x - (200 + 0.35x) &\geq 460 \\ 2x - 200 - 0.35x &\geq 460 \\ 1.65x - 200 &\geq 460 \\ +200 \quad & \quad +200 \\ 1.65x &\geq 660 \\ x &\geq \frac{660}{1.65} \\ x &\geq 400 \end{aligned}$$

Kelly needs to sell at least 400 corn dogs in order for her daily profit to be at least \$460.

3. Let $x =$ number of snow cones sold per month
$1.50x =$ revenue
$600 + 0.25x =$ overhead costs + production costs = total cost

Revenue > Cost

$$1.50x > 600 + 0.25x$$
$$-0.25x \qquad -0.25x$$
$$1.25x > 600$$
$$x > \frac{600}{1.25}$$
$$x > 480$$

The owner should sell more than 480 snow cones per month to make a profit.

4. Let $x =$ number of tee shirts sold per month
$18x =$ revenue
$1500 + 8x =$ overhead costs + production costs = total cost
$18x - (1500 + 8x) =$ profit

$$\text{Profit} \geq 3500$$
$$18x - (1500 + 8x) \geq 3500$$
$$18x - 1500 - 8x \geq 3500$$
$$10x - 1500 \geq 3500$$
$$+1500 \quad +1500$$
$$10x \geq 5000$$
$$x \geq \frac{5000}{10}$$
$$x \geq 500$$

At least 500 tee shirts would need to be sold each month to make a monthly profit of at least \$3500.

5. Let $x =$ number of daily miles
The \$24 option costs $24 + 0.80x$ per day.

$$24 + 0.80x \leq 40$$
$$-24 \qquad\qquad -24$$
$$0.80x \leq 16$$
$$x \leq \frac{16}{0.80}$$
$$x \leq 200$$

The most a customer could drive is 200 miles per day in order for the \$24 plan to cost no more than the \$40 plan.

6. The first 50 hours are free under the \$12 plan, so let x represent the number of hours used beyond 50 hours. Each hour beyond 50 costs $0.65x$.

$$25 < 12 + 0.65x$$
$$-12 \quad -12$$
$$13 < 0.65x$$
$$\frac{13}{0.65} < x$$
$$20 < x \text{ (or } x > 20)$$

A family would need to use more than 20 hours per month beyond 50 hours (or more than 70 hours per month) in order for the unlimited plan to cost less than the limited hour plan.

7. Let $x =$ number of times Sharon uses her skates
The cost to rent skates is $3x$.

$$60 < 3x$$
$$\frac{60}{3} < x$$
$$20 < x \text{ (or } x > 20)$$

Sharon would need to use her skates more than 20 times to justify purchasing them instead of renting them.

8. Let $x =$ number of kilowatt-hours used per month
$28 + 0.07x =$ monthly bill

$$
\begin{aligned}
28 + 0.07x &\leq 210 \\
-28 \qquad &\quad -28 \\
0.07x &\leq 182 \\
x &\leq \frac{182}{0.07} \\
x &\leq 2600
\end{aligned}
$$

The James family can use no more than 2600 kilowatt-hours per month in order to keep their electricity costs within their budget.

9. Let $x =$ amount spent at the store annually
   $0.05x =$ extra 5% purchase charge per year

$$
\begin{aligned}
40 &\leq 0.05x \\
\frac{40}{0.05} &\leq x \\
800 &\leq x \text{ (or } x \geq 800\text{)}
\end{aligned}
$$

A shopper would need to spend at least \$800 per year to justify the \$40 annual fee.

10. Let $x =$ annual sales level
    $0.08x =$ annual commission
    $15{,}000 + 0.08x =$ annual salary plus commission

$$
\begin{aligned}
15{,}000 + 0.08x &\geq 25{,}000 \\
-15{,}000 \qquad &\quad -15{,}000 \\
0.08x &\geq 10{,}000 \\
x &\geq \frac{10{,}000}{0.08} \\
x &\geq 125{,}000
\end{aligned}
$$

The sales clerk would need an annual sales level of \$125,000 or more in order for the salary plus commission option to be at least as attractive as the straight salary option.

Double Inequalities

Double inequalities represent bounded regions on the number line. The double inequality $a < x < b$ means all real numbers between a and b, where

a is the smaller number and b is the larger number. All double inequalities are of the form $a < x < b$ where one or both of the "$<$" signs might be replaced by "$\leq$." Keep in mind, though, that "$a < x < b$" is the same as "$b > x > a$." An inequality such as $10 < x < 5$ is *never* true because no number x is both larger than 10 and smaller than 5. In other words an inequality in the form "larger number $< x <$ smaller number" is meaningless.

The following table shows the number line region and interval notation for each type of double inequality.

Inequality	Region on the Number Line	Verbal Description	Interval
$a < x < b$	a b	All real numbers between a and b but not including a and b	(a, b)
$a \leq x \leq b$	a b	All real numbers between a and b including a and b	$[a, b]$
$a < x \leq b$	a b	All real numbers between a and b including b but not including a	$(a, b]$
$a \leq x < b$	a b	All real numbers between a and b including a but not including b	$[a, b)$

Examples

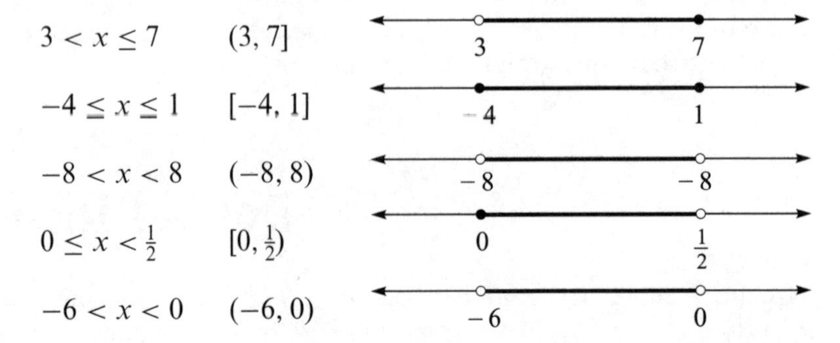

Practice

Give the interval notation and shade the region on the number line for the double inequality.

1. $6 < x < 8$
2. $-4 \leq x < 5$
3. $-2 \leq x < 2$
4. $0 \leq x \leq 10$
5. $9 < x \leq 11$
6. $\frac{1}{4} \leq x \leq \frac{1}{2}$
7. $904 < x < 1100$

Solutions

Inequality	Interval	Number line endpoints
1. $6 < x < 8$	$(6, 8)$	6, 8
2. $-4 \leq x < 5$	$[-4, 5)$	-4, 5
3. $-2 \leq x < 2$	$[-2, 2)$	-2, 2
4. $0 \leq x \leq 10$	$[0, 10]$	0, 10
5. $9 < x \leq 11$	$(9, 11]$	9, 11
6. $\frac{1}{4} \leq x \leq \frac{1}{2}$	$[\frac{1}{4}, \frac{1}{2}]$	$\frac{1}{4}$, $\frac{1}{2}$
7. $904 < x < 1100$	$(904,1100)$	904, 1100

Double inequalities are solved the same way as other inequalities except that there are three "sides" to the inequality instead of two.

Examples

$4 \leq 2x \leq 12$

$$\frac{4}{2} \le \frac{2}{2}x \le \frac{12}{2}$$

$2 \le x \le 6$ [2,6]

$$\begin{array}{lll} 6 \le 4x - 2 \le 10 \\ +2 \quad\quad +2 \quad +2 \end{array}$$

$8 \le 4x \le 12$

$$\frac{8}{4} \le \frac{4}{4}x \le \frac{12}{4}$$

$2 \le x \le 3$ [2,3]

$$\begin{array}{l} -6 < -2x + 3 < 1 \\ -3 \quad\quad\quad -3 \quad -3 \end{array}$$

$-9 < -2x < -2$

$$\frac{-9}{-2} > \frac{-2}{-2}x > \frac{-2}{-2}$$

$\frac{9}{2} > x > 1$ or $1 < x < \frac{9}{2}$ $\left(1, \frac{9}{2}\right)$

$$\begin{array}{l} 7 \le 3x + 7 < 4 \\ -7 \quad\quad -7 \quad -7 \end{array}$$

$0 \le 3x < -3$

$$\frac{0}{3} \le \frac{3}{3}x < \frac{-3}{3}$$

$0 \le x < -1$ [0, −1)

$16 < 4(2x - 1) \le 20$

$$\begin{array}{l} 16 < 8x - 4 \le 20 \\ +4 \quad\quad +4 \quad +4 \end{array}$$

$20 < 8x \le 24$

$$\frac{20}{8} < \frac{8}{8}x \leq \frac{24}{8}$$

$$\frac{5}{2} < x \leq 3 \qquad \left(\frac{5}{2}, 3\right]$$

$$-2 < \frac{-1}{2}(4x - 5) < 2$$

$$-2 < -2x + \frac{5}{2} < 2$$
$$-\frac{5}{2} \qquad -\frac{5}{2} \qquad -\frac{5}{2}$$

$$-\frac{9}{2} < -2x < -\frac{1}{2}$$

$$\frac{-1}{2} \cdot \frac{-9}{2} > \frac{-1}{2}(-2x) > \frac{-1}{2} \cdot \frac{-1}{2}$$

$$\frac{9}{4} > x > \frac{1}{4} \quad \text{or} \quad \frac{1}{4} < x < \frac{9}{4} \qquad \left(\frac{1}{4}, \frac{9}{4}\right)$$

$$2 < \frac{5x - 1}{4} < 6$$

$$4(2) < \frac{4}{1} \cdot \frac{5x - 1}{4} < 4(6)$$

$$8 < 5x - 1 < 24$$
$$+1 \qquad +1 \quad +1$$

$$9 < 5x < 25$$

$$\frac{9}{5} < \frac{5}{5}x < \frac{25}{5}$$

$$\frac{9}{5} < x < 5 \qquad \left(\frac{9}{5}, 5\right)$$

Practice

Give your solution in interval notation.

1. $14 < 2x < 20$

2. $5 \le 3x - 1 \le 8$

3. $-2 \le 3x - 4 \le 5$

4. $-4 < -2x + 6 < 4$

5. $0.12 \le 4x - 1 \le 1.8$

6. $4 < 3(-2x + 1) \le 7$

7. $-1 < -6x + 11 < 1$

8. $\frac{7}{8} < \frac{1}{4}x \le 2$

9. $8 \le 4.5x - 1 \le 11$

10. $-6 \le \frac{2}{3}x + 4 < 0$

11. $-1 < \frac{2x - 5}{3} < 1$

Solutions

1. $14 < 2x < 20$

$$\frac{14}{2} < \frac{2}{2}x < \frac{20}{2}$$

$$7 < x < 10 \qquad (7, 10)$$

2. $5 \le 3x - 1 \le 8$

$$+1 \qquad +1 \quad +1$$

$$6 \le 3x \le 9$$

$$\frac{6}{3} \le \frac{3}{3}x \le \frac{9}{3}$$

$$2 \le x \le 3 \qquad [2, 3]$$

3. $-2 \le 3x - 4 \le 5$

$$\begin{aligned}
+4 \qquad\quad +4 \quad +4 \\
2 \le 3x \le 9 \\
\frac{2}{3} \le \frac{3}{3}x \le \frac{9}{3} \\
\frac{2}{3} \le x \le 3 \qquad\qquad \left[\frac{2}{3}, 3\right]
\end{aligned}$$

4. $-4 < -2x + 6 < 4$

$$\begin{aligned}
-6 \qquad\quad -6 \quad -6 \\
-10 < -2x < -2 \\
\frac{-10}{-2} > \frac{-2}{-2}x > \frac{-2}{-2} \\
5 > x > 1 \text{ or } 1 < x < 5 \qquad\qquad (1, 5)
\end{aligned}$$

5. $0.12 \le 4x - 1 \le 1.8$

$$\begin{aligned}
+1.00 \qquad\quad +1 \quad +1.0 \\
1.12 \le 4x \le 2.8 \\
\frac{1.12}{4} \le \frac{4}{4}x \le \frac{2.8}{4} \\
0.28 \le x \le 0.7 \qquad\qquad [0.28, 0.7]
\end{aligned}$$

6. $4 < 3(-2x + 1) \le 7$

$$\begin{aligned}
4 < -6x + 3 \le 7 \\
-3 \qquad\quad -3 \quad -3 \\
1 < -6x \le 4 \\
\frac{1}{-6} > \frac{-6}{-6}x \ge \frac{4}{-6} \\
-\frac{1}{6} > x \ge -\frac{2}{3} \quad \text{or} \quad -\frac{2}{3} \le x < -\frac{1}{6} \qquad\qquad \left[-\frac{2}{3}, -\frac{1}{6}\right)
\end{aligned}$$

7. $-1 < -6x + 11 < 1$

$$\begin{aligned}
-11 \qquad\quad -11 \quad -11 \\
-12 < -6x < -10 \\
\frac{-12}{-6} > \frac{-6}{-6}x > \frac{-10}{-6} \\
2 > x > \frac{5}{3} \quad \text{or} \quad \frac{5}{3} < x < 2 \qquad\qquad \left(\frac{5}{3}, 2\right)
\end{aligned}$$

8.
$$\frac{7}{8} < \frac{1}{4}x \le 2$$
$$\frac{4}{1}\cdot\frac{7}{8} < \frac{4}{1}\cdot\frac{1}{4}x \le 4(2)$$
$$\frac{7}{2} < x \le 8 \qquad \left(\frac{7}{2}, 8\right]$$

9.
$$8 \le 4.5x - 1 \le 11$$
$$+1 \qquad +1 \quad +1$$
$$9 \le 4.5x \le 12$$
$$\frac{9}{4.5} \le \frac{4.5}{4.5}x \le \frac{12}{4.5}$$
$$2 \le x \le \frac{8}{3} \qquad \left[2, \frac{8}{3}\right]$$

10.
$$-6 \le \frac{2}{3}x + 4 < 0$$
$$-4 \qquad -4 - 4$$
$$-10 \le \frac{2}{3}x < -4$$
$$\frac{3}{2}\cdot\frac{-10}{1} \le \frac{3}{2}\cdot\frac{2}{3}x < \frac{3}{2}\cdot\frac{-4}{1}$$
$$-15 \le x < -6 \qquad [-15, -6)$$

11.
$$-1 < \frac{2x-5}{3} < 1$$
$$(3)(-1) < \frac{3}{1}\cdot\frac{2x-5}{3} < 3(1)$$
$$-3 < 2x - 5 < 3$$
$$+5 \qquad +5 \quad +5$$
$$2 < 2x < 8$$
$$\frac{2}{2} < \frac{2}{2}x < \frac{8}{2}$$
$$1 < x < 4 \qquad (1, 4)$$

Double inequalities are used to solve word problems where the solution is a limited range of values. Usually there are two variables and you are given the range of one of them and asked to find the range of the other.

Examples

$y = 3x - 2$

If $7 \le y \le 10$, what is the corresponding interval for x?

Because $y = 3x - 2$, replace "y" with "$3x - 2$."

"$7 \le y \le 10$" becomes "$7 \le 3x - 2 \le 10$"

$$7 \le 3x - 2 \le 10$$
$$+2 \qquad +2 \quad +2$$
$$9 \le 3x \le 12$$
$$\frac{9}{3} \le \frac{3}{3}x \le \frac{12}{3}$$
$$3 \le x \le 4$$

$y = 4x + 1$

If $-3 < y < 3$, the corresponding interval for x can be found by solving $-3 < 4x + 1 < 3$.

$$-3 < 4x + 1 < 3$$
$$-1 \qquad -1 \quad -1$$
$$-4 < 4x < 2$$
$$\frac{-4}{4} < \frac{4}{4}x < \frac{2}{4}$$
$$-1 < x < \frac{1}{2}$$

$y = 3 - x$

If $0 \le y < 4$, the corresponding interval for x can be found by solving $0 \le 3 - x < 4$.

$$0 \le 3 - x < 4$$
$$-3 \quad -3 \qquad -3$$
$$-3 \le -x < 1$$
$$-(-3) \ge -(-x) > -1$$
$$3 \ge x > -1 \quad \text{or} \quad -1 < x \le 3$$

Practice

Give the corresponding interval for x.

1. $y = x - 4 \qquad -5 < y < 5$

2. $y = 4x - 3$ $\quad 0 < y \le 2$

3. $y = 7 - 2x$ $\quad 4 \le y \le 10$

4. $y = 8x + 1$ $\quad -5 \le y \le -1$

5. $y = \frac{2}{3}x - 8$ $\quad 4 < y < 6$

6. $y = \dfrac{x-4}{2}$ $\quad -1 \le y < 3$

Solutions

1. $y = x - 4$ $\quad -5 < y < 5$

$$\begin{array}{l} -5 < x - 4 < 5 \\ +4 \quad\quad +4 \quad +4 \\ -1 < x < 9 \end{array}$$

2. $y = 4x - 3$ $\quad 0 < y \le 2$

$$\begin{array}{l} 0 < 4x - 3 \le 2 \\ +3 \quad\quad +3 \quad +3 \\ 3 < 4x \le 5 \\ \frac{3}{4} < \frac{4}{4}x \le \frac{5}{4} \\ \frac{3}{4} < x \le \frac{5}{4} \end{array}$$

3. $y = 7 - 2x$ $\quad 4 \le y \le 10$

$$\begin{array}{l} 4 \le 7 - 2x \le 10 \\ -7 \quad -7 \quad\quad -7 \\ -3 \le -2x \le 3 \\ \frac{-3}{-2} \ge \frac{-2}{-2}x \ge \frac{3}{-2} \\ \frac{3}{2} \ge x \ge \frac{-3}{2} \quad \text{or} \quad \frac{-3}{2} \le x \le \frac{3}{2} \end{array}$$

4. $y = 8x + 1 \qquad -5 \le y \le -1$

$$-5 \le 8x + 1 \le -1$$
$$-1 \qquad -1 \quad -1$$

$$-6 \le 8x \le -2$$

$$\frac{-6}{8} \le \frac{8}{8}x \le \frac{-2}{8}$$

$$\frac{-3}{4} \le x \le \frac{-1}{4}$$

5. $y = \frac{2}{3}x - 8 \qquad 4 < y < 6$

$$4 < \frac{2}{3}x - 8 < 6$$

$$+8 \qquad +8 \quad +8$$

$$12 < \frac{2}{3}x < 14$$

$$\frac{3}{2} \cdot \frac{12}{1} < \frac{3}{2} \cdot \frac{2}{3}x < \frac{3}{2} \cdot \frac{14}{1}$$

$$18 < x < 21$$

6. $y = \frac{x-4}{2} \qquad -1 \le y < 3$

$$-1 \le \frac{x-4}{2} < 3$$

$$2(-1) \le \frac{2}{1} \cdot \frac{x-4}{2} < 2(3)$$

$$-2 \le x - 4 < 6$$
$$+4 \qquad +4 \quad +4$$

$$2 \le x < 10$$

The applied problems in this section are similar to problems earlier in this chapter. The only difference is that you are given a range for one value and you are asked to find the range for the other.

Examples

A high school student earns \$8 per hour in her summer job. She hopes to earn between \$120 and \$200 per week. What range of hours will she need to work so that her pay is in this range?

Let x represent the number of hours worked per week. Represent her weekly pay by $p = 8x$. The student wants $120 \leq p \leq 200$. The inequality to solve is $120 \leq 8x \leq 200$.

$$120 \leq 8x \leq 200$$

$$\frac{120}{8} \leq \frac{8}{8}x \leq \frac{200}{8}$$

$$15 \leq x \leq 25$$

The student would need to work between 15 and 25 hours per week for her pay to range from \$120 to \$200 per week.

A manufacturing plant produces pencils. It has monthly overhead costs of \$60,000. Each gross (144) of pencils costs \$3.60 to manufacture. The company wants to keep total costs between \$96,000 and \$150,000 per month. How many gross of pencils should the plant produce to keep its costs in this range?

Let x represent the number of gross of pencils manufactured monthly. Production cost is represented by $3.60x$. Represent the total cost by $c = 60{,}000 + 3.60x$. The manufacturer wants $96{,}000 \leq c \leq 150{,}000$. The inequality to be solved is $96{,}000 \leq 60{,}000 + 3.60x \leq 150{,}000$.

$$\begin{array}{rrr} 96{,}000 \leq & 60{,}000 \quad +3.60x \leq & 150{,}000 \\ -60{,}000 & -60{,}000 \qquad\qquad & -60{,}000 \end{array}$$

$$36{,}000 \leq 3.60x \leq 90{,}000$$

$$\frac{36{,}000}{3.60} \leq \frac{3.60}{3.60}x \leq \frac{90{,}000}{3.60}$$

$$10{,}000 \leq x \leq 25{,}000$$

The manufacturing plant should produce between 10,000 and 25,000 gross per month to keep its monthly costs between \$96,000 and \$150,000.

Practice

1. According to Hooke's Law, the force, F (in pounds), required to stretch a certain spring x inches beyond its natural length is $F = 4.2x$. If $7 \le F \le 14$, what is the corresponding range for x?

2. Recall that the relationship between the Fahrenheit and Celsius temperature scales is given by $\mathrm{F} = \frac{9}{5}\mathrm{C} + 32$. If $5 \le \mathrm{F} \le 23$, what is the corresponding range for C?

3. A saleswoman's salary is a combination of an annual base salary of \$15,000 plus a 10% commission on sales. What level of sales does she need to maintain in order that her annual salary range from \$25,000 to \$40,000?

4. The Smith's electric bills consist of a base charge of \$20 plus 6 cents per kilowatt-hour. If the Smiths want to keep their electric bill in the \$80 to \$110 range, what range of kilowatt-hours do they need to maintain?

5. A particular collect call costs \$2.10 plus 75 cents per minute. (The company bills in two-second intervals.) How many minutes would a call need to last to keep a charge between \$4.50 and \$6.00?

Solutions

1. $7 \le F \le 14 \quad \text{and} \quad F = 4.2x.$

$$7 \le 4.2x \le 14$$

$$\frac{7}{4.2} \le \frac{4.2}{4.2}x \le \frac{14}{4.2}$$

$$\frac{5}{3} \le x \le \frac{10}{3}$$

If the force is to be kept between 7 and 14 pounds, the spring will stretch between $\frac{5}{3}$ and $\frac{10}{3}$ inches beyond its natural length.

2. $5 \le \mathrm{F} \le 23 \text{ and } \mathrm{F} = \frac{9}{5}\mathrm{C} + 32.$

$$5 \le \frac{9}{5}\mathrm{C} + 32 \le 23$$

$$-32 \qquad\qquad -32 \quad -32$$

$$-27 \le \frac{9}{5}\text{C} \le -9$$

$$\frac{5}{9} \cdot \frac{-27}{1} \le \frac{5}{9} \cdot \frac{9}{5}\text{C} \le \frac{5}{9} \cdot \frac{-9}{1}$$

$$-15 \le \text{C} \le -5$$

3. Let x represent the saleswoman's annual sales. Let $s = 15,000 + 0.10x$ represent her annual salary. She wants $25{,}000 \le s \le 40{,}000$.

$$\begin{array}{rcr} 25{,}000 \le & 15{,}000 + 0.10x \le & 40{,}000 \\ -15{,}000 & -15{,}000 & -15{,}000 \end{array}$$

$$10{,}000 \le 0.10x \le 25{,}000$$

$$\frac{10{,}000}{0.10} \le \frac{0.10}{0.10}x \le \frac{25{,}000}{0.10}$$

$$100{,}000 \le x \le 250{,}000$$

She needs to have her annual sales range from \$100,000 to \$250,000 in order to maintain her annual salary between \$25,000 and \$40,000.

4. Let x represent the number of kilowatt-hours the Smiths use per month. Then $c = 20 + 0.06x$ represents their monthly electric bill.

$$80 \le c \le 110$$

$$\begin{array}{rcr} 80 \le & 20 + 0.06x \le & 110 \\ -20 & -20 & -20 \end{array}$$

$$60 \le 0.06x \le 90$$

$$\frac{60}{0.06} \le \frac{0.06}{0.06}x \le \frac{90}{0.06}$$

$$1000 \le x \le 1500$$

The Smiths would need to keep their monthly usage between 1000 and 1500 kilowatt-hours monthly to keep their monthly bills in the \$80 to \$110 range.

5. Let x represent the number of minutes the call lasts. Let $c = 2.10 + 0.75x$ represent the total cost of the call.

$$4.50 \le c \le 6.00$$

$$\begin{array}{rcl} 4.50 \le & 2.10 + 0.75x & \le 6.00 \\ -2.10 & -2.10 & -2.10 \end{array}$$

$$2.40 \le 0.75x \le 3.90$$

$$\frac{2.40}{0.75} \le \frac{0.75}{0.75}x \le \frac{3.90}{0.75}$$

$$3.2 \le x \le 5.2$$

3.2 minutes is three minutes 12 seconds and 5.2 minutes is five minutes 12 seconds because 0.20 minutes is 0.20(60) seconds = 12 seconds

A call would need to last between three minutes 12 seconds and 5 minutes 12 seconds in order to cost between \$4.50 and \$6.00.

Chapter Review

1. $2x - 3 \le 7 + x$

(a) $x \le 10$ (b) $x < 10$ (c) $x \le -10$ (c) $x < -10$

2. $x < 5$ is represented by

(a) 5 (b) 5

(c) 5 (d) 5

3. $x \ge -6$ is represented by

(a) −6 (b) −6

(c) −6 (d) −6

4. $4x + 12 > 10x$

(a) $x > 2$ (b) $x > -2$ (c) $x < 2$ (d) $x < -2$

5. The interval notation for $x > 1$ is

(a) $(-\infty, 1)$ (b) $(1, \infty)$ (c) $(-\infty, 1]$ (d) $[1, \infty)$

6. A financial officer is splitting \$500,000 between two bonds, one paying 6% interest and the other paying 8% interest (this bond carries more risk). A minimum of \$36,500 of interest payments per year is required. How much can she spend on the 6% bond?

(a) She can spend at most \$175,000 on the 6% bond.
(b) She can spend at least \$175,000 on the 6% bond.
(c) She can spend at most \$325,000 on the 6% bond.
(d) She can spend at least \$325,000 on the 6% bond.

7. The interval notation for $3 < x \leq 10$ is

(a) $(3, 10]$ (b) $[3, 10)$ (c) $[10, 3)$ (d) $(10, 3]$

8. $-6 < x < 6$ is represented on the number line by

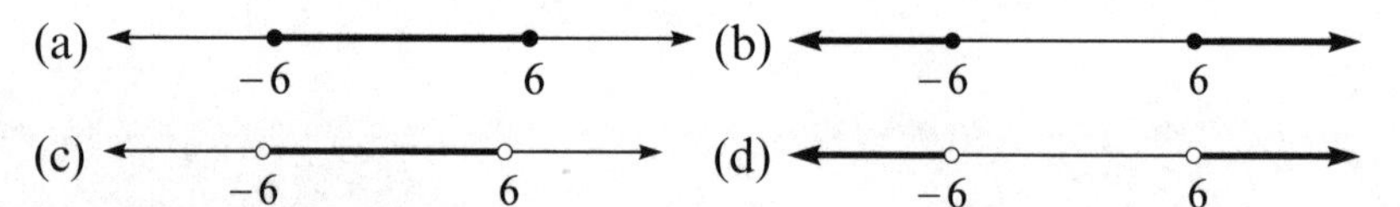

9. The interval notation for $8 \leq x \leq 2$ is

(a) $[8, 2]$ (b) $[2, 8]$ (c) not an interval of numbers

10. $-4 < 2x - 6 < 8$

(a) $-4 < x < -2$ (b) $-1 < x < 7$ (c) $-2 < x < 4$
(d) $1 < x < 7$

11. The Martinez family budgets \$60–85 for its monthly electric bill. The electric company charges a customer charge of \$15 per month plus \$0.08 per kilowatt-hour. What range of kilowatt-hours can they use each month to keep their electric bill in this range?

(a) At least 562.5 kilowatt-hours but no more than 875 kilowatt-hours
(b) More than 562.5 kilowatt-hours but less than 875 kilowatt-hours
(c) At least 750 kilowatt-hours but no more than 1062.5 kilowatt-hours
(d) More than 750 kilowatt-hours but less than 1062.5 kilowatt-hours

12. Joel wants to invest \$10,000. Some will be deposited into an account earning 6% interest and the rest into an account earning $7\frac{1}{4}$% interest. If he wants at least \$650 interest each year, how much can he invest at $7\frac{1}{4}$%?

(a) At least \$6000 at $7\frac{1}{4}$%
(b) More than \$6000 at $7\frac{1}{4}$%
(c) At least \$4000 at $7\frac{1}{4}$%
(d) More than \$4000 at $7\frac{1}{4}$%

13. $-2 < 4 - 3x < 2$

(a) $-2 < x < -\frac{2}{3}$ (b) $\frac{2}{3} < x < 2$ (c) $2 < x < \frac{2}{3}$
(d) $\frac{2}{3} > x > 2$

14. $3 < \dfrac{2x - 1}{5} < 7$

(a) $2 < x < 4$ (b) $10 < x < 20$ (c) $7 < x < 17$
(d) $8 < x < 18$

15. The interval notation for $x \leq 6$ is

(a) $[-\infty, 6]$ (b) $(-\infty, 6)$ (c) $(-\infty, 6]$ (d) $[-\infty, 6)$

Solutions

1. (a) **2.** (c) **3.** (b) **4.** (c)
5. (b) **6.** (a) **7.** (a) **8.** (c)
9. (c) **10.** (d) **11.** (a) **12.** (c)
13. (b) **14.** (d) **15.** (c)

Quadratic Equations

A quadratic equation is one that can be put in the form $ax^2 + bx + c = 0$ where a, b, and c are numbers and a is not zero (b and/or c might be zero). For instance $3x^2 + 7x = 4$ is a quadratic equation.

$$\begin{aligned} 3x^2 + 7x &= 4 \\ -4 &\quad -4 \\ 3x^2 + 7x - 4 &= 0 \end{aligned}$$

In this example $a = 3$, $b = 7$, and $c = -4$.

There are two main approaches to solving these equations. One approach uses the fact that if the product of two numbers is zero, at least one of the numbers must be zero. In other words, $wz = 0$ implies $w = 0$ or $z = 0$ (or both $w = 0$ and $z = 0$.) To use this fact on a quadratic equation first make sure that one side of the equation is zero and factor the other side. Set each factor equal to zero then solve for x.

Examples

$x^2 + 2x - 3 = 0$

$x^2 + 2x - 3$ can be factored as $(x + 3)(x - 1)$

$x^2 + 2x - 3 = 0$ becomes $(x + 3)(x - 1) = 0$

Now set each factor equal to zero and solve for x.

$$\begin{array}{rcl} x+3=0 & \qquad & x-1=0 \\ -3\quad -3 & & +1\quad +1 \\ x=-3 & & x=1 \end{array}$$

You can check your solutions by substituting them into the original equation.

$$x^2+2x-3=0$$

$$x=-3\text{: } (-3)^2+2(-3)-3=9-6-3=0\checkmark$$

$$x=1\text{: } 1^2+2(1)-3=1+2-3=0\checkmark$$

$x^2+5x+6=0$ becomes $(x+2)(x+3)=0$

$$\begin{array}{rcl} x+2=0 & \qquad & x+3=0 \\ -2\quad -2 & & -3\quad -3 \\ x=-2 & & x=-3 \end{array}$$

$$\begin{array}{r} x^2+7x=8 \\ -8\quad -8 \end{array}$$

$x^2+7x-8=0$ becomes $(x+8)(x-1)=0$

$$\begin{array}{rcl} x+8=0 & \qquad & x-1=0 \\ -8\quad -8 & & +1\quad +1 \\ x=-8 & & x=1 \end{array}$$

$x^2-16=0$ becomes $(x-4)(x+4)=0$

$$\begin{array}{rcl} x-4=0 & \qquad & x+4=0 \\ +4\quad +4 & & -4\quad -4 \\ x=4 & & x=-4 \end{array}$$

$3x^2-9x-30=0$ becomes $3(x^2-3x-10)=0$ which becomes $3(x-5)(x+2)=0$

$$\begin{array}{rcl} x-5=0 & \qquad & x+2=0 \\ +5\quad +5 & & -2\quad -2 \\ x=5 & & x=-2 \end{array}$$

The factor 3 was not set equal to zero because "$3=0$" does not lead to any solution.

Practice

1. $x^2 - x - 12 = 0$
2. $x^2 + 7x + 12 = 0$
3. $x^2 + 8x = -15$
4. $x^2 - 10x = -21$
5. $3x^2 - x - 2 = 0$
6. $4x^2 - 8x = 5$
7. $x^2 - 25 = 0$
8. $9x^2 - 16 = 0$
9. $x^2 = 100$
10. $x^2 + 6x + 9 = 0$
11. $x^2 = 0$
12. $5x^2 = 0$
13. $x^2 - \frac{1}{9} = 0$

Solutions

1. $x^2 - x - 12 = 0$

   $(x - 4)(x + 3) = 0$

   $x - 4 = 0$ $\qquad$ $x + 3 = 0$

   $+4 \quad +4$ $\qquad$ $-3 \quad -3$

   $x = 4$ $\qquad$ $x = -3$

2. $x^2 + 7x + 12 = 0$

   $(x + 3)(x + 4) = 0$

$$x + 3 = 0 \qquad x + 4 = 0$$
$$-3 \quad -3 \qquad -4 \quad -4$$
$$x = -3 \qquad x = -4$$

3. $x^2 + 8x = -15$

$$+15 \quad +15$$
$$x^2 + 8x + 15 = 0$$
$$(x + 3)(x + 5) = 0$$
$$x + 3 = 0 \qquad x + 5 = 0$$
$$-3 \quad -3 \qquad -5 \quad -5$$
$$x = -3 \qquad x = -5$$

4. $x^2 - 10x = -21$

$$+21 \quad +21$$
$$x^2 - 10x + 21 = 0$$
$$(x - 3)(x - 7) = 0$$
$$x - 3 = 0 \qquad x - 7 = 0$$
$$+3 \quad +3 \qquad +7 \quad +7$$
$$x = 3 \qquad x = 7$$

5. $3x^2 - x - 2 = 0$

$$(3x + 2)(x - 1) = 0$$
$$3x + 2 = 0 \qquad x - 1 = 0$$
$$-2 \quad -2 \qquad +1 \quad +1$$
$$3x = -2 \qquad x = 1$$
$$x = \frac{-2}{3}$$

6. $4x^2 - 8x = 5$

$$-5 \quad -5$$
$$4x^2 - 8x - 5 = 0$$
$$(2x + 1)(2x - 5) = 0$$
$$2x + 1 = 0 \qquad 2x - 5 = 0$$
$$-1 \quad -1 \qquad +5 \quad +5$$
$$2x = -1 \qquad 2x = 5$$
$$x = \frac{-1}{2} \qquad x = \frac{5}{2}$$

7. $x^2 - 25 = 0$

$(x-5)(x+5) = 0$

$x - 5 = 0$ $x + 5 = 0$

$+5 \quad +5$ $-5 \quad -5$

$x = 5$ $x = -5$

8. $9x^2 - 16 = 0$

$(3x-4)(3x+4) = 0$

$3x - 4 = 0$ $3x + 4 = 0$

$+4 \quad +4$ $-4 \quad -4$

$3x = 4$ $3x = -4$

$x = \frac{4}{3}$ $x = \frac{-4}{3}$

9. $x^2 = 100$

$-100 - 100$

$x^2 - 100 = 0$

$(x-10)(x+10) = 0$

$x - 10 = 0$ $x + 10 = 0$

$+10 \quad +10$ $-10 \quad -10$

$x = 10$ $x = -10$

10. $x^2 + 6x + 9 = 0$

$(x+3)(x+3) = 0$

$x + 3 = 0$

$-3 \quad -3$

$x = -3$

11. $x^2 = 0$

$(x)(x) = 0$

$x = 0$

12. $5x^2 = 0$

$5(x)(x) = 0$

$x = 0$

13. $x^2 - \frac{1}{9} = 0$

$$\left(x - \frac{1}{3}\right)\left(x + \frac{1}{3}\right) = 0$$

$$x - \frac{1}{3} = 0 \qquad\qquad x + \frac{1}{3} = 0$$

$$+\frac{1}{3} + \frac{1}{3} \qquad\qquad -\frac{1}{3} - \frac{1}{3}$$

$$x = \frac{1}{3} \qquad\qquad x = -\frac{1}{3}$$

Not all quadratic expressions will be as easy to factor as the previous examples and problems were. Sometimes you will need to multiply or divide both sides of the equation by a number. Because zero multiplied or divided by any nonzero number is still zero, only one side of the equation will change. Keep in mind that not all quadratic expressions can be factored using rational numbers (fractions) or even real numbers. Fortunately there is another way of solving quadratic equations, which bypasses the factoring method.

Examples

The equation $-x^2 + 4x - 3 = 0$ is awkward to factor because of the negative sign in front of x^2. Multiply both sides of the equation by -1 then factor.

$$-1(-x^2 + 4x - 3) = -1(0)$$

$$x^2 - 4x + 3 = 0$$

$$(x - 3)(x - 1) = 0$$

$$x - 3 = 0 \qquad\qquad x - 1 = 0$$

$$+3 \quad +3 \qquad\qquad +1 \quad +1$$

$$x = 3 \qquad\qquad x = 1$$

Decimals and fractions in a quadratic equation can be eliminated in the same way. Multiply both sides of the equation by a power of 10 to eliminate decimal points. Multiply both sides of the equation by the LCD to eliminate fractions.

$$0.1x^2 - 1.5x + 5.6 = 0$$

Multiply both sides of the equation by 10 to clear the decimal.

$$10(0.1x^2 - 1.5x + 5.6) = 10(0)$$

$$x^2 - 15x + 56 = 0$$

$$(x - 8)(x - 7) = 0$$

$$x - 8 = 0 \qquad x - 7 = 0$$

$$+8 \quad +8 \qquad +7 \quad +7$$

$$x = 8 \qquad x = 7$$

$$\frac{3}{4}x^2 + \frac{1}{2}x - \frac{1}{4} = 0$$

Clear the fraction by multiplying both sides of the equation by 4 (the LCD).

$$4\left(\frac{3}{4}x^2 + \frac{1}{2}x - \frac{1}{4}\right) = 4(0)$$

$$3x^2 + 2x - 1 = 0$$

$$(3x - 1)(x + 1) = 0$$

$$3x - 1 = 0 \qquad x + 1 = 0$$

$$+1 \quad +1 \qquad -1 \quad -1$$

$$3x = 1 \qquad x = -1$$

$$x = \frac{1}{3}$$

$$\frac{1}{2}x^2 - 3x + 4 = 0$$

$$2\left(\frac{1}{2}x^2 - 3x + 4\right) = 2(0)$$

$$x^2 - 6x + 8 = 0$$

$$(x - 4)(x - 2) = 0$$

$$x - 4 = 0 \qquad x - 2 = 0$$

$$+4 \quad +4 \qquad +2 \quad +2$$

$$x = 4 \qquad x = 2$$

$$-2x^2 - 18x - 28 = 0$$
$$-(-2x^2 - 18x - 28) = -0$$
$$2x^2 + 18x + 28 = 0$$
$$\frac{1}{2}(2x^2 + 18x + 28) = \frac{1}{2}(0)$$
$$x^2 + 9x + 14 = 0$$
$$(x + 7)(x + 2) = 0$$

$$x + 7 = 0 \qquad\qquad x + 2 = 0$$
$$-7 \quad -7 \qquad\qquad -2 \quad -2$$
$$x = -7 \qquad\qquad x = -2$$

Multiplying both sides of the equation by $-\frac{1}{2}$ would have combined two steps.

Practice

1. $-x^2 - x + 30 = 0$
2. $-9x^2 + 25 = 0$
3. $0.01x^2 + 0.14x + 0.13 = 0$
4. $-0.1x^2 + 1.1x - 2.8 = 0$
5. $\frac{1}{5}x^2 + \frac{1}{5}x - 6 = 0$
6. $\frac{1}{6}x^2 - \frac{2}{3}x - \frac{16}{3} = 0$
7. $x^2 - \frac{1}{2}x - 3 = 0$
8. $-\frac{3}{2}x^2 + \frac{1}{2}x + 1 = 0$
9. $6x^2 + 18x - 24 = 0$
10. $-10x^2 - 34x - 12 = 0$

Solutions

1. $-x^2 - x + 30 = 0$

$$-(-x^2 - x + 30) = -0$$
$$x^2 + x - 30 = 0$$
$$(x + 6)(x - 5) = 0$$

$$\begin{array}{ll} x + 6 = 0 & x - 5 = 0 \\ \quad -6 \quad -6 & \quad +5 \quad +5 \\ \quad x = -6 & \quad x = 5 \end{array}$$

2. $-9x^2 + 25 = 0$

$$-(-9x^2 + 25) = -0$$
$$9x^2 - 25 = 0$$
$$(3x - 5)(3x + 5) = 0$$

$$\begin{array}{ll} 3x - 5 = 0 & 3x + 5 = 0 \\ \quad +5 \quad +5 & \quad -5 \quad -5 \\ \quad 3x = 5 & \quad 3x = -5 \\ \quad x = \frac{5}{3} & \quad x = \frac{-5}{3} \end{array}$$

3. $0.01x^2 + 0.14x + 0.13 = 0$

$$100(0.01x^2 + 0.14x + 0.13) = 100(0)$$
$$x^2 + 14x + 13 = 0$$
$$(x + 13)(x + 1) = 0$$

$$\begin{array}{ll} x + 13 = 0 & x + 1 = 0 \\ \quad -13 \quad -13 & \quad -1 \quad -1 \\ \quad x = -13 & \quad x = -1 \end{array}$$

4. $-0.1x^2 + 1.1x - 2.8 = 0$

$$-10(-0.1x^2 + 1.1x - 2.8) = -10(0)$$
$$x^2 - 11x + 28 = 0$$
$$(x - 7)(x - 4) = 0$$

$$\begin{aligned} x - 7 &= 0 \\ +7 \quad & +7 \\ x &= 7 \end{aligned} \qquad \begin{aligned} x - 4 &= 0 \\ +4 \quad & +4 \\ x &= 4 \end{aligned}$$

5. $\frac{1}{5}x^2 + \frac{1}{5}x - 6 = 0$

$$5\left(\frac{1}{5}x^2 + \frac{1}{5}x - 6\right) = 5(0)$$

$$x^2 + x - 30 = 0$$

$$(x + 6)(x - 5) = 0$$

$$\begin{aligned} x + 6 &= 0 \\ -6 \quad & -6 \\ x &= -6 \end{aligned} \qquad \begin{aligned} x - 5 &= 0 \\ +5 \quad & +5 \\ x &= 5 \end{aligned}$$

6. $\frac{1}{6}x^2 - \frac{2}{3}x - \frac{16}{3} = 0$

$$6\left(\frac{1}{6}x^2 - \frac{2}{3}x - \frac{16}{3}\right) = 6(0)$$

$$x^2 - 4x - 32 = 0$$

$$(x - 8)(x + 4) = 0$$

$$\begin{aligned} x - 8 &= 0 \\ +8 \quad & +8 \\ x &= 8 \end{aligned} \qquad \begin{aligned} x + 4 &= 0 \\ -4 \quad & -4 \\ x &= -4 \end{aligned}$$

7. $x^2 - \frac{1}{2}x - 3 = 0$

$$2\left(x^2 - \frac{1}{2}x - 3\right) = 2(0)$$

$$2x^2 - x - 6 = 0$$

$$(2x + 3)(x - 2) = 0$$

$$2x + 3 = 0 \qquad x - 2 = 0$$
$$-3 \quad -3 \qquad +2 \quad +2$$
$$2x = -3 \qquad x = 2$$
$$x = \frac{-3}{2}$$

8. $-\frac{3}{2}x^2 + \frac{1}{2}x + 1 = 0$

$$-2\left(-\frac{3}{2}x^2 + \frac{1}{2}x + 1\right) = -2(0)$$
$$3x^2 - x - 2 = 0$$
$$(3x + 2)(x - 1) = 0$$
$$3x + 2 = 0 \qquad x - 1 = 0$$
$$-2 \quad -2 \qquad +1 \quad +1$$
$$3x = -2 \qquad x = 1$$
$$\frac{3}{3}x = \frac{-2}{3}$$
$$x = \frac{-2}{3}$$

9. $6x^2 + 18x - 24 = 0$

$$\frac{1}{6}(6x^2 + 18x - 24) = \frac{1}{6}(0)$$
$$x^2 + 3x - 4 = 0$$
$$(x + 4)(x - 1) = 0$$
$$x + 4 = 0 \qquad x - 1 = 0$$
$$-4 \quad -4 \qquad +1 \quad +1$$
$$x = -4 \qquad x = 1$$

10. $-10x^2 - 34x - 12 = 0$

$$\frac{-1}{2}(-10x^2 - 34x - 12) = \frac{-1}{2}(0)$$
$$5x^2 + 17x + 6 = 0$$
$$(5x + 2)(x + 3) = 0$$

$$\begin{aligned} 5x + 2 &= \ \ 0 \\ -2 & \quad -2 \\ 5x &= -2 \\ x &= \frac{-2}{5} \end{aligned} \qquad\qquad \begin{aligned} x + 3 &= \ \ 0 \\ -3 & \quad -3 \\ x &= -3 \end{aligned}$$

Sometimes using the fact that $x^2 = k$ implies $x = \pm\sqrt{k}$ can be used to solve quadratic equations. For instance, if $x^2 = 9$, then $x = 3$ or -3 because $3^2 = 9$ and $(-3)^2 = 9$. This method works if the equation can be put in the form $ax^2 - c = 0$, where a and c are not negative.

Examples

$$x^2 = 16$$

$$x = \pm\sqrt{16}$$

$$x = \pm 4$$

$$\begin{aligned} 3x^2 &= 27 \\ x^2 &= 9 \\ x &= \pm 3 \end{aligned}$$

$$\begin{aligned} 25 - x^2 &= 0 \\ 25 &= x^2 \\ \pm 5 &= x \end{aligned}$$

$$\begin{aligned} 4x^2 &= 49 \\ x^2 &= \frac{49}{4} \\ x &= \pm\sqrt{\frac{49}{4}} \\ x &= \pm\frac{7}{2} \end{aligned}$$

$$3x^2 = 36$$
$$x^2 = 12$$
$$x = \pm\sqrt{12}$$
$$x = \pm\sqrt{4 \cdot 3} = \pm 2\sqrt{3}$$

Practice

1. $x^2 - 81 = 0$
2. $64 - x^2 = 0$
3. $4x^2 = 100$
4. $2x^2 = 3$
5. $-6x^2 = -80$

Solutions

1. $x^2 - 81 = 0$
$$x^2 = 81$$
$$x = \pm 9$$

2. $64 - x^2 = 0$
$$64 = x^2$$
$$\pm 8 = x$$

3. $4x^2 = 100$
$$x^2 = \frac{100}{4}$$
$$x^2 = 25$$
$$x = \pm 5$$

4. $$2x^2 = 3$$
$$x^2 = \frac{3}{2}$$
$$x = \pm\sqrt{\frac{3}{2}}$$
$$x = \pm\frac{\sqrt{3}}{\sqrt{2}} \cdot \frac{\sqrt{2}}{\sqrt{2}}$$
$$x = \pm\frac{\sqrt{6}}{2}$$

5. $$-6x^2 = -80$$
$$x^2 = \frac{-80}{-6}$$
$$x^2 = \frac{40}{3}$$
$$x = \pm\sqrt{\frac{40}{3}}$$
$$x = \pm\frac{\sqrt{40}}{\sqrt{3}} \cdot \frac{\sqrt{3}}{\sqrt{3}} = \pm\frac{\sqrt{120}}{3}$$
$$x = \pm\frac{\sqrt{4 \cdot 30}}{3} = \pm\frac{2\sqrt{30}}{3}$$

The Quadratic Formula

The other main approach to solving quadratic equations comes from the fact that $x^2 = k$ implies $x = \sqrt{k}, -\sqrt{k}$ and a technique called *completing the square*. The solutions to $ax^2 + bx + c = 0$ are

$$x = \frac{-b + \sqrt{b^2 - 4ac}}{2a}, \frac{-b - \sqrt{b^2 - 4ac}}{2a}.$$

These solutions are abbreviated as

$$x = \frac{-b \pm \sqrt{b^2 - 4ac}}{2a}.$$

This formula is called the *quadratic formula*. It will solve every quadratic equation. The quadratic formula is very important in algebra and is worth memorizing. You might wonder why we bother factoring quadratic expressions to solve quadratic equations when the quadratic formula will work. There are two reasons. One, factoring is an important skill in algebra and calculus. Two, factoring is often easier and faster than computing the quadratic formula. The quadratic formula is normally used to solve quadratic equations where the factoring is difficult.

Before the formula can be used, the quadratic equation must be in the form $ax^2 + bx + c = 0$. Once a, b, and c are identified, applying the quadratic formula is simply a matter of performing arithmetic.

$2x^2 - x - 7 = 0$	$a = 2, b = -1, c = -7$
$10x^2 - 4 = 0$ is equivalent to $10x^2 + 0x - 4 = 0$	$a = 10, b = 0, c = -4$
$3x^2 + x = 0$ is equivalent to $3x^2 + x + 0 = 0$	$a = 3, b = 1, c = 0$
$4x^2 = 0$ is equivalent to $4x^2 + 0x + 0 = 0$	$a = 4, b = 0, c = 0$
$x^2 + 3x = 4$ is equivalent to $x^2 + 3x - 4 = 0$	$a = 1, b = 3, c = -4$
$-8x^2 = -64$ is equivalent to $8x^2 + 0x - 64 = 0$	$a = 8, b = 0, c = -64$

Practice

Identify a, b, and c for $ax^2 + bx + c = 0$.

1. $2x^2 + 9x + 3 = 0$
2. $-3x^2 + x + 5 = 0$
3. $x^2 - x - 6 = 0$
4. $x^2 - 9 = 0$

5. $2x^2 = 32$

6. $x^2 + x = 0$

7. $x^2 - x = 0$

8. $9x^2 = 10x$

9. $8x^2 + 20x = 9$

10. $4x - 5 - 3x^2 = 0$

Solutions

1. $2x^2 + 9x + 3 = 0$ $a = 2$ $b = 9$ $c = 3$

2. $-3x^2 + x + 5 = 0$ $a = -3$ $b = 1$ $c = 5$

3. $x^2 - x - 6 = 0$ $a = 1$ $b = -1$ $c = -6$

4. $x^2 - 9 = 0$ $a = 1$ $b = 0$ $c = -9$

5. $2x^2 = 32$ Rewritten: $2x^2 - 32 = 0$ $a = 2$ $b = 0$ $c = -32$

6. $x^2 + x = 0$ $a = 1$ $b = 1$ $c = 0$

7. $x^2 - x = 0$ $a = 1$ $b = -1$ $c = 0$

8. $9x^2 = 10x$ Rewritten: $9x^2 - 10x = 0$
   $a = 9$ $b = -10$ $c = 0$

9. $8x^2 + 20x = 9$ Rewritten: $8x^2 + 20x - 9 = 0$
   $a = 8$ $b = 20$ $c = -9$

10. $4x - 5 - 3x^2 = 0$ Rewritten: $-3x^2 + 4x - 5 = 0$
    $a = -3$ $b = 4$ $c = -5$

The quadratic formula can be messy to compute when any of a, b, or c are fractions or decimals. You can get around this by multiplying both sides of the equation by the least common denominator or some power of ten.

$$ax^2 + bx + c = 0 \qquad x = \frac{-b \pm \sqrt{b^2 - 4ac}}{2a}$$

Example

$$\frac{1}{2}x^2 - \frac{1}{2}x - 1 = 0 \qquad a = \frac{1}{2} \qquad b = -\frac{1}{2} \qquad c = -1$$

$$x = \frac{-\left(-\frac{1}{2}\right) \pm \sqrt{\left(-\frac{1}{2}\right)^2 - 4\left(\frac{1}{2}\right)(-1)}}{2\left(\frac{1}{2}\right)}$$

The fractions in the formula could be eliminated if we multiplied both sides of the equation by 2.

$$2\left(\frac{1}{2}x^2 - \frac{1}{2}x - 1\right) = 2(0)$$

$$x^2 - x - 2 = 0 \qquad a = 1 \qquad b = -1 \qquad c = -2$$

$$x = \frac{-(-1) \pm \sqrt{(-1)^2 - 4(1)(-2)}}{2(1)}$$

Sometimes the solutions to a quadratic equation need to be simplified.

Examples

$$\frac{8 \pm \sqrt{24}}{2}$$

$$\sqrt{24} = \sqrt{4 \cdot 6} = 2\sqrt{6}$$

$$\frac{8 \pm \sqrt{24}}{2} = \frac{8 \pm 2\sqrt{6}}{2}$$

The denominator is divisible by 2 and each term in the numerator is divisible by 2, so factor 2 from each term in the numerator. Next use this 2 to cancel the 2 in the denominator.

$$\frac{8 \pm 2\sqrt{6}}{2} = \frac{2(4 \pm \sqrt{6})}{2} = 4 \pm \sqrt{6}$$

$$\frac{-3 \pm \sqrt{18}}{6} = \frac{-3 \pm \sqrt{9 \cdot 2}}{6} = \frac{-3 \pm 3\sqrt{2}}{6} = \frac{3(-1 \pm \sqrt{2})}{6} = \frac{-1 \pm \sqrt{2}}{2}$$

$$\frac{15 \pm \sqrt{50}}{10} = \frac{15 \pm \sqrt{25 \cdot 2}}{10} = \frac{15 \pm 5\sqrt{2}}{10} = \frac{5(3 \pm \sqrt{2})}{10} = \frac{3 \pm \sqrt{2}}{2}$$

Practice

Simplify.

1. $\frac{6 \pm \sqrt{12}}{2} =$

2. $\frac{12 \pm \sqrt{27}}{6} =$

3. $\frac{2 \pm \sqrt{48}}{4} =$

4. $\frac{20 \pm \sqrt{300}}{10} =$

5. $\frac{-6 \pm \sqrt{20}}{-2} =$

Solutions

1. $\frac{6 \pm \sqrt{12}}{2} = \frac{6 \pm \sqrt{4 \cdot 3}}{2} = \frac{6 \pm 2\sqrt{3}}{2} = \frac{2(3 \pm \sqrt{3})}{2} = 3 \pm \sqrt{3}$

2. $\frac{12 \pm \sqrt{27}}{6} = \frac{12 \pm \sqrt{9 \cdot 3}}{6} = \frac{12 \pm 3\sqrt{3}}{6} = \frac{3(4 \pm \sqrt{3})}{6} = \frac{4 \pm \sqrt{3}}{2}$

3. $\frac{2 \pm \sqrt{48}}{4} = \frac{2 \pm \sqrt{16 \cdot 3}}{4} = \frac{2 \pm 4\sqrt{3}}{4} = \frac{2(1 \pm 2\sqrt{3})}{4} = \frac{1 \pm 2\sqrt{3}}{2}$

4. $\frac{20 \pm \sqrt{300}}{10} = \frac{20 \pm \sqrt{100 \cdot 3}}{10} = \frac{20 \pm 10\sqrt{3}}{10} = \frac{10(2 \pm \sqrt{3})}{10} = 2 \pm \sqrt{3}$

5. $\dfrac{-6 \pm \sqrt{20}}{-2} = \dfrac{-6 \pm \sqrt{4 \cdot 5}}{-2} = \dfrac{-6 \pm 2\sqrt{5}}{-2} = \dfrac{-2(3 \pm \sqrt{5})}{-2} = 3 \pm \sqrt{5}$

(The negative of $\pm\sqrt{5}$ is still $\pm\sqrt{5}$.)

Now that we can identify a, b, and c in the quadratic formula and can simplify the solutions, we are ready to solve quadratic equations using the formula.

Examples

$2x^2 + 3x + 1 = 0 \qquad a = 2 \qquad b = 3 \qquad c = 1$

$$x = \frac{-3 \pm \sqrt{(3)^2 - 4(2)(1)}}{2(2)} = \frac{-3 \pm \sqrt{9-8}}{4} = \frac{-3 \pm \sqrt{1}}{4}$$

$$= \frac{-3+1}{4}, \frac{-3-1}{4} = \frac{-2}{4}, \frac{-4}{4} = -\frac{1}{2}, -1$$

$x^2 - x - 1 = 0 \qquad a = 1 \qquad b = -1 \qquad c = -1$

$$x = \frac{-(-1) \pm \sqrt{(-1)^2 - 4(1)(-1)}}{2(1)} = \frac{1 \pm \sqrt{1-(-4)}}{2} = \frac{1 \pm \sqrt{1+4}}{2}$$

$$= \frac{1 \pm \sqrt{5}}{2} = \frac{1+\sqrt{5}}{2}, \frac{1-\sqrt{5}}{2}$$

$x^2 - 18 = 0 \qquad a = 1 \qquad b = 0 \qquad c = -18$

$$x = \frac{-0 \pm \sqrt{0^2 - 4(1)(-18)}}{2(1)} = \frac{\pm\sqrt{0-(-72)}}{2} = \frac{\pm\sqrt{72}}{2} = \frac{\pm\sqrt{36 \cdot 2}}{2}$$

$$= \frac{\pm 6\sqrt{2}}{2} = \pm 3\sqrt{2} = 3\sqrt{2}, -3\sqrt{2}$$

$2x^2 + 6x = 5$ Rewrite as $2x^2 + 6x - 5 = 0$

$a = 2 \qquad b = 6 \qquad c = -5$

$$x = \frac{-6 \pm \sqrt{6^2 - 4(2)(-5)}}{2(2)} = \frac{-6 \pm \sqrt{36-(-40)}}{4} = \frac{-6 \pm \sqrt{76}}{4}$$

$$= \frac{-6 \pm \sqrt{4 \cdot 19}}{4} = \frac{-6 \pm 2\sqrt{19}}{4} = \frac{2(-3 \pm \sqrt{19})}{4} = \frac{-3 \pm \sqrt{19}}{2}$$

$$= \frac{-3 + \sqrt{19}}{2}, \frac{-3 - \sqrt{19}}{2}$$

$\frac{1}{3}x^2 + x - 2 = 0$

Multiply both sides of the equation by 3 to eliminate the fraction.

$$3\left(\frac{1}{3}x^2 + x - 2\right) = 3(0)$$

$$x^2 + 3x - 6 = 0 \qquad a = 1 \qquad b = 3 \qquad c = -6$$

$$x = \frac{-3 \pm \sqrt{3^2 - 4(1)(-6)}}{2(1)} = \frac{-3 \pm \sqrt{9 - (-24)}}{2} = \frac{-3 \pm \sqrt{33}}{2}$$

$$= \frac{-3 + \sqrt{33}}{2}, \frac{-3 - \sqrt{33}}{2}$$

$0.1x^2 - 0.8x + 0.21 = 0$

Multiply both sides of the equation by 100 to eliminate the decimal.

$$100(0.1x^2 - 0.8x + 0.21) = 0$$

$$10x^2 - 80x + 21 = 0 \qquad a = 10 \qquad b = -80 \qquad c = 21$$

$$x = \frac{-(-80) \pm \sqrt{(-80)^2 - 4(10)(21)}}{2(10)} = \frac{80 \pm \sqrt{6400 - 840}}{20}$$

$$= \frac{80 \pm \sqrt{5560}}{20} = \frac{80 \pm \sqrt{4 \cdot 1390}}{20} = \frac{80 \pm 2\sqrt{1390}}{20} = \frac{2(40 \pm \sqrt{1390})}{20}$$

$$= \frac{40 \pm \sqrt{1390}}{10} = \frac{40 + \sqrt{1390}}{10}, \frac{40 - \sqrt{1390}}{10}$$

Practice

1. $x^2 - 5x + 3 = 0$
2. $4x^2 + x - 6 = 0$

3. $7x^2 + 3x = 2$

4. $2x^2 = 9$

5. $-3x^2 + 4x + 1 = 0$

6. $9x^2 - x = 10$

7. $10x^2 - 5x = 0$

8. $0.1x^2 - 0.11x - 1 = 0$

9. $\frac{1}{3}x^2 + \frac{1}{6}x - \frac{1}{8} = 0$

10. $80x^2 - 16x - 32 = 0$

11. $18x^2 + 39x + 20 = 0$

12. $x^2 + 10x + 25 = 0$

Solutions

1. $x^2 - 5x + 3 = 0$

$$x = \frac{-(-5) \pm \sqrt{(-5)^2 - 4(1)(3)}}{2(1)} = \frac{5 \pm \sqrt{25 - 12}}{2} = \frac{5 \pm \sqrt{13}}{2}$$

2. $4x^2 + x - 6 = 0$

$$x = \frac{-1 \pm \sqrt{1^2 - 4(4)(-6)}}{2(4)} = \frac{-1 \pm \sqrt{1 - (-96)}}{8} = \frac{-1 \pm \sqrt{97}}{8}$$

3. $7x^2 + 3x = 2$ (Equivalent to $7x^2 + 3x - 2 = 0$)

$$x = \frac{-3 \pm \sqrt{3^2 - 4(7)(-2)}}{2(7)} = \frac{-3 \pm \sqrt{9 - (-56)}}{14} = \frac{-3 \pm \sqrt{65}}{14}$$

4. $2x^2 = 9$ (Equivalent to $2x^2 - 9 = 0$)

$$x = \frac{-0 \pm \sqrt{0^2 - 4(2)(-9)}}{2(2)} = \frac{0 \pm \sqrt{-(-72)}}{4} = \frac{\pm\sqrt{72}}{4} = \frac{\pm\sqrt{36 \cdot 2}}{4}$$

$$= \frac{\pm 6\sqrt{2}}{4} = \frac{\pm 3\sqrt{2}}{2}$$

5. $-3x^2 + 4x + 1 = 0$

$$x = \frac{-4 \pm \sqrt{4^2 - 4(-3)(1)}}{2(-3)} = \frac{-4 \pm \sqrt{16 - (-12)}}{-6} = \frac{-4 \pm \sqrt{28}}{-6}$$

$$= \frac{-4 \pm \sqrt{4 \cdot 7}}{-6} = \frac{-4 \pm 2\sqrt{7}}{-6} = \frac{-2(2 \pm \sqrt{7})}{-6} = \frac{2 \pm \sqrt{7}}{3}$$

or $-1(-3x^2 + 4x + 1) = -1(0)$

$$3x^2 - 4x - 1 = 0$$

$$x = \frac{-(-4) \pm \sqrt{(-4)^2 - 4(3)(-1)}}{2(3)} = \frac{4 \pm \sqrt{16 - (-12)}}{6}$$

$$= \frac{4 \pm \sqrt{28}}{6} = \frac{4 \pm \sqrt{4 \cdot 7}}{6} = \frac{4 \pm 2\sqrt{7}}{6} = \frac{2(2 \pm \sqrt{7})}{6} = \frac{2 \pm \sqrt{7}}{3}$$

6. $9x^2 - x = 10$ (Equivalent to $9x^2 - x - 10 = 0$)

$$x = \frac{-(-1) \pm \sqrt{(-1)^2 - 4(9)(-10)}}{2(9)} = \frac{1 \pm \sqrt{1 - (-360)}}{18}$$

$$= \frac{1 \pm \sqrt{361}}{18} = \frac{1 \pm 19}{18} = \frac{1 + 19}{18}, \frac{1 - 19}{18} = \frac{20}{18}, \frac{-18}{18} = \frac{10}{9}, -1$$

7. $10x^2 - 5x = 0$

$$x = \frac{-(-5) \pm \sqrt{(-5)^2 - 4(10)(0)}}{2(10)} = \frac{5 \pm \sqrt{25 - 0}}{20} = \frac{5 \pm \sqrt{25}}{20}$$

$$= \frac{5 \pm 5}{20} = \frac{10}{20}, \frac{0}{20} = \frac{1}{2}, 0$$

8. $0.1x^2 - 0.11x - 1 = 0$

$100(0.1x^2 - 0.11x - 1) = 100(0)$

$10x^2 - 11x - 100 = 0$

$$x = \frac{-(-11) \pm \sqrt{(-11)^2 - 4(10)(-100)}}{2(10)} = \frac{11 \pm \sqrt{121 - (-4000)}}{20}$$

$$= \frac{11 \pm \sqrt{4121}}{20}$$

9. $\frac{1}{3}x^2 + \frac{1}{6}x - \frac{1}{8} = 0$

$$24\left(\frac{1}{3}x^2 + \frac{1}{6}x - \frac{1}{8}\right) = 24(0)$$

$$8x^2 + 4x - 3 = 0$$

$$x = \frac{-4 \pm \sqrt{4^2 - 4(8)(-3)}}{2(8)} = \frac{-4 \pm \sqrt{16 - (-96)}}{16} = \frac{-4 \pm \sqrt{112}}{16}$$

$$= \frac{-4 \pm \sqrt{16 \cdot 7}}{16} = \frac{-4 \pm 4\sqrt{7}}{16} = \frac{4(-1 \pm \sqrt{7})}{16} = \frac{-1 \pm \sqrt{7}}{4}$$

10. $80x^2 - 16x - 32 = 0$

$$\frac{1}{16}(80x^2 - 16x - 32) = \frac{1}{16}(0)$$

$$5x^2 - x - 2 = 0$$

$$x = \frac{-(-1) \pm \sqrt{(-1)^2 - 4(5)(-2)}}{2(5)} = \frac{1 \pm \sqrt{1 - (-40)}}{10} = \frac{1 \pm \sqrt{41}}{10}$$

11. $18x^2 + 39x + 20 = 0$

$$x = \frac{-39 \pm \sqrt{39^2 - 4(18)(20)}}{2(18)} = \frac{-39 \pm \sqrt{1521 - 1440}}{36}$$

$$= \frac{-39 \pm \sqrt{81}}{36} = \frac{-39 \pm 9}{36} = \frac{-39 + 9}{36}, \frac{-39 - 9}{36} = \frac{-30}{36}, \frac{-48}{36}$$

$$= \frac{-5}{6}, \frac{-4}{3}$$

12. $x^2 + 10x + 25 = 0$

$$x = \frac{-10 \pm \sqrt{10^2 - 4(1)(25)}}{2(1)} = \frac{-10 \pm \sqrt{100 - 100}}{2} = \frac{-10 \pm 0}{2}$$

$$= \frac{-10}{2} = -5$$

Rational Equations

Some rational equations (an equation with one or more fractions as terms) become quadratic equations once each term has been multiplied by the least common denominator. Remember, you *must* be sure that any solutions do not lead to a zero in any denominator in the original equation.

There are two main approaches to clearing the denominator(s) in a rational equation. If the equation is in the form of "fraction = fraction," cross multiply. If the equation is not in this form, find the least common denominator (LCD). Finding the least common denominator often means you need to factor each denominator completely. We learned in Chapter 7 to multiply both sides of the LCD, then to distribute the LCD. In this chapter we will simply multiply each term on each side of the equation by the LCD. If the new equation is a quadratic equation, collect all terms on one side of the equal sign and leave a zero on the other. In the examples and practice problems below, the solutions that lead to a zero in a denominator will be stated.

Examples

$\frac{x-1}{x^2-2x-8} = \frac{-2}{5}$ This is in the form "fraction = fraction" so we will cross multiply.

$$5(x-1) = -2(x^2-2x-8)$$

$$5x-5 = -2x^2+4x+16$$
$$+2x^2-4x-16 \quad +2x^2-4x-16$$

$$2x^2+x-21=0$$

$$(2x+7)(x-3)=0$$

$$2x+7=0 \qquad x-3=0$$
$$-7 \quad -7 \qquad +3 \quad +3$$

$$2x=-7 \qquad x=3$$

$$x=\frac{-7}{2}$$

$$\frac{4}{x^2-2x}+\frac{x+1}{x}=\frac{-4}{x-2}$$

First factor each denominator, second find the LCD, and third multiply all three terms by the LCD.

$$\frac{4}{x(x-2)}+\frac{x+1}{x}=\frac{-4}{x-2} \qquad \text{LCD} = x(x-2)$$

$$\frac{x(x-2)}{1}\cdot\frac{4}{x(x-2)}+\frac{x(x-2)}{1}\cdot\frac{x+1}{x}=\frac{x(x-2)}{1}\cdot\frac{-4}{x-2}$$

$$\begin{aligned}4+(x-2)(x+1)&=-4x\\ 4+x^2-x-2&=-4x\\ x^2-x+2&=-4x\\ +4x\quad&\quad +4x\\ x^2+3x+2&=0\\ (x+2)(x+1)&=0\end{aligned}$$

$$\begin{aligned}x+2&=0 \qquad\qquad & x+1&=0\\ -2\quad&\ -2 & -1\quad&\ -1\\ x&=-2 & x&=-1\end{aligned}$$

$$\frac{3}{x-4}=\frac{-6x+24}{x^2-2x-8} \qquad \text{Cross multiply.}$$

$$\begin{aligned}3(x^2-2x-8)&=(x-4)(-6x+24)\\ 3x^2-6x-24&=-6x^2+48x-96\\ +6x^2-48x+96\quad&\quad +6x^2-48x+96\\ 9x^2-54x+72&=0\\ \frac{1}{9}(9x^2-54x+72)&=\frac{1}{9}(0)\\ x^2-6x+8&=0\\ (x-4)(x-2)&=0\end{aligned}$$

$$\begin{aligned}x-4&=0 \qquad\qquad & x-2&=0\\ +4\quad&\ +4 & +2\quad&\ +2\\ x&=4 & x&=2\end{aligned}$$

We cannot let x be 4 because $x = 4$ leads to a zero in the denominator of $\frac{3}{x-4}$. The only solution is $x = 2$.

Practice

Because all of these problems factor, factoring is used in the solutions. If factoring takes too long on some of these, you may use the quadratic formula.

1. $\frac{3x}{x-4} = \frac{-3x}{2}$

2. $\frac{x-1}{2x+3} = \frac{6}{x-2}$

3. $\frac{x+2}{x-3} + \frac{2x+1}{x^2-9} = \frac{12x+3}{x+3}$

4. $\frac{2x-1}{x+1} - \frac{3x}{x-2} = \frac{-8x-7}{x^2-x-2}$

5. $\frac{4x+1}{x-1} + \frac{x-5}{1-x} = \frac{24}{x}$

6. $\frac{2x}{x+1} + \frac{3}{x} = \frac{2}{x^2+x}$

7. $\frac{2x}{2x+1} - \frac{3}{x} = \frac{-3x-4}{3x}$

8. $\frac{2}{x-5} + \frac{1}{3x} = \frac{-8}{3x+15}$

9. $\frac{1}{x-4} + \frac{2}{x+1} - \frac{3}{x+3} = \frac{x-3}{x^2-3x-4}$

10. $\frac{4}{x-1} + \frac{3}{x+1} = \frac{1}{2x} - \frac{6}{x^2-1}$

Solutions

1. $\frac{3x}{x-4} = \frac{-3x}{2}$

$$3x(2) = (x-4)(-3x)$$

$$6x = -3x^2 + 12x$$
$$+3x^2 - 12x \quad +3x^2 - 12x$$

$$3x^2 - 6x = 0$$

$$3x(x-2) = 0$$

$$3x = 0 \qquad x - 2 = 0$$

$$x = \frac{0}{3} \qquad +2 \quad +2$$

$$x = 0 \qquad x = 2$$

2. $\dfrac{x-1}{2x+3} = \dfrac{6}{x-2}$

$$(x-1)(x-2) = 6(2x+3)$$

$$x^2 - 3x + 2 = 12x + 18$$
$$-12x - 18 \quad -12x - 18$$

$$x^2 - 15x - 16 = 0$$

$$(x-16)(x+1) = 0$$

$$x - 16 = 0 \qquad x + 1 = 0$$
$$+16 \quad +16 \qquad -1 \quad -1$$

$$x = 16 \qquad x = -1$$

3. $\dfrac{x+2}{x-3} + \dfrac{2x+1}{x^2-9} = \dfrac{12x+3}{x+3}$

Denominator factored: $\dfrac{x+2}{x-3} + \dfrac{2x+1}{(x-3)(x+3)} = \dfrac{12x+3}{x+3}$

LCD $= (x-3)(x+3)$

$$(x-3)(x+3)\cdot\frac{x+2}{x-3} + (x-3)(x+3)\cdot\frac{2x+1}{(x-3)(x+3)}$$

$$= (x-3)(x+3)\cdot\frac{12x+3}{x+3}$$

$$(x+3)(x+2)+2x+1=(x-3)(12x+3)$$
$$x^2+5x+6+2x+1=12x^2-33x-9$$
$$x^2+7x+7=12x^2-33x-9$$
$$-x^2-7x-7 \quad -x^2 \quad -7x-7$$
$$0=11x^2-40x-16$$
$$0=(11x+4)(x-4)$$

$$11x+4=0 \qquad\qquad x-4=0$$
$$-4 \quad -4 \qquad\qquad +4 \quad +4$$
$$11x=-4 \qquad\qquad x=4$$
$$x=\frac{-4}{11}$$

4. $$\frac{2x-1}{x+1}-\frac{3x}{x-2}=\frac{-8x-7}{x^2-x-2}$$

Denominator factored: $\frac{2x-1}{x+1}-\frac{3x}{x-2}=\frac{-8x-7}{(x-2)(x+1)}$

LCD $=(x+1)(x-2)$

$$(x+1)(x-2)\cdot\frac{2x-1}{x+1}-(x+1)(x-2)\cdot\frac{3x}{x-2}$$
$$=(x+1)(x-2)\cdot\frac{-8x-7}{(x-2)(x+1)}$$

$$(x-2)(2x-1)-3x(x+1)=-8x-7$$
$$2x^2-5x+2-3x^2-3x=-8x-7$$
$$-x^2-8x+2=-8x-7$$
$$+x^2+8x-2 \quad +x^2+8x-2$$
$$0=x^2-9$$
$$0=(x-3)(x+3)$$

$$x-3=0 \qquad\qquad x+3=0$$
$$+3+3 \qquad\qquad -3-3$$
$$x=3 \qquad\qquad x=-3$$

5. $\dfrac{4x+1}{x-1}+\dfrac{x-5}{1-x}=\dfrac{24}{x}$

Using the fact that $1-x=-(1-x)$ allows us to write the second denominator the same as the first.

$$\frac{4x+1}{x-1}+\frac{x-5}{-(x-1)}=\frac{24}{x}$$

$$\frac{4x+1}{x-1}+\frac{-(x-5)}{x-1}=\frac{24}{x} \qquad \text{LCD}=x(x-1)$$

$$x(x-1)\cdot\frac{4x+1}{x-1}+x(x-1)\cdot\frac{-(x-5)}{x-1}=x(x-1)\cdot\frac{24}{x}$$

$$x(4x+1)+-x(x-5)=24(x-1)$$

$$x(4x+1)-x(x-5)=24(x-1)$$

$$4x^2+x-x^2+5x=24x-24$$

$$3x^2+6x \qquad =24x-24$$

$$-24x+24 \quad -24x+24$$

$$3x^2-18x+24=0$$

$$\frac{1}{3}(3x^2-18x+24)=\frac{1}{3}(0)$$

$$x^2-6x+8=0$$

$$(x-4)(x-2)=0$$

$$x-4=0 \qquad\qquad x-2=0$$

$$+4 \quad +4 \qquad\qquad +2 \quad +2$$

$$x=4 \qquad\qquad x=2$$

6. $\dfrac{2x}{x+1}+\dfrac{3}{x}=\dfrac{2}{x^2+x}$ Denominator factored: $\dfrac{2x}{x+1}+\dfrac{3}{x}=\dfrac{2}{x(x+1)}$

$\text{LCD}=x(x+1)$

$$x(x+1)\cdot\frac{2x}{x+1}+x(x+1)\cdot\frac{3}{x}=x(x+1)\cdot\frac{2}{x(x+1)}$$

$$2x^2+3(x+1)=2$$

$$2x^2+3x+3=2$$

$$-2 \quad -2$$

$$2x^2 + 3x + 1 = 0$$
$$(2x + 1)(x + 1) = 0$$

$$\begin{aligned} 2x + 1 &= 0 \\ -1 \quad & -1 \\ 2x &= -1 \\ x &= \frac{-1}{2} \end{aligned}$$

$$\begin{aligned} x + 1 &= 0 \\ -1 \quad & -1 \\ x &= -1 \end{aligned}$$

But $x = -1$ leads to a zero in a denominator, so $x = -1$ is *not* a solution.

7. $\dfrac{2x}{2x+1} - \dfrac{3}{x} = \dfrac{-3x-4}{3x}$ LCD $= 3x(2x + 1)$

$$3x(2x+1) \cdot \frac{2x}{2x+1} - 3x(2x+1) \cdot \frac{3}{x} = 3x(2x+1) \cdot \frac{-3x-4}{3x}$$
$$3x(2x) - 3(2x + 1)3 = (2x + 1)(-3x - 4)$$
$$6x^2 - 9(2x + 1) = -6x^2 - 11x - 4$$
$$6x^2 - 18x - 9 = -6x^2 - 11x - 4$$
$$+6x^2 + 11x + 4 \quad +6x^2 + 11x + 4$$

$$12x^2 - 7x - 5 = 0$$
$$(12x + 5)(x - 1) = 0$$

$$\begin{aligned} 12x + 5 &= 0 \\ -5 \quad & -5 \\ 12x &= -5 \\ x &= \frac{-5}{12} \end{aligned}$$

$$\begin{aligned} x - 1 &= 0 \\ +1 \quad & +1 \\ x &= 1 \end{aligned}$$

8. $\dfrac{2}{x-5} + \dfrac{1}{3x} = \dfrac{-8}{3x+15}$

Denominator factored: $\dfrac{2}{x-5} + \dfrac{1}{3x} = \dfrac{-8}{3(x+5)}$

LCD $= 3x(x - 5)(x + 5)$

$$3x(x-5)(x+5) \cdot \frac{2}{x-5} + 3x(x-5)(x+5) \cdot \frac{1}{3x}$$

$$= 3x(x-5)(x+5)\cdot\frac{-8}{3(x+5)}$$

$$3x(x+5)2+(x-5)(x+5)=x(x-5)(-8)$$
$$6x(x+5)+(x-5)(x+5)=-8x(x-5)$$
$$6x^2+30x+x^2-25=-8x^2+40x$$

$$7x^2+30x-25=-8x^2+40x$$
$$+8x^2-40x \qquad\qquad +8x^2-40x$$
$$15x^2-10x-25=0$$

$$\frac{1}{5}(15x^2-10x-25)=\frac{1}{5}(0)$$
$$3x^2-2x-5=0$$
$$(3x-5)(x+1)=0$$

$$3x-5=0 \qquad\qquad x+1=0$$
$$+5 \quad +5 \qquad\qquad -1 \quad -1$$
$$3x=5 \qquad\qquad x=-1$$
$$x=\frac{5}{3}$$

9. $$\frac{1}{x-4}+\frac{2}{x+1}-\frac{3}{x+3}=\frac{x-3}{x^2-3x-4}$$

Denominator factored: $\frac{1}{x-4}+\frac{2}{x+1}-\frac{3}{x+3}=\frac{x-3}{(x-4)(x+1)}$

LCD $=(x-4)(x+1)(x+3)$

$$(x-4)(x+1)(x+3)\cdot\frac{1}{x-4}+(x-4)(x+1)(x+3)\cdot\frac{2}{x+1}$$
$$-(x-4)(x+1)(x+3)\cdot\frac{3}{x+3}$$
$$=(x-4)(x+1)(x+3)\cdot\frac{x-3}{(x-4)(x+1)}$$

$$(x+1)(x+3)+2[(x-4)(x+3)]-3[(x-4)(x+1)]=(x+3)(x-3)$$
$$x^2+4x+3+2(x^2-x-12)-3(x^2-3x-4)=x^2-9$$
$$x^2+4x+3+2x^2-2x-24-3x^2+9x+12=x^2-9$$

$$\begin{aligned} 11x - 9 &= x^2 - 9 \\ -11x + 9 & \quad -11x + 9 \\ 0 &= x^2 - 11x \\ 0 &= x(x - 11) \end{aligned}$$

$x = 0$

$$\begin{aligned} x - 11 &= 0 \\ +11 & \quad +11 \\ x &= 11 \end{aligned}$$

10. $\dfrac{4}{x-1} + \dfrac{3}{x+1} = \dfrac{1}{2x} - \dfrac{6}{x^2 - 1}$

Denominator factored: $\dfrac{4}{x-1} + \dfrac{3}{x+1} = \dfrac{1}{2x} - \dfrac{6}{(x-1)(x+1)}$

LCD $= 2x(x-1)(x+1)$

$$\begin{aligned} &2x(x-1)(x+1) \cdot \frac{4}{x-1} + 2x(x-1)(x+1)\frac{3}{x+1} \\ &\quad = 2x(x-1)(x+1) \cdot \frac{1}{2x} - 2x(x-1)(x+1)\frac{6}{(x-1)(x+1)} \end{aligned}$$

$$\begin{aligned} 2x(x+1)(4) + 2x(x-1)(3) &= (x-1)(x+1) - 2x(6) \\ 8x(x+1) + 6x(x-1) &= x^2 - 1 - 12x \\ 8x^2 + 8x + 6x^2 - 6x &= x^2 - 12x - 1 \\ 14x^2 + 2x &= x^2 - 12x - 1 \\ -x^2 + 12x + 1 & \quad -x^2 + 12x + 1 \end{aligned}$$

$$\begin{aligned} 13x^2 + 14x + 1 &= 0 \\ (13x + 1)(x + 1) &= 0 \end{aligned}$$

$$\begin{aligned} 13x + 1 &= 0 \\ -1 & \quad -1 \\ 13x &= -1 \\ x &= \frac{-1}{13} \end{aligned}$$

$$\begin{aligned} x + 1 &= 0 \\ -1 & \quad -1 \\ x &= -1 \end{aligned}$$

But $x = -1$ leads to a zero in a denominator, so $x = -1$ is *not* a solution.

Chapter Review

1. If $(x+1)(x-5)=0$, then the solutions are

   (a) $x=1,-5$ (b) $x=1,5$ (c) $x=-1,-5$
   (d) $x=-1,5$

2. If $x^2-x-1=0$, then $x=$

   a) $\dfrac{-1\pm\sqrt{5}}{2}$ (b) $-1\pm\dfrac{\sqrt{5}}{2}$ (c) $\dfrac{1\pm\sqrt{5}}{2}$ (d) $1\pm\dfrac{\sqrt{5}}{2}$

3. $\dfrac{2\pm\sqrt{24}}{2}$ in simplified form is

   (a) $1\pm\sqrt{24}$ (b) $1\pm\sqrt{6}$ (c) $2\pm\sqrt{6}$
   (d) cannot be simplified

4. To apply the quadratic formula to $2x^2-x=3$,

   (a) $a=2, b=-1, c=3$ (b) $a=2, b=1, c=3$
   (c) $a=2, b=-1, c=-3$ (d) $a=2, b=-1, c=0$

5. If $x^2-3x-4=0$, then the solutions are

   (a) $x=4,-1$ (b) $x=-4,1$ (c) $x=4,1$
   (d) $x=-4,-1$

6. If $2x^2+4x-9=0$, the solutions are

   (a) $x=\dfrac{2\pm\sqrt{22}}{2}$ (b) $x=\dfrac{-2\pm\sqrt{22}}{2}$
   (c) $x=2\pm\dfrac{\sqrt{22}}{2}$ (d) $x=-2\pm\dfrac{\sqrt{22}}{2}$

7. If $x^2-\frac{1}{4}=0$, the solutions are

   (a) $x=\pm\dfrac{1}{2}$ (b) $x=\pm\dfrac{1}{4}$ (c) $x=\pm\dfrac{1}{8}$ (d) $x=\pm\dfrac{1}{16}$

8. If $\dfrac{x}{2x+6} = \dfrac{x-6}{x-1}$, the solutions are

(a) $x = 9, -4$ (b) $x = -9, 4$ (c) $x = \dfrac{3 \pm \sqrt{77}}{2}$

(d) $x = 3 \pm \dfrac{\sqrt{77}}{2}$

9. If $\dfrac{x+1}{x-2} + \dfrac{1}{x} = \dfrac{13}{x^2 - 2x}$, the solutions are

(a) $x = 11$ only (b) $x = \dfrac{-1 \pm \sqrt{57}}{2}$ (c) $x = -5, 3$

(d) $x = -1 \pm \dfrac{\sqrt{57}}{2}$

10. If $\dfrac{2x}{x^2 + x - 2} = \dfrac{3}{x+2}$, the solutions are

(a) $x = -3, 2$ (b) $x = 3, -2$ (c) $x = -3$ only

(d) $x = 3$ only

Solutions

1. (d)	**2.** (c)	**3.** (b)	**4.** (c)
5. (a)	**6.** (b)	**7.** (a)	**8.** (a)
9. (c)	**10.** (d)		

CHAPTER 11

Quadratic Applications

Most of the problems in this chapter are not much different from the word problems in previous chapters. The only difference is that quadratic equations are used to solve them. Because quadratic equations usually have two solutions, some of these applied problems will have two solutions. Most will have only one—one of the "solutions" will be invalid. More often than not, the invalid solutions are easy to recognize.

Examples

The product of two consecutive positive numbers is 240. Find the numbers.

Let x represent the first number. Because the numbers are consecutive, the next number is one more than the first: $x + 1$ represents the next number. The product of these two numbers is $x(x + 1)$, which equals 240.

$$x(x + 1) = 240$$

$$x^2 + x = 240$$

$$x^2 + x - 240 = 0$$

$$(x - 15)(x + 16) = 0$$

$$\begin{aligned} x - 15 &= \ \ 0 \qquad (x + 16 = 0 \text{ leads to a negative solution}) \\ +15 & \quad +15 \\ x &= 15 \end{aligned}$$

The consecutive positive numbers are 15 and 16.

(This problem could have been set up with x representing the first number and $x - 1$ representing the second number.)

The product of two consecutive even numbers is 528. What are the numbers?

Let x represent the first number. Consecutive even numbers (and consecutive odd numbers) differ by two, so let $x + 2$ represent the second number. Their product is $x(x + 2)$.

$$x(x + 2) = 528$$

$$x^2 + 2x = 528$$

$$x^2 + 2x - 528 = 0$$

$$(x - 22)(x + 24) = 0$$

$$\begin{aligned} x - 22 &= \ \ 0 \\ +22 & \quad +22 \\ x &= 22 \end{aligned} \qquad \begin{aligned} x + 24 &= \ \ 0 \\ -24 & \quad -24 \\ x &= -24 \end{aligned}$$

The two solutions are 22 and 24, and -24 and -22.

Two positive numbers differ by five. Their product is 104. Find the two numbers.

Let x represent the first number. If x differs from the other number by five, then the other number could either be $x + 5$ or $x - 5$; it does not matter which representation you use. We will work this problem with both representations.

Let $x + 5$ represent the other number	Let $x - 5$ represent the other number
$x(x + 5) = 104$	$x(x - 5) = 104$
$x^2 + 5x = 104$	$x^2 - 5x = 104$
$x^2 + 5x - 104 = 0$ $(x + 13)(x - 8) = 0$ $x - 8 = 0$ ($x + 13 = 0$ leads to a $+8 \quad +8$ negative solution) $x = 8$ The numbers are 8 and $8 + 5 = 13$.	$x^2 - 5x - 104 = 0$ $(x - 13)(x + 8) = 0$ $x - 13 = 0$ ($x + 8 = 0$ leads to a $+13 \quad +13$ negative solution) $x = 13$ The numbers are 13 and $13 - 5 = 8$.

Practice

1. The product of two consecutive odd numbers is 399. Find the numbers.
2. The product of two consecutive numbers is 380. Find the numbers.
3. The product of two consecutive numbers is 650. Find the numbers.
4. The product of two consecutive even numbers is 288. What are the numbers?
5. Two numbers differ by 7. Their product is 228. What are the numbers?

Solution

1. Let $x =$ first number $\quad x + 2 =$ second number

$$x(x+2) = 399$$

$$x^2 + 2x = 399$$

$$x^2 + 2x - 399 = 0$$

$$(x - 19)(x + 21) = 0$$

$$x - 19 = 0 \qquad x + 21 = 0$$

$$x = 19 \qquad x = -21$$

$$x + 2 = 21 \qquad x + 2 = -19$$

There are two solutions: 19 and 21, and -21 and -19.

2. Let $x =$ first number $\quad x + 1 =$ second number

$$x(x+1) = 380$$

$$x^2 + x = 380$$

$$x^2 + x - 380 = 0$$

$$(x + 20)(x - 19) = 0$$

$x + 20 = 0$ $\quad$ $x - 19 = 0$

$x = -20$ $\quad$ $x = 19$

$x + 1 = -19$ $\quad$ $x + 1 = 20$

There are two solutions: 19 and 20, and −19 and −20.

3. Let x = first number $\quad$ $x + 1$ = second number

$$x(x+1) = 650$$
$$x^2 + x = 650$$
$$x^2 + x - 650 = 0$$
$$(x - 25)(x + 26) = 0$$

$x - 25 = 0$ $\quad$ $x + 26 = 0$

$x = 25$ $\quad$ $x = -26$

$x + 1 = 26$ $\quad$ $x + 1 = -25$

There are two solutions: 25 and 26, and −25 and −26.

4. Let x = first number $\quad$ $x + 2$ = second number

$$x(x+2) = 288$$
$$x^2 + 2x = 288$$
$$x^2 + 2x - 288 = 0$$
$$(x - 16)(x + 18) = 0$$

$x - 16 = 0$ $\quad$ $x + 18 = 0$

$x = 16$ $\quad$ $x = -18$

$x + 2 = 18$ $\quad$ $x + 2 = -16$

There are two solutions: 16 and 18, and −16 and −18.

5. Let x = first number $\quad$ $x + 7$ = second number

$$x(x+7) = 228$$
$$x^2 + 7x = 228$$
$$x^2 + 7x - 228 = 0$$
$$(x - 12)(x + 19) = 0$$

$$x - 12 = 0 \qquad x + 19 = 0$$
$$x = 12 \qquad x = -19$$
$$x + 7 = 19 \qquad x + 7 = -12$$

There are two solutions: 12 and 19, −12 and −19.

Revenue

A common business application of quadratic equations occurs when raising a price results in lower sales or lowering a price results in higher sales. The obvious question is what to charge to bring in the most revenue. This problem is addressed in Algebra II and Calculus. The problem addressed here is finding a price that would bring in a particular revenue.

The problem involves raising (or lowering) a price by a certain number of increments and sales decreasing (or increasing) by a certain amount for each incremental change in the price. For instance, suppose for each increase of \$10 in the price, two customers are lost. The price and sales level both depend on the *number* of \$10 increases. If the price is increased by \$10, two customers are lost. If the price is increased by \$20, $2(2) = 4$ customers will be lost. If the price is increased by \$30, $2(3) = 6$ customers will be lost. If the price does not change, $2(0) = 0$ customers will be lost. The variable will represent the *number* of incremental increases (or decreases) of the price.

The revenue formula is $R = PQ$ where R represents the revenue, P represents the price, and Q represents the number sold. If the price is increased, then P will equal the current price plus the variable times the increment. If the price is decreased, then P will equal the current price minus the variable times the increment. If the sales level is decreased, then Q will equal the current sales level minus the variable times the incremental loss. If the sales level is increased, then Q will equal the current sales level plus the variable times the incremental gain.

Examples

A department store sells 20 portable stereos per week at \$80 each. The manager believes that for each decrease of \$5 in the price, six more stereos will be sold.

Let x represent the number of \$5 decreases in the price. Then the price will decrease by $5x$:

$P = 80 - 5x.$

The sales level will increase by six for each \$5 decrease in the price—the sales level will increase by $6x$:

$Q = 20 + 6x.$
$R = PQ$ becomes $R = (80 - 5x)(20 + 6x).$

A rental company manages an office complex with 16 offices. Each office can be rented if the monthly rent is \$1000. For each \$200 increase in the rent, one tenant will be lost.

Let x represent the number of \$200 increases in the rent.

$P = 1000 + 200x \qquad Q = 16 - 1x \qquad R = (1000 + 200x)(16 - x)$

A grocery store sells 300 pounds of bananas each day when they are priced at 45 cents per pound. The produce manager observes that for each 5-cent decrease in the price per pound of bananas, an additional 50 pounds are sold.

Let x represent the number of 5-cent decreases in the rent.

$P = 45 - 5x \qquad Q = 300 + 50x \qquad R = (45 - 5x)(300 + 5x)$

(The revenue will be in cents instead of dollars.)

A music storeowner sells 60 newly released CDs per day when the price is \$12 per CD. For each \$1.50 decrease in the price, the store will sell an additional 16 CDs each week.

Let x represent the number of \$1.50 decreases in the price.

$P = 12.00 - 1.50x \quad Q = 60 + 16x \qquad R = (12.00 - 1.50x)(60 + 16x)$

Practice

Let x represent the number of increases/decreases in the price.

1. The owner of an apartment complex knows he can rent all 50 apartments when the monthly rent is \$400. He thinks that for each \$25 increase in the rent, he will lose two tenants.

   $P =$ ____________

   $Q =$ ____________ $R =$ ____________

2. A grocery store sells 4000 gallons of milk per week when the price is \$2.80 per gallon. Customer research indicates that for each \$0.10 decrease in the price, 200 more gallons of milk will be sold.

   $P =$ ____________

   $Q =$ ____________ $R =$ ____________

3. A movie theater's concession stand sells an average of 500 buckets of popcorn each weekend when the price is \$4 per bucket. The manager knows from experience that for every \$0.05 decrease in the price, 20 more buckets of popcorn will be sold each weekend.
$P =$ ____________
$Q =$ ____________ $R =$ ____________

4. An automobile repair shop performs 40 oil changes per day when the price is \$30. Industry research indicates that the shop will lose 5 customers for each \$2 increase in the price.
$P =$ ____________
$Q =$ ____________ $R =$ ____________

5. A fast food restaurant sells an average of 250 orders of onion rings each week when the price is \$1.50 per order. The manager believes that for each \$0.05 decrease in the price, 10 more orders will be sold.
$P =$ ____________
$Q =$ ____________ $R =$ ____________

6. A shoe store sells a certain athletic shoe for \$40 per pair. The store averages sales of 80 pairs each week. The store owner's past experience leads him to believe that for each \$2 increase in the price of the shoe, one less pair would be sold each week.
$P =$ ____________
$Q =$ ____________ $R =$ ____________

Solutions

1. $P = 400 + 25x$ $Q = 50 - 2x$
$R = (400 + 25x)(50 - 2x)$

2. $P = 2.80 - 0.10x$ $Q = 4000 + 200x$
$R = (2.80 - 0.10x)(4000 + 200x)$

3. $P = 4 - 0.05x$ $Q = 500 + 20x$
$R = (4 - 0.05x)(500 + 20x)$

4. $P = 30 + 2x$ $Q = 40 - 5x$
$R = (30 + 2x)(40 - 5x)$

5. $P = 1.50 - 0.05x \qquad Q = 250 + 10x$
   $R = (1.50 - 0.05x)(250 + 10x)$

6. $P = 40 + 2x \qquad Q = 80 - 1x$
   $R = (40 + 2x)(80 - x)$

Now that we can set up these problems, we are ready to solve them. For each of the previous examples and problems, a desired revenue will be given. We will set that revenue equal to the revenue equation. This will be a quadratic equation. Some of these equations will be solved by factoring, others by the quadratic formula. Some problems will have more than one solution.

Examples

A department store sells 20 portable stereos per week at \$80 each. The manager believes that for each decrease of \$5 in the price, six more stereos will be sold.

Let x represent the number of \$5 decreases in the price.

$$P = 80 - 5x \qquad Q = 20 + 6x \qquad R = (80 - 5x)(20 + 6x).$$

What price should be charged if the revenue needs to be \$2240?

$$R = (80 - 5x)(20 + 6x) \text{ becomes } 2240 = (80 - 5x)(20 + 6x)$$

$$2240 = (80 - 5x)(20 + 6x)$$

$$2240 = 1600 + 380x - 30x^2$$

$$30x^2 - 380x + 640 = 0$$

$$\frac{1}{10}\left(30x^2 - 380x + 640\right) = \frac{1}{10}(0)$$

$$3x^2 - 38x + 64 = 0$$

$$(3x - 32)(x - 2) = 0$$

$$3x - 32 = 0 \qquad x - 2 = 0$$

$$3x = 32 \qquad x = 2$$

$$x = \frac{32}{3}$$

If $x = \frac{32}{3}$, the price for each stereo will be $P = 80 - 5\left(\frac{32}{3}\right) = \$26.67$.

If $x = 2$, the price for each stereo will be $P = 80 - 5(2) = \$70$.

A rental company manages an office complex with 16 offices. Each office can be rented if the monthly rent is \$1000. For each \$200 increase in the rent, one tenant will be lost.

Let x represent the number of \$200 increases in the rent.

$P = 1000 + 200x \qquad Q = 16 - 1x \qquad R = (1000 + 200x)(16 - x)$

What should the monthly rent be if the rental company needs \$20,800 each month in revenue?

$$R = (1000 + 200x)(16 - x)$$

$$20{,}800 = (1000 + 200x)(16 - x)$$

$$20{,}800 = 16{,}000 + 2200x - 200x^2$$

$$200x^2 - 2200x + 4800 = 0$$

$$\frac{1}{200}\left(200x^2 - 2200x + 4800\right) = \frac{1}{200}(0)$$

$$x^2 - 11x + 24 = 0$$

$$(x - 3)(x - 8) = 0$$

$$x - 3 = 0 \qquad x - 8 = 0$$

$$x = 3 \qquad x = 8$$

If $x = 3$, the monthly rent will be $1000 + 200(3) = \$1600$. If $x = 8$, the monthly rent will be $1000 + 200(8) = \$2600$.

A grocery store sells 300 pounds of bananas each day when they are priced at 45 cents per pound. The produce manager observes that for each 5-cent decrease in the price per pound of bananas, an additional 50 pounds are sold.

Let x represent the number of 5-cent decreases in the price.

$P = 45 - 5x \qquad Q = 300 + 50x \qquad R = (45 - 5x)(300 + 5x)$

What should the price of bananas be for weekly sales to be \$140? How many bananas (in pounds) will be sold at this price (these prices)?

(The revenue will be in terms of cents, so \$140 becomes 14,000 cents.)

$$R = (45 - 5x)(300 + 50x)$$

$$14{,}000 = (45 - 5x)(300 + 50x)$$

$$14{,}000 = 13{,}500 + 750x - 250x^2$$

$$250x^2 - 750x + 500 = 0$$

$$\frac{1}{250}(250x^2 - 750x + 500) = \frac{1}{250}(0)$$

$$x^2 - 3x + 2 = 0$$

$$(x - 2)(x - 1) = 0$$

$$x - 2 = 0 \qquad x - 1 = 0$$

$$x = 2 \qquad x = 1$$

If $x = 2$, the price per pound will be $45 - 5(2) = 35$ cents. The number of pounds sold each week will be $300 + 50(2) = 400$. If $x = 1$, the price per pound will be $45 - 5(1) = 40$ cents and the number of pounds sold each week will be $300 + 50(1) = 350$.

A music storeowner sells 60 newly released CDs per week when the cost is \$12 per CD. For each \$1.50 decrease in the price, the store will sell an additional 16 CDs per week.

Let x represent the number of \$1.50 decreases in the price.

$$P = 12.00 - 1.50x \qquad Q = 60 + 16x \qquad R = (12.00 - 1.50x)(60 + 16x)$$

What should the price be if the storeowner needs revenue of \$810 per week for the sale of these CDs? How many will be sold at this price (these prices)?

$$R = (12.00 - 1.50x)(60 + 16x)$$

$$810 = (12.00 - 1.50x)(60 + 16x)$$

$$810 = 720 + 102x - 24x^2$$

$$24x^2 - 102x + 90 = 0$$

$$\frac{1}{6}(24x^2 - 102x + 90) = \frac{1}{6}(0)$$

$$4x^2 - 17x + 15 = 0$$

$$(4x - 5)(x - 3) = 0$$

$$4x - 5 = 0 \qquad x - 3 = 0$$

$$4x = 5 \qquad x = 3$$

$$x = \frac{5}{4} = 1.25$$

When $x = 1.25$, the price should be $12 - 1.50(1.25) = \$10.13$ and the number sold would be $60 + 16(1.25) = 80$. If $x = 3$, the price should be $12 - 1.50(3) = \$7.50$ and the number sold would be $60 + 16(3) = 108$.

Practice

1. The owner of an apartment complex knows he can rent all 50 apartments when the monthly rent is \$400. He thinks that for each \$25 increase in the rent, he will lose two tenants. What should the rent be for the revenue to be \$20,400?

2. A grocery store sells 4000 gallons of milk per week when the price is \$2.80 per gallon. Customer research indicates that for each \$0.10 decrease in the price, 200 more gallons of milk will be sold. What does the price need to be so that weekly milk sales reach \$11,475?

3. A movie theater's concession stand sells an average of 500 buckets of popcorn each weekend when the price is \$4 per bucket. The manager knows from experience that for every \$0.05 decrease in the price, 20 more buckets of popcorn will be sold each weekend. What should the price be so that \$2450 worth of popcorn is sold? How many buckets will be sold at this price (these prices)?

4. An automobile repair shop performs 40 oil changes per day when the price is \$30. Industry research indicates that the shop will lose 5 customers for each \$2 increase in the price. What would the shop have to charge in order for the daily revenue from oil changes to be \$1120? How many oil changes will the shop perform each day?

5. A fast food restaurant sells an average of 250 orders of onion rings each week when the price is \$1.50 per order. The manager believes that for each \$0.05 decrease in the price, 10 more orders are sold. If

the manager wants \$378 weekly revenue from onion ring sales, what should she charge for onion rings?

6. A shoe store sells a certain athletic shoe for \$40 per pair. The store averages sales of 80 pairs each week. The store owner's past experience leads him to believe that for each \$2 increase in the price of the shoe, one less pair would be sold each week. What price would result in \$3648 weekly sales?

Solutions

1. $P = 400 + 25x \qquad Q = 50 - 2x \qquad R = (400 + 25x)(50 - 2x)$

$$20{,}400 = (400 + 25x)(50 - 2x)$$

$$20{,}400 = 20{,}000 + 450x - 50x^2$$

$$50x^2 - 450x + 400 = 0$$

$$\frac{1}{50}(50x^2 - 450x + 400) = \frac{1}{50}(0)$$

$$x^2 - 9x + 8 = 0$$

$$(x - 8)(x - 1) = 0$$

$$x - 8 = 0 \qquad x - 1 = 0$$

$$x = 8 \qquad x = 1$$

If $x = 1$, the rent should be $400 + 25(1) = \$425$. If $x = 8$, the rent should be $400 + 25(8) = \$600$.

2. $P = 2.80 - 0.10x \qquad Q = 4000 + 200x$

$R = (2.80 - 0.10x)(4000 + 200x)$

$$11{,}475 = (2.80 - 0.10x)(4000 + 200x)$$

$$11{,}475 = 11{,}200 + 160x - 20x^2$$

$$20x^2 - 160x + 275 = 0$$

$$\frac{1}{5}(20x^2 - 160x + 275) = \frac{1}{5}(0)$$

$4x^2 - 32x + 55 = 0$

$(2x - 5)(2x - 11) = 0$

$$\begin{aligned} 2x - 5 &= 0 & 2x - 11 &= 0 \\ 2x &= 5 & 2x &= 11 \\ x &= \frac{5}{2} & x &= \frac{11}{2} \\ x &= 2.5 & x &= 5.5 \end{aligned}$$

If $x = 2.50$, the price should be $2.80 - 0.10(2.5) = \$2.55$. If $x = 5.5$, the price should be $2.80 - 0.10(5.50) = \$2.25$.

3. $P = 4 - 0.05x \qquad Q = 500 + 20x \qquad R = (4 - 0.05x)(500 + 20x)$

$2450 = (4 - 0.05x)(500 + 20x)$

$2450 = 2000 + 55x - x^2$

$x^2 - 55x + 450 = 0$

$(x - 45)(x - 10) = 0$

$$\begin{aligned} x - 45 &= 0 & x - 10 &= 0 \\ x &= 45 & x &= 10 \end{aligned}$$

If $x = 45$, the price should be $4 - 0.05(45) = \$1.75$ and $500 + 20(45) = 1400$ buckets would be sold. If $x = 10$, the price should be $4 - 0.05(10) = \$3.50$ and $500 + 20(10) = 700$ buckets would be sold.

4. $P = 30 + 2x \qquad Q = 40 - 5x \qquad R = (30 + 2x)(40 - 5x)$

$1120 = (30 + 2x)(40 - 5x)$

$1120 = 1200 - 70x - 10x^2$

$10x^2 + 70x - 80 = 0$

$\frac{1}{10}\left(10x^2 + 70x - 80\right) = \frac{1}{10}(0)$

$x^2 + 7x - 8 = 0$

$(x-1)(x+8)=0$

$x-1=0 \qquad x+8=0$

$x=1 \qquad x=-8 \quad (x=-8 \text{ is not a solution})$

The price should be $30+2(1)=\$32$. There would be $40-5(1)=35$ oil changes performed each day.

5. $P=1.50-0.05x \qquad Q=250+10x$

$R=(1.50-0.05x)(250+10x)$

$378=(1.50-0.05x)(250+10x)$

$378=375+2.5x-0.5x^2$

$0.5x^2-2.5x+3=0$

$2(0.5x^2-2.5x+3)=2(0)$

$x^2-5x+6=0$

$(x-2)(x-3)=0$

$x-2=0 \qquad x-3=0$

$x=2 \qquad x=3$

If $x=2$, the price should be $1.50-0.05(2)=\$1.40$. If $x=3$, the price should be $1.50-0.05(3)=\$1.35$.

6. $P=40+2x \qquad Q=80-1x \qquad R=(40+2x)(80-x)$

$3648=(40+2x)(80-x)$

$3648=3200+120x-2x^2$

$2x^2-120x+448=0$

$\frac{1}{2}\left(2x^2-120x+448\right)=\frac{1}{2}(0)$

$x^2-60x+224=0$

$$x = \frac{-(-60) \pm \sqrt{(-60)^2 - 4(1)(224)}}{2(1)} = \frac{60 \pm \sqrt{3600 - 896}}{2}$$

$$= \frac{60 \pm \sqrt{2704}}{2} = \frac{60 \pm 52}{2} = 4, 56$$

($x = 56$ is not likely to be a solution—the price would be \$152!) If the price of the shoes are $40 + 2(4) = \$48$ per pair, the revenue will be \$3648.

Work Problems

REVIEW

Solve work problems by filling in the table below. In the work formula $Q = rt$ (Q = quantity, r = rate, and t = time), Q is usually "1." Usually the equation to solve is

Worker 1's Rate + Worker 2's Rate = Together Rate.

The information given in the problem is usually the time one or both workers need to complete the job. We want the *rates* not the *times*. We can solve for r in $Q = rt$ to get the rates.

$$Q = rt$$

$$\frac{Q}{t} = r$$

Because Q is usually "1,"

$$\frac{1}{t} = r.$$

The equation to solve is usually

$$\frac{1}{\text{Worker 1's time}} + \frac{1}{\text{Worker 2's time}} = \frac{1}{\text{Together time}}.$$

Worker	Quantity	Rate	Time
Worker 1	1	$\dfrac{1}{\text{Worker 1's time}}$	Worker 1's time
Worker 2	1	$\dfrac{1}{\text{Worker 2's time}}$	Worker 2's time
Together	1	$\dfrac{1}{\text{Together time}}$	Together time

Example

Together John and Michael can paint a wall in 18 minutes. Alone John needs 15 minutes longer to paint the wall than Michael needs. How much time does John and Michael each need to paint the wall by himself?

Let t represent the number of minutes Michael needs to paint the wall. Then $t+15$ represents the number of minutes John needs to paint the wall.

Worker	Quantity	Rate	Time
Michael	1	$\dfrac{1}{t}$	t
John	1	$\dfrac{1}{t+15}$	$t+15$
Together	1	$\dfrac{1}{18}$	18

The equation to solve is $\frac{1}{t}+\frac{1}{t+15}=\frac{1}{18}$. The LCD is $18t(t+15)$.

$$\frac{1}{t}+\frac{1}{t+15}=\frac{1}{18}$$

$$18t(t+15)\cdot\frac{1}{t}+18t(t+15)\cdot\frac{1}{t+15}=18t(t+15)\cdot\frac{1}{18}$$

$$18(t+15)+18t=t(t+15)$$

$$18t + 270 + 18t = t^2 + 15t$$

$$36t + 270 = t^2 + 15t$$

$$0 = t^2 - 21t - 270$$

$$0 = (t - 30)(t + 9)$$

$t - 30 = 0$ $\quad$ $t + 9 = 0$ (This does not lead to a solution.)

$t = 30$

Michael needs 30 minutes to paint the wall by himself and John needs $30 + 15 = 45$ minutes.

Practice

1. Alex and Tina working together can peel a bag of potatoes in six minutes. By herself Tina needs five minutes more than Alex to peel the potatoes. How long would each need to peel the potatoes if he or she were to work alone?

2. Together Rachel and Jared can wash a car in 16 minutes. Working alone Rachel needs 24 minutes longer than Jared does to wash the car. How long would it take for each Rachel and Jared to wash the car?

3. Two printing presses working together can print a magazine order in six hours. Printing Press I can complete the job alone in five fewer hours than Printing Press II. How long would each press need to print the run by itself?

4. Together two pipes can fill a small reservoir in two hours. Working alone Pipe I can fill the reservoir in one hour forty minutes less time than Pipe II can. How long would each pipe need to fill the reservoir by itself?

5. John and Gary together can unload a truck in 1 hour 20 minutes. Working alone John needs 36 minutes more to unload the truck than Gary needs. How long would each John and Gary need to unload the truck by himself?

Solutions

1. Let t represent the number of minutes Alex needs to peel the potatoes. Tina needs $t+5$ minutes to complete the job alone.

Worker	Quantity	Rate	Time
Alex	1	$\frac{1}{t}$	t
Tina	1	$\frac{1}{t+5}$	$t+5$
Together	1	$\frac{1}{6}$	6

The equation to solve is $\frac{1}{t}+\frac{1}{t+5}=\frac{1}{6}$. The LCD is $6t(t+5)$.

$$\frac{1}{t}+\frac{1}{t+5}=\frac{1}{6}$$

$$6t(t+5)\cdot\frac{1}{t}+6t(t+5)\cdot\frac{1}{t+5}=6t(t+5)\cdot\frac{1}{6}$$

$$6(t+5)+6t=t(t+5)$$

$$6t+30+6t=t^2+5t$$

$$12t+30=t^2+5t$$

$$0=t^2-7t-30$$

$$0=(t-10)(t+3)$$

$t-10=0 \qquad t+3=0$ (This does not lead to a solution.)

$t=10$

Alex can peel the potatoes in 10 minutes and Tina can peel them in $10+5=15$ minutes.

2. Let t represent the number of minutes Jared needs to wash the car by himself. The time Rachel needs to wash the car by herself is $t+24$.

Worker	Quantity	Rate	Time
Jared	1	$\frac{1}{t}$	t
Rachel	1	$\frac{1}{t+24}$	$t+24$
Together	1	$\frac{1}{16}$	16

The equation to solve is $\frac{1}{t}+\frac{1}{t+24}=\frac{1}{16}$. The LCD is $16t(t+24)$.

$$\frac{1}{t}+\frac{1}{t+24}=\frac{1}{16}$$

$$16t(t+24)\cdot\frac{1}{t}+16t(t+24)\cdot\frac{1}{t+24}=16t(t+24)\cdot\frac{1}{16}$$

$$16(t+24)+16t=t(t+24)$$

$$16t+384+16t=t^2+24t$$

$$32t+384=t^2+24t$$

$$0=t^2-8t-384$$

$$0=(t-24)(t+16)$$

$t-24=0$ $\quad$ $t+16=0$ (This does not lead to a solution.)

$t=24$

Jared needs 24 minutes to wash the car alone and Rachel needs $24+24=48$ minutes.

3. Let t represent the number of hours Printing Press II needs to print the run by itself. Because Printing Press I needs five fewer hours than Printing Press II, $t-5$ represents the number of hours Printing Press I needs to complete the run by itself.

Worker	Quantity	Rate	Time
Press I	1	$\frac{1}{t-5}$	$t-5$
Press II	1	$\frac{1}{t}$	t
Together	1	$\frac{1}{6}$	6

The equation to solve is $\frac{1}{t-5}+\frac{1}{t}=\frac{1}{6}$. The LCD is $6t(t-5)$.

$$\frac{1}{t-5}+\frac{1}{t}=\frac{1}{6}$$

$$6t(t-5)\cdot\frac{1}{t-5}+6t(t-5)\cdot\frac{1}{t}=6t(t-5)\cdot\frac{1}{6}$$

$$6t+6(t-5)=t(t-5)$$

$$6t+6t-30=t^2-5t$$

$$12t-30=t^2-5t$$

$$0=t^2-17t+30$$

$$0=(t-15)(t-2)$$

$t-15=0$ $\quad$ $t-2=0$ (This cannot be a solution because
$t=15$ $\quad$ $t=2$ $\quad$ $2-5$ is negative.)

Printing Press II can print the run alone in 15 hours and Printing Press I needs $15-5=10$ hours.

4. Let t represent the number of hours Pipe II needs to fill the reservoir alone. Pipe I needs one hour forty minutes less to do the job, so $t-1\frac{40}{60}=t-1\frac{2}{3}=t-\frac{5}{3}$ represents the time Pipe I needs to fill the reservoir by itself.

Worker	Quantity	Rate	Time
Pipe I	1	$\frac{1}{t-\frac{5}{3}}$	$t-\frac{5}{3}$
Pipe II	1	$\frac{1}{t}$	t
Together	1	$\frac{1}{2}$	2

The equation to solve is $\frac{1}{t-\frac{5}{3}}+\frac{1}{t}=\frac{1}{2}$. The LCD is $2t(t-\frac{5}{3})$.

$$\frac{1}{t-\frac{5}{3}}+\frac{1}{t}=\frac{1}{2}$$

$$2t\left(t-\frac{5}{3}\right)\cdot\frac{1}{t-\frac{5}{3}}+2t\left(t-\frac{5}{3}\right)\cdot\frac{1}{t}=2t\left(t-\frac{5}{3}\right)\cdot\frac{1}{2}$$

$$2t+2\left(t-\frac{5}{3}\right)=t\left(t-\frac{5}{3}\right)$$

$$2t+2t-\frac{10}{3}=t^2-\frac{5}{3}t$$

$$4t-\frac{10}{3}=t^2-\frac{5}{3}t$$

$$3\left(4t-\frac{10}{3}\right)=3\left(t^2-\frac{5}{3}t\right)$$

$$12t-10=3t^2-5t$$

$$0=3t^2-17t+10$$

$$0=(t-5)(3t-2)$$

$$t-5=0 \qquad 3t-2=0$$

$$t=5 \qquad 3t=2$$

$$t=\frac{2}{3}$$

($t=\frac{2}{3}$ cannot be a solution because $t-\frac{5}{3}$ would be negative)

Pipe II can fill the reservoir in 5 hours and Pipe I can fill it in $5-\frac{5}{3}=\frac{5}{1}\cdot\frac{3}{3}-\frac{5}{3}=\frac{15}{3}-\frac{5}{3}=\frac{10}{3}=3\frac{1}{3}$ hours or 3 hours 20 minutes.

5. Let t represent the number of hours Gary needs to unload the truck by himself. John needs 36 minutes more than Gary needs to unload the truck by himself, so John needs $\frac{36}{60}$ more hours or $\frac{3}{5}$ more hours. The number of hours John needs to unload the truck by himself is $t+\frac{3}{5}$.

   Together they can unload the truck in 1 hour 20 minutes, which is $1\frac{1}{3}=\frac{4}{3}$ hours. This means that the Together rate is $\frac{1}{\frac{4}{3}}=1\div\frac{4}{3}=1\cdot\frac{3}{4}=\frac{3}{4}$.

Worker	Quantity	Rate	Time
John	1	$\frac{1}{t+\frac{3}{5}}$	$t+\frac{3}{5}$
Gary	1	$\frac{1}{t}$	t
Together	1	$\frac{3}{4}$	$\frac{4}{3}$

The equation to solve is $\frac{1}{t+\frac{3}{5}}+\frac{1}{t}=\frac{3}{4}$. The LCD is $4t\left(t+\frac{3}{5}\right)$.

$$\frac{1}{t+\frac{3}{5}}+\frac{1}{t}=\frac{3}{4}$$

$$4t\left(t+\frac{3}{5}\right)\cdot\frac{1}{t+\frac{3}{5}}+4t\left(t+\frac{3}{5}\right)\cdot\frac{1}{t}=4t\left(t+\frac{3}{5}\right)\cdot\frac{3}{4}$$

$$4t+4\left(t+\frac{3}{5}\right)=3t\left(t+\frac{3}{5}\right)$$

$$4t+4t+\frac{12}{5}=3t^2+\frac{9}{5}t$$

$$8t+\frac{12}{5}=3t^2+\frac{9}{5}t$$

$$5\left(8t+\frac{12}{5}\right)=5\left(3t^2+\frac{9}{5}t\right)$$

$$40t+12=15t^2+9t$$

$$0=15t^2-31t-12$$

$$0=(5t-12)(3t+1)$$

$5t-12=0$ $\quad$ $3t+1=0$ (This does not lead to a solution.)

$5t=12$

$t=\frac{12}{5}=2\frac{2}{5}$

Gary needs $2\frac{2}{5}$ hours or 2 hours 24 minutes to unload the truck. John needs 2 hours 24 minutes + 36 minutes = 3 hours to unload the truck.

The Height of a Falling Object

The height of an object dropped, thrown or fired can be computed using quadratic equations. The general formula is $h = -16t^2 + v_0 t + h_0$, where h is the object's height (in feet), t is time (in seconds), h_0 is the object's initial height (that is, its height at $t = 0$ seconds) and v_0 is the object's initial velocity (that is, its speed at $t = 0$ seconds) in feet per second. If the object is tossed, thrown, or fired upward, v_0 is positive. If the object is thrown downward, v_0 is negative. If the object is dropped, v_0 is zero. The object reaches the ground when $h = 0$. (The effect of air resistance is ignored.)

Typical questions are:

When will the object be ___ feet high?
When will the object reach the ground?
What is the object's height after ____ seconds?

Examples

An object is dropped from a height of 1600 feet. How long will it take for the object to hit the ground?

Because the object is dropped, the initial velocity, v_0, is zero: $v_0 = 0$. The object is dropped from a height of 1600 feet, so $h_0 = 1600$. The formula $h = -16t^2 + v_0 t + h_0$ becomes $h = -16t^2 + 1600$. The object hits the ground when $h = 0$, so $h = -16t^2 + 1600$ becomes $0 = -16t^2 + 1600$.

$$0 = -16t^2 + 1600$$

$$16t^2 = 1600$$

$$t^2 = \frac{1600}{16}$$

$$t^2 = 100$$

$$t = \sqrt{100} \quad (t = -\sqrt{100} \text{ is not a solution})$$

$$t = 10$$

The object will hit the ground 10 seconds after it is dropped.

A ball is dropped from the top of a four-story building. The building is 48 feet tall. How long will it take for the ball to reach the ground?

Because the object is dropped, the initial velocity, v_0, is zero: $v_0 = 0$. The object is dropped from a height of 48 feet, so $h_0 = 48$. The formula $h = -16t^2 + v_0 t + h_0$ becomes $h = -16t^2 + 48$. The object hits the ground when $h = 0$.

$$h = -16t^2 + 48$$

$$0 = -16t^2 + 48$$

$$16t^2 = 48$$

$$t^2 = \frac{48}{16}$$

$$t^2 = 3$$

$$t = \sqrt{3} \quad (t = -\sqrt{3} \text{ is not a solution})$$

$$t \approx 1.73$$

The ball will reach the ground in about 1.73 seconds.

Practice

1. An object is dropped from a 56-foot bridge over a bay. How long will it take for the object to reach the water?

2. An object is dropped from the top of a 240-foot tall observation tower. How long will it take for the object to reach the ground?

3. A ball is dropped from a sixth-floor window at a height of 70 feet. When will the ball hit the ground?

4. An object falls from the top of a 100-foot communications tower. After how much time will the object hit the ground?

Solutions

For all of these problems, both a negative t and a positive t will be solutions for the quadratic equations. Only the positive t will be a solution to the problem.

1. For the formula $h = -16t^2 + v_0 t + h_0$, $h_0 = 56$ and $v_0 = 0$ (because the object is being dropped). The object reaches the ground when $h = 0$.

$$h = -16t^2 + 56$$

$$0 = -16t^2 + 56$$

$$16t^2 = 56$$

$$t^2 = \frac{56}{16}$$

$$t^2 = \frac{7}{2}$$

$$t = \sqrt{\frac{7}{2}}$$

$$t \approx 1.87$$

The object will reach the water in about 1.87 seconds.

2. For the formula $h = -16t^2 + v_0 t + h_0$, $h_0 = 240$ and $v_0 = 0$ (because the object is being dropped). The object reaches the ground when $h = 0$.

$$h = -16t^2 + 240$$

$$0 = -16t^2 + 240$$

$$16t^2 = 240$$

$$t^2 = \frac{240}{16}$$

$$t^2 = 15$$

$$t = \sqrt{15}$$

$$t \approx 3.87$$

The object will reach the ground in about 3.87 seconds.

3. For the formula $h = -16t^2 + v_0 t + h_0$, $h_0 = 70$ and $v_0 = 0$ (because the object is being dropped). The object reaches the ground when $h = 0$.

$$h = -16t^2 + 70$$

$$0 = -16t^2 + 70$$

$$16t^2 = 70$$

$$t^2 = \frac{70}{16}$$

$$t^2 = \frac{35}{8}$$

$$t = \sqrt{\frac{35}{8}}$$

$$t \approx 2.09$$

The ball will hit the ground in about 2.09 seconds.

4. For the formula $h = -16t^2 + v_0t + h_0$, $h_0 = 100$ and $v_0 = 0$ (because the object is being dropped). The object reaches the ground when $h = 0$.

$$h = -16t^2 + 100$$

$$0 = -16t^2 + 100$$

$$16t^2 = 100$$

$$t^2 = \frac{100}{16}$$

$$t^2 = \frac{25}{4}$$

$$t = \frac{5}{2}$$

$$t = 2.5$$

The object will hit the ground after 2.5 seconds.

Example

An object is dropped from the roof of a 60-foot building. How long must it fall to reach a height of 28 feet?

In the formula $h = -16t^2 + v_0t + h_0$, h_0 is 60 and v_0 is zero (because the object is dropped). The object reaches a height of 28 feet when $h = 28$.

$$h = -16t^2 + 60$$

$$28 = -16t^2 + 60$$

$$16t^2 = 32$$

$$t^2 = \frac{32}{16}$$

$$t^2 = 2$$

$$t = \sqrt{2} \quad (t = -\sqrt{2} \text{ is not a solution})$$

$$t \approx 1.41$$

The object will reach a height of 28 feet after about 1.41 seconds.

Practice

1. A ball is dropped from a height of 50 feet. How long after it is dropped will it reach a height of 18 feet?

2. A small object falls from a height of 200 feet. How long will it take to reach a height of 88 feet?

3. A small object is dropped from a tenth-floor window (at a height of 110 feet). How long will it take for the object to pass the third-floor window (at a height of 35 feet)?

4. An object is dropped from 120 feet. How long will it take for the object to fall 100 feet? (*Hint*: the height the object has reached after it has fallen 100 feet is 120 – 100 = 20 feet.)

Solutions

Negative values of t will not be solutions.

1. In the formula $h = -16t^2 + v_0t + h_0$, $h_0 = 50$ and $v_0 = 0$.

   $h = -16t^2 + 50$

   We want to find t when $h = 18$.

$$18 = -16t^2 + 50$$
$$16t^2 = 32$$
$$t^2 = \frac{32}{16}$$
$$t = \sqrt{2}$$
$$t \approx 1.41$$

The ball reaches a height of 18 feet about 1.41 seconds after it is dropped.

2. In the formula $h = -16t^2 + v_0t + h_0$, $h_0 = 200$ and $v_0 = 0$.

   $h = -16t^2 + 200$

   We want to find t when $h = 88$.

$$88 = -16t^2 + 200$$
$$16t^2 = 112$$
$$t^2 = \frac{112}{16}$$
$$t^2 = 7$$
$$t = \sqrt{7}$$
$$t \approx 2.65$$

   The object will reach a height of 88 feet after about 2.65 seconds.

3. In the formula $h = -16t^2 + v_0t + h_0$, $h_0 = 110$ and $v_0 = 0$.
   $h = -16t^2 + 110$

   We want to find t when $h = 35$.

$$35 = -16t^2 + 110$$
$$16t^2 = 75$$
$$t^2 = \frac{75}{16}$$
$$t = \sqrt{\frac{75}{16}}$$
$$t \approx 2.17$$

   The object will pass the third floor window after about 2.17 seconds.

4. In the formula $h = -16t^2 + v_0 t + h_0$, $h_0 = 120$ and $v_0 = 0$.

   $h = -16t^2 + 120$

   The object has fallen 100 feet when the height is $120 - 100 = 20$ feet, so we want to find t when $h = 20$.

   $$20 = -16t^2 + 120$$
   $$16t^2 = 100$$
   $$t^2 = \frac{100}{16}$$
   $$t = \sqrt{\frac{100}{16}}$$
   $$t = \frac{10}{4}$$
   $$t = 2.5$$

   The object will have fallen 100 feet 2.5 seconds after it is dropped.

Examples

An object is tossed up in the air at the rate of 40 feet per second. How long will it take for the object to hit the ground?

In the formula $h = -16t^2 + v_0 t + h_0$, $v_0 = 40$ and $h_0 = 0$.

$h = -16t^2 + 40t$

We want to find t when $h = 0$.

$$0 = -16t^2 + 40t$$
$$0 = t(-16t + 40)$$
$$-16t + 40 = 0 \qquad t = 0 \text{ (This is when the object is tossed.)}$$
$$40 = 16t$$
$$\frac{40}{16} = t$$
$$\frac{5}{2} = t$$
$$2.5 = t$$

The object will hit the ground after 2.5 seconds.

A projectile is fired upward from the ground at an initial velocity of 60 feet per second. When will the projectile be 44 feet above the ground?

In the formula $h = -16t^2 + v_0 t + h_0$, $v_0 = 60$ and $h_0 = 0$.

$$h = -16t^2 + 60t$$

We want to find t when $h = 44$.

$$44 = -16t^2 + 60t$$

$$16t^2 - 60t + 44 = 0$$

$$\frac{1}{4}(16t^2 - 60t + 44) = \frac{1}{4}(0)$$

$$4t^2 - 15t + 11 = 0$$

$$(t - 1)(4t - 11) = 0$$

$$t - 1 = 0 \qquad 4t - 11 = 0$$

$$t = 1 \qquad 4t = 11$$

$$t = \frac{11}{4}$$

$$t = 2.75$$

The projectile will be 44 feet off the ground at 1 second (on the way up) and again at 2.75 seconds (on the way down).

Practice

1. An object on the ground is thrown upward at the rate of 25 feet per second. After how much time will the object hit the ground?

2. A projectile is fired upward from the ground at the rate of 150 feet per second. How long will it take the projectile to fall back to the ground?

3. An object is thrown upward from the top of a 50-foot building. Its initial velocity is 20 feet per second. When will the object be 55 feet off the ground?

4. A projectile is fired upward from the top of a 36-foot building. Its initial velocity is 80 feet per second. When will it be 90 feet above the ground?

Solutions

1. In the formula $h = -16t^2 + v_0 t + h_0$, $v_0 = 25$ and $h_0 = 0$.

   $h = -16t^2 + 25t$

   We want to find t when $h = 0$.

   $0 = -16t^2 + 25t$

   $0 = t(-16t^2 + 25)$

   $-16t + 25 = 0$ $\qquad t = 0$ (This is when the object is thrown.)

   $-16t = -25$

   $t = \dfrac{-25}{-16}$

   $t = \dfrac{25}{16}$

   $t = 1.5625$

   The object will hit the ground after 1.5625 seconds.

2. In the formula $h = -16t^2 + v_0 t + h_0$, $v_0 = 150$ and $h_0 = 0$.

   $h = -16t^2 + 150t$

   We want to find t when $h = 0$.

   $0 = -16t^2 + 150t$

   $0 = t(-16t + 150)$

   $-16t + 150 = 0$ $\qquad t = 0$ (This is when the projectile is fired.)

   $-16t = -150$

   $t = \dfrac{-150}{-16}$

   $t = \dfrac{75}{8}$

   $t = 9.375$

   The object will fall back to the ground after 9.375 seconds.

3. In the formula $h = -16t^2 + v_0 t + h_0$, $v_0 = 20$ and $h_0 = 50$.

   $h = -16t^2 + 20t + 50$

We want to find t when $h = 55$.

$$55 = -16t^2 + 20t + 50$$
$$0 = -16t^2 + 20t - 5$$

$$t = \frac{-20 \pm \sqrt{(20)^2 - 4(-16)(-5)}}{2(-16)} = \frac{-20 \pm \sqrt{400 - 320}}{-32}$$
$$= \frac{-20 \pm \sqrt{80}}{-32}$$
$$\approx \frac{-20 \pm 8.94}{-32} \approx 0.35, 0.90$$

The object will reach a height of 55 feet at about 0.35 seconds (on its way up) and again at about 0.90 seconds (on its way down).

4. In the formula $h = -16t^2 + v_0 t + h_0$, $v_0 = 80$ and $h_0 = 36$.

$$h = -16t^2 + 80t + 36$$

We want to find t when $h = 90$.

$$90 = -16t^2 + 80t + 36$$
$$0 = -16t^2 + 80t - 54$$
$$\frac{-1}{2}(0) = \frac{-1}{2}(-16t^2 + 80t - 54)$$
$$0 = 8t^2 - 40t + 27 = 0$$

$$t = \frac{-(-40) \pm \sqrt{(-40)^2 - 4(8)(27)}}{2(8)} = \frac{40 \pm \sqrt{1600 - 864}}{16}$$
$$= \frac{40 \pm \sqrt{736}}{16}$$
$$\approx \frac{40 \pm 27.13}{16} \approx 0.80, 4.20$$

The object will reach a height of 90 feet after about 0.80 seconds (on its way up) and again at about 4.20 seconds (on its way down).

Geometric Problems

To solve word problems involving geometric shapes, write down the formula or formulas referred to in the problem. For example, after reading "The perimeter of a rectangular room . . . " write $P = 2L + 2W$. Then fill in the information given about the formula. For example, after reading "The perimeter of the room is 50 feet . . . " write $P = 50$ and $50 = 2L + 2W$. "Its width is two-thirds its length." Write $W = \frac{2}{3}L$ and $50 = 2L + 2W$ becomes $50 = 2L + 2(\frac{2}{3}L)$.

The formulas you will need in this section are listed below.

RECTANGLE FORMULAS

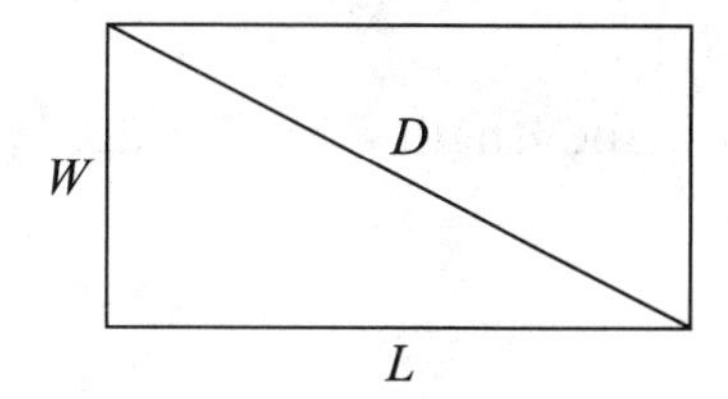

- Area $A = LW$
- Perimeter (the length around its sides) $P = 2L + 2W$
- Diagonal $D^2 = L^2 + W^2$

TRIANGLE FORMULAS

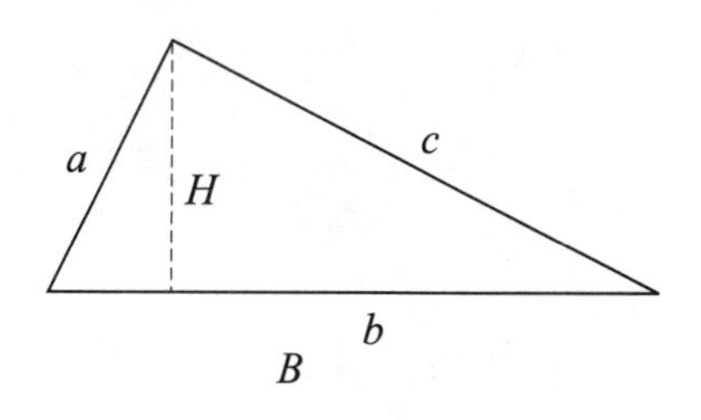

Right triangle

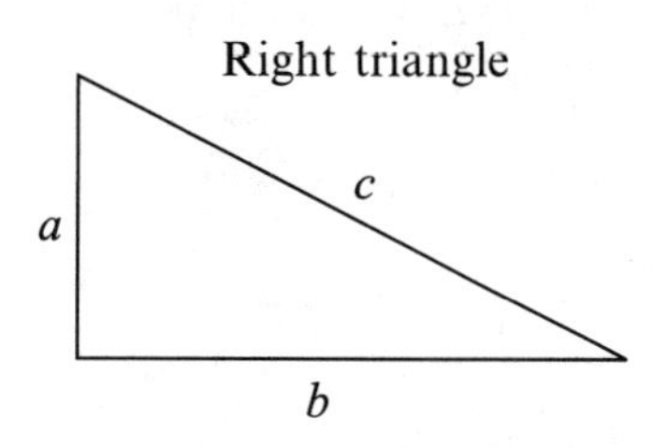

- Area $A = \frac{1}{2}BH$
- Perimeter $P = a + b + c$ (for any triangle)
- Pythagorean Theorem $a^2 + b^2 = c^2$ (for right triangles only)

MISCELLANEOUS SHAPES

- Volume of a right circular cylinder $V = \pi r^2 h$, where r is the cylinder's radius, h is the cylinder's height
- Surface area of a sphere (ball) is $SA = 4\pi r^2$, where r is the sphere's radius
- Area of a circle $A = \pi r^2$, where r is the circle's radius
- Volume of a rectangular box $V = LWH$, where L is the box's length, W is the box's width, and H is the box's height

In many of the examples and practice problems, there will be two solutions to the equation but only one solution to the geometric problem. The extra solutions come from solving quadratic equations.

Examples

A square has a diameter of 50 cm. What is the length of each side?
Let x represent the length of each side.

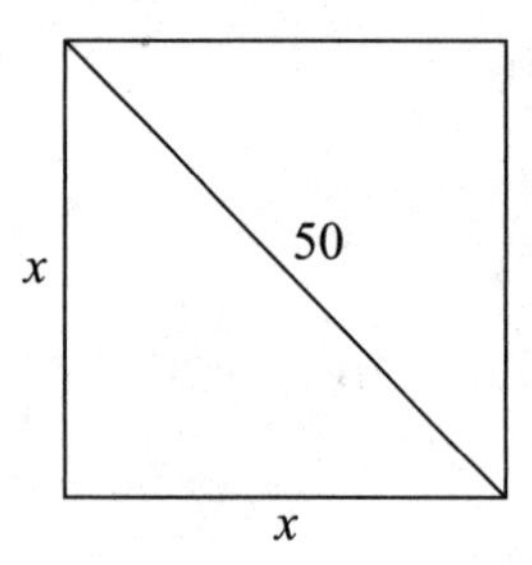

The diagonal formula for a rectangle is $D^2 = L^2 + W^2$. In this example, $D = 50$, $L = x$, and $W = x$. $D^2 = L^2 + W^2$ becomes $x^2 + x^2 = 50^2$.

$$x^2 + x^2 = 50^2$$

$$2x^2 = 2500$$

$$x^2 = \frac{2500}{2}$$

$$x^2 = 1250$$

$$x = \sqrt{1250} \quad (x = -\sqrt{1250} \text{ is not a solution})$$

$$x = \sqrt{25^2 \cdot 2}$$

$$x = 25\sqrt{2}$$

Each side is $25\sqrt{2}$ cm long.

A rectangle is one inch longer than it is wide. Its diameter is five inches. What are its dimensions?

$L = W + 1$

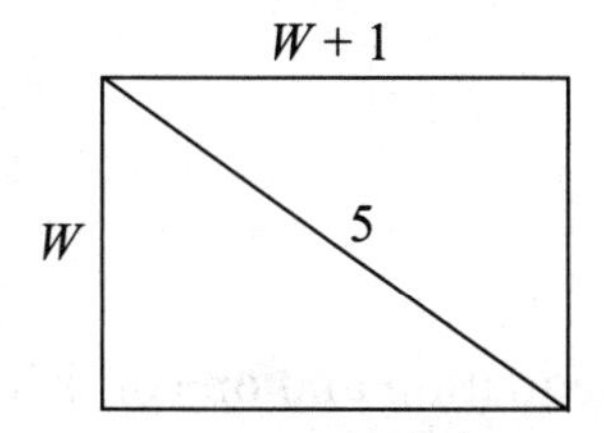

The diagonal formula for a rectangle is $D^2 = L^2 + W^2$. In this example, $D = 5$ and $L = W + 1$. $D^2 = L^2 + W^2$ becomes $5^2 = (W + 1)^2 + W^2$.

$$5^2 = (W + 1)^2 + W^2$$
$$25 = (W + 1)(W + 1) + W^2$$
$$25 = W^2 + 2W + 1 + W^2$$
$$25 = 2W^2 + 2W + 1$$
$$0 = 2W^2 + 2W - 24$$
$$\frac{1}{2}(0) = \frac{1}{2}(2W^2 + 2W - 24)$$
$$0 = W^2 + W - 12$$
$$0 = (W - 3)(W + 4)$$

$W - 3 = 0$ $\qquad$ $W + 4 = 0$

$W = 3$ $\qquad$ $W = -4$ (not a solution)

The width is 3 inches and the length is $3 + 1 = 4$ inches.

Practice

1. The diameter of a square is 60 feet. What is the length of its sides?

2. A rectangle has one side 14 cm longer than the other. Its diameter is 34 cm. What are its dimensions?

3. The length of a rectangle is 7 inches more than its width. The diagonal is 17 inches. What are its dimensions?

4. The width of a rectangle is three-fourths its length. The diagonal is 10 inches. What are its dimensions?

5. The diameter of a rectangular classroom is 34 feet. The room's length is 14 feet longer than its width. How wide and long is the classroom?

Solutions

When there is more than one solution to an equation and one of them is not valid, only the valid solution will be given.

1. Let x represent the length of each side (in feet). The diagonal is 60 feet, so $D = 60$. The formula $D^2 = L^2 + W^2$ becomes $60^2 = x^2 + x^2$.

$$60^2 = x^2 + x^2$$
$$3600 = 2x^2$$
$$\frac{3600}{2} = x^2$$
$$1800 = x^2$$
$$\sqrt{1800} = x$$
$$\sqrt{30^2 \cdot 2} = x$$
$$30\sqrt{2} = x$$

The length of the square's sides is $30\sqrt{2}$ feet, or approximately 42.4 feet.

2. The length is 14 cm more than the width, so $L = W + 14$. The diameter is 34 cm, so $D = 34$. The formula $D^2 = L^2 + W^2$ becomes $34^2 = (W + 14)^2 + W^2$.

$$34^2 = (W + 14)^2 + W^2$$
$$1156 = (W + 14)(W + 14) + W^2$$
$$1156 = W^2 + 28W + 196 + W^2$$
$$1156 = 2W^2 + 28W + 196$$
$$0 = 2W^2 + 28W - 960$$

$$\frac{1}{2}(0) = \frac{1}{2}(2W^2 + 28W - 960)$$
$$0 = W^2 + 14W - 480$$
$$0 = (W - 16)(W + 30)$$

$W - 16 = 0 \quad W + 30 = 0$ (This does not lead to a solution.)
$W = 16$

The width is 16 cm and the length is $16 + 14 = 30$ cm.

3. The length is 7 inches more than the width, so $L = W + 7$. The diagonal is 17 inches. The formula $D^2 = L^2 + W^2$ becomes $17^2 = (W + 7)^2 + W^2$.

$$17^2 = (W + 7)^2 + W^2$$
$$289 = (W + 7)(W + 7) + W^2$$
$$289 = W^2 + 14W + 49 + W^2$$
$$289 = 2W^2 + 14W + 49$$
$$0 = 2W^2 + 14W - 240$$
$$\frac{1}{2}(0) = \frac{1}{2}(2W^2 + 14W - 240)$$
$$0 = W^2 + 7W - 120$$
$$0 = (W - 8)(W + 15)$$

$W - 8 = 0 \quad W + 15 = 0$ (This does not lead to a solution.)
$W = 8$

The rectangle's width is 8 inches and its length is $8 + 7 = 15$ inches.

4. The width is three-fourths its length, so $W = \frac{3}{4}L$. The diagonal is 10 inches, so the formula $D^2 = L^2 + W^2$ becomes $10^2 = L^2 + (\frac{3}{4}L)^2$.

$$10^2 = L^2 + \left(\frac{3}{4}\right)^2 L^2$$
$$100 = L^2 + \frac{9}{16}L^2$$
$$100 = L^2\left(1 + \frac{9}{16}\right)$$
$$100 = L^2\left(\frac{16}{16} + \frac{9}{16}\right)$$

$$100 = \frac{25}{16}L^2$$

$$\frac{16}{25}(100) = L^2$$

$$64 = L^2$$

$$\sqrt{64} = L$$

$$8 = L$$

The rectangle's length is 8 inches and its width is $8(\frac{3}{4}) = 6$ inches.

5. The classroom's length is 14 feet more than its width, so $L = W + 14$. The diameter is 34 feet. The formula $D^2 = L^2 + W^2$ becomes $34^2 = (W + 14)^2 + W^2$.

$$34^2 = (W + 14)^2 + W^2$$

$$1156 = (W + 14)(W + 14) + W^2$$

$$1156 = W^2 + 28W + 196 + W^2$$

$$1156 = 2W^2 + 28W + 196$$

$$0 = 2W^2 + 28W - 960$$

$$\frac{1}{2}(0) = \frac{1}{2}(2W^2 + 28W - 960)$$

$$0 = W^2 + 14W - 480$$

$$0 = (W - 16)(W + 30)$$

$W - 16 = 0$ $\quad$ $W + 30 = 0$ (This does not lead to a solution.)

$W = 16$

The classroom is 16 feet wide and $16 + 14 = 30$ feet long.

Examples

The area of a triangle is 40 in^2. Its height is four-fifths the length of its base. What are its base and height?

The area is 40 and $H = \frac{4}{5}B$ so the formula $A = \frac{1}{2}BH$ becomes $40 = \frac{1}{2}B(\frac{4}{5}B)$.

$$40 = \frac{1}{2}B\left(\frac{4}{5}B\right)$$

$$40 = \frac{1}{2}\cdot\frac{4}{5}B^2$$

$$40 = \frac{2}{5}B^2$$

$$\frac{5}{2}\cdot 40 = B^2$$

$$100 = B^2$$

$$10 = B$$

The triangle's base is 10 inches long and its height is $(\frac{4}{5})(10) = 8$ inches.

The hypotenuse of a right triangle is 34 feet. The sum of the lengths of the two legs is 46 feet. Find the lengths of the legs.

The sum of the lengths of the legs is 46 feet, so if a and b are the lengths of the legs, $a + b = 46$, so $a = 46 - b$. The hypotenuse is 34 feet so if c is the length of the hypotenuse, then the formula $a^2 + b^2 = c^2$ becomes $(46 - b)^2 + b^2 = 34^2$.

$$(46 - b)^2 + b^2 = 34^2$$

$$(46 - b)(46 - b) + b^2 = 1156$$

$$2116 - 92b + b^2 + b^2 = 1156$$

$$2b^2 - 92b + 2116 = 1156$$

$$2b^2 - 92b + 960 = 0$$

$$\frac{1}{2}(2b^2 - 92b + 960) = \frac{1}{2}(0)$$

$$b^2 - 46b + 480 = 0$$

$$(b - 30)(b - 16) = 0$$

$$b - 30 = 0 \qquad b - 16 = 0$$

$$b = 30 \qquad b = 16$$

One leg is 30 feet long and the other is $46 - 30 = 16$ feet long.

A can's height is four inches and its volume is 28 cubic inches. What is the can's radius?

The volume formula for a right circular cylinder is $V = \pi r^2 h$. The can's volume is 28 cubic inches and its height is 4 inches, so $V = \pi r^2 h$ becomes $28 = \pi r^2(4)$.

$$28 = \pi r^2(4)$$
$$\frac{28}{4\pi} = r^2$$
$$\sqrt{\frac{28}{4\pi}} = r$$
$$1.493 \approx r$$

The can's radius is about 1.493 inches.

The volume of a box is 72 cm^3. Its height is 3 cm. Its length is 1.5 times its width. What are the length and width of the box?

The formula for the volume of the box is $V = LWH$. The volume is 72, the height is 3 and the length is 1.5 times the width ($L = 1.5W$) so the formula becomes $72 = (1.5W)W(3)$.

$$72 = (1.5W)W(3)$$
$$72 = 4.5W^2$$
$$\frac{72}{4.5} = W^2$$
$$16 = W^2$$
$$4 = W$$

The box's width is 4 cm and its length is $(1.5)(4) = 6$ cm.

The surface area of a ball is 314 square inches. What is the ball's diameter?

The formula for the surface area of a sphere is $SA = 4\pi r^2$. The area is 314, so the formula becomes $314 = 4\pi r^2$.

$$314 = 4\pi r^2$$
$$\frac{314}{4\pi} = r^2$$
$$\sqrt{\frac{314}{4\pi}} = r$$
$$5 \approx r$$

The radius of the ball is approximately 5 inches. The diameter is twice the radius, so the diameter is approximately 10 inches.

The manufacturer of a six-inch drinking cup is considering increasing its radius. The cup has straight sides (the top is the same size as the bottom). If the radius is increased by one inch, the new volume would be 169.6 cubic inches. What is the cup's current radius?

The formula for the volume of a right circular cylinder is $V = \pi r^2 h$. The cup's height is 6. If the cup's radius is increased, the volume would be 169.6. Let x represent the cup's current radius. Then the radius of the new cup would be $x + 1$. The volume formula becomes $169.6 = \pi(x+1)^2 6$.

$$169.6 = \pi(x+1)^2 6$$

$$\frac{169.6}{6\pi} = (x+1)^2 \qquad \left(\frac{169.6}{6\pi} \approx 9\right)$$

$$9 = (x+1)^2$$

$$9 = (x+1)(x+1)$$

$$9 = x^2 + 2x + 1$$

$$0 = x^2 + 2x - 8$$

$$0 = (x-2)(x+4)$$

$$x - 2 = 0 \qquad x + 4 = 0 \text{ (This does not lead to a solution.)}$$

$$x = 2$$

The cup's current radius is approximately 2 inches.

Practice

1. The area of a triangle is 12 in^2. The length of its base is two-thirds its height. What are the base and height?

2. The area of a triangle is 20 cm^2. The height is 3 cm more than its base. What are the base and height?

3. The sum of the base and height of a triangle is 14 inches. The area is 20 in^2. What are the base and height?

4. The hypotenuse of a right triangle is 85 cm long. One leg is 71 cm longer than the other. What are the lengths of its legs?

5. The manufacturer of a food can wants to increase the capacity of one of its cans. The can is 5 inches tall and its diameter is 6 inches. The manufacturer wants to increase the can's capacity by 50% and wants the can's height to remain 5 inches. How much does the diameter need to increase?

6. A pizza restaurant advertises that its large pizza is 20% larger than the competition's large pizza. The restaurant's large pizza is 16 inches in diameter. What is the diameter of the competition's large pizza?

Solutions

1. The area formula for a triangle is $A = \frac{1}{2}BH$. The area is 12. The length of its base is two-thirds its height, so $B = \frac{2}{3}H$. The formula becomes $12 = \frac{1}{2}\left(\frac{2}{3}H\right)H$.

$$12 = \frac{1}{2}\left(\frac{2}{3}H\right)H$$
$$12 = \frac{1}{2} \cdot \frac{2}{3}H^2$$
$$12 = \frac{1}{3}H^2$$
$$3(12) = H^2$$
$$36 = H^2$$
$$6 = H$$

The height of the triangle is 6 inches. Its base is $(\frac{2}{3})6 = 4$ inches.

2. The formula for the area of a triangle is $A = \frac{1}{2}BH$. The area is 20. The height is 3 cm more than the base, so $H = B + 3$. The formula becomes $20 = \frac{1}{2}B(B + 3)$.

$$20 = \frac{1}{2}B(B + 3)$$
$$2(20) = B(B + 3)$$
$$40 = B(B + 3)$$
$$40 = B^2 + 3B$$
$$0 = B^2 + 3B - 40$$
$$0 = (B - 5)(B + 8)$$

$B - 5 = 0$ $\quad\quad$ $B + 8 = 0$ (This does not lead to a solution.)

$B = 5$

The triangle's base is 5 cm and its height is $5 + 3 = 8$ cm.

3. The formula for the area of a triangle is $A = \frac{1}{2}BH$. The area is 20. $B + H = 14$ so $H = 14 - B$. The formula becomes $20 = \frac{1}{2}B(14 - B)$.

$$20 = \frac{1}{2}B(14 - B)$$

$$40 = B(14 - B)$$

$$40 = 14B - B^2$$

$$0 = 14B - B^2 - 40$$

$$-(0) = -(14B - B^2 - 40)$$

$$0 = -14B + B^2 + 40$$

$$0 = B^2 - 14B + 40$$

$$0 = (B - 10)(B - 4)$$

$$B - 10 = 0 \qquad B - 4 = 0$$

$$B = 10 \qquad B = 4$$

There are two triangles that satisfy the conditions. If the base is 10 inches, the height is $14 - 10 = 4$ inches. If the base is 4 inches, the height is $14 - 4 = 10$ inches.

4. By the Pythagorean theorem, $a^2 + b^2 = c^2$. The hypotenuse is c, so $c = 85$. One leg is 71 longer than the other so $a = b + 71$ ($b = a + 71$ also works). The Pythagorean theorem becomes $85^2 = (b + 71)^2 + b^2$.

$$85^2 = (b + 71)^2 + b^2$$

$$7225 = (b + 71)(b + 71) + b^2$$

$$7225 = b^2 + 142b + 5041 + b^2$$

$$7225 = 2b^2 + 142b + 5041$$

$$0 = 2b^2 + 142b - 2184$$

$$\frac{1}{2}(0) = \frac{1}{2}(2b^2 + 142b - 2184)$$

$$0 = b^2 + 71b - 1092$$

$$0 = (b - 13)(b + 84)$$

$b - 13 = 0$ $\quad b + 84 = 0$ (This does not lead to a solution.)

$b = 13$

The shorter leg is 13 cm and the longer leg is $13 + 71 = 84$ cm.

5. Because the can's diameter is 6, the radius is 3. Let x represent the increase in the can's radius. The radius of the new can is $3 + x$. The volume of the current can is $V = \pi r^2 h = \pi(3)^2 5 = 45\pi$. To increase the volume by 50% means to add half of 45π to itself; the new volume would be $45\pi + \frac{1}{2}45\pi = \frac{90\pi}{2} + \frac{45\pi}{2} = \frac{135\pi}{2}$. The volume formula for the new can becomes $\frac{135\pi}{2} = \pi(3 + x)^2 5$.

$$\frac{135\pi}{2} = \pi(3 + x)^2 5$$

$$\frac{1}{5\pi} \cdot \frac{135\pi}{2} = (3 + x)^2$$

$$\frac{27}{2} = (3 + x)^2$$

$$\frac{27}{2} = (3 + x)(3 + x)$$

$$\frac{27}{2} = 9 + 6x + x^2$$

$$2\left(\frac{27}{2}\right) = 2(9 + 6x + x^2)$$

$$27 = 18 + 12x + 2x^2$$

$$27 = 2x^2 + 12x + 18$$

$$0 = 2x^2 + 12x - 9$$

$$x = \frac{-12 \pm \sqrt{12^2 - 4(2)(-9)}}{2(2)} = \frac{-12 \pm \sqrt{144 + 72}}{4}$$

$$= \frac{-12 \pm \sqrt{216}}{4} = \frac{-12 \pm \sqrt{6^2 \cdot 6}}{4} = \frac{-12 \pm 6\sqrt{6}}{4}$$

$$= \frac{2(-6 \pm 3\sqrt{6})}{4} = \frac{-6 \pm 3\sqrt{6}}{2} \approx 0.674.$$

(The other solution is negative.)

The manufacturer should increase the can's radius by about 0.674 inches. Because the diameter is twice the radius, the manufacturer should increase the can's diameter by about $2(0.674) = 1.348$ inches.

6. A pizza's shape is circular so we need the area formula for a circle which is $A = \pi r^2$. The radius is half the diameter, so the restaurant's large pizza has a radius of 8 inches. The area of the restaurant's large pizza is $\pi(8)^2 = 64\pi \approx 201$. The restaurant's large pizza is 20% larger than the competition's large pizza. Let A represent the area of the competition's large pizza. Then 201 is 20% more than A:

$$201 = A + 0.20A = A(1 + 0.20) = 1.20A$$
$$201 = 1.20A$$
$$\frac{201}{1.20} = A$$
$$167.5 = A$$
$$A = \pi r^2$$
$$167.5 = \pi r^2$$
$$\frac{167.5}{\pi} = r^2$$
$$\sqrt{\frac{167.5}{\pi}} = r$$
$$7.3 \approx r$$

The competition's radius is approximately 7.3 inches, so its diameter is approximately $2(7.3) = 14.6$ inches.

Distance Problems

There are several distance problems that quadratic equations can solve. One of these types is "stream" problems: a vehicle travels the same distance up and back where in one direction, the "stream's" average speed is added to the vehicle's speed and in the other, the "stream's" average speed is subtracted from the vehicle's speed. Another type involves two bodies moving away from each other where their paths form a right angle (for instance, one travels north and the other west). Finally, the last type is where a vehicle makes a round trip that takes longer in one direction than in the other. In all of these types, the formula $D = RT$ is key.

"Stream" distance problems usually involve boats (traveling upstream or downstream) and planes (traveling against a headwind or with a tailwind). The boat or plane generally travels in one direction then turns around and travels in the opposite direction. The distance upstream and downstream is usually the same. If r represents the boat's or plane's average speed traveling without the "stream," then $r +$ stream's speed represents the boat's or plane's average speed traveling with the stream, and $r -$ stream's speed represents the boat's or plane's average speed traveling against the stream.

Example

Miami and Pittsburgh are 1000 miles apart. A plane flew into a 50-mph headwind from Miami to Pittsburgh. On the return flight the 50-mph wind became a tailwind. The plane was in the air a total of $4\frac{1}{2}$ hours for the round trip. What would have been the plane's average speed without the wind?

Let r represent the plane's average speed (in mph) without the wind. The plane's average speed against the wind is $r - 50$ (from Miami to Pittsburgh) and the plane's average speed with the wind is $r + 50$ (from Pittsburgh to Miami). The distance from Miami to Pittsburgh is 1000 miles. With this information we can use $T = \frac{D}{R}$ to compute the time in the air in each direction. The time in the air from Miami to Pittsburgh is $\frac{1000}{r-50}$. The time in the air from Pittsburgh to Miami is $\frac{1000}{r+50}$. The time in the air from Miami to Pittsburgh plus the time in the air from Pittsburgh to Miami is $4\frac{1}{2} = 4.5$ hours. The equation to solve is $\frac{1000}{r-50} + \frac{1000}{r+50} = 4.5$. The LCD is $(r - 50)(r + 50)$.

$$(r-50)(r+50)\frac{1000}{r-50} +$$

$$(r-50)(r+50)\frac{1000}{r+50} = (r-50)(r+50)(4.5)$$

$$1000(r+50) + 1000(r-50) = 4.5[(r-50)(r+50)]$$

$$1000r + 50{,}000 + 1000r - 50{,}000 = 4.5(r^2 - 2500)$$

$$2000r = 4.5r^2 - 11{,}250$$

$$0 = 4.5r^2 - 2000r - 11{,}250$$

$$r = \frac{-(-2000) \pm \sqrt{(-2000)^2 - 4(4.5)(-11,250)}}{2(4.5)}$$

$$= \frac{2000 \pm \sqrt{4,000,000 + 202,500}}{9} = \frac{2000 \pm \sqrt{4,202,500}}{9}$$

$$= \frac{2000 \pm 2050}{9} = \frac{2000 + 2050}{9}, \frac{2000 - 2050}{9} = 450$$

($-\frac{50}{9}$ is not a solution.) The plane's average speed without the wind is 450 mph.

Practice

1. A flight from Dallas to Chicago is 800 miles. A plane flew with a 40-mph tailwind from Dallas to Chicago. On the return trip, the plane flew against the same 40-mph wind. The plane was in the air a total of 5.08 hours for the flight from Dallas to Chicago and the return flight. What would have been the plane's speed without the wind?

2. A flight from Houston to New Orleans faced a 50-mph headwind, which became a 50-mph tailwind on the return flight. The total time in the air was $1\frac{3}{4}$ hours. The distance between Houston and New Orleans is 300 miles. How long was the plane in flight from Houston to New Orleans?

3. A small motorboat traveled 15 miles downstream then turned around and traveled 15 miles back. The total trip took 2 hours. The stream's speed is 4 mph. How fast would the boat have traveled in still water?

4. A plane on a flight from Denver to Indianapolis flew with a 20-mph tailwind. On the return flight, the plane flew into a 20-mph headwind. The distance between Denver and Indianapolis is 1000 miles and the plane was in the air a total of $5\frac{1}{2}$ hours. What would have been the plane's average speed without the wind?

5. A plane flew from Minneapolis to Atlanta, a distance of 900 miles, against a 30 mph-headwind. On the return flight, the 30-mph wind became a tailwind. The plane was in the air for a total of $5\frac{1}{2}$ hours. What would the plane's average speed have been without the wind?

Solutions

1. Let r represent the plane's average speed (in mph) with no wind. Then the average speed from Dallas to Chicago (with the tailwind) is $r+40$, and the average speed from Chicago to Dallas is $r-40$ (against the headwind). The distance between Dallas and Chicago is 800 miles. The time in the air from Dallas to Chicago plus the time in the air from Chicago to Dallas is 5.08 hours. The time in the air from Dallas to Chicago is $\frac{800}{r+40}$. The time in the air from Chicago to Dallas is $\frac{800}{r-40}$. The equation to solve is $\frac{800}{r+40}+\frac{800}{r-40}=5.08.$ The LCD is $(r+40)(r-40)$.

$$\begin{aligned}
(r+40)(r-40)\frac{800}{r+40}&+\\
(r+40)(r-40)\frac{800}{r-40}&=(r+40)(r-40)(5.08)\\
800(r-40)+800(r+40)&=5.08[(r-40)(r+40)]\\
800r-32{,}000+800r+32{,}000&=5.08(r^2-1600)\\
1600r&=5.08r^2-8128\\
0&=5.08r^2-1600r-8128
\end{aligned}$$

$$\begin{aligned}
r&=\frac{-(-1600)\pm\sqrt{(-1600)^2-4(5.08)(-8128)}}{2(5.08)}\\
&=\frac{1600\pm\sqrt{2{,}560{,}000+165{,}160.96}}{10.16}\\
&=\frac{1600\pm\sqrt{2{,}725{,}160.96}}{10.16}\approx\frac{1600\pm1650.806155}{10.16}\\
&\approx 320\left(\frac{1600-1650.806155}{10.16}\text{ is negative}\right)
\end{aligned}$$

The plane's average speed without the wind would have been about 320 mph.

2. Let r represent the plane's average speed without the wind. The average speed from Houston to New Orleans (against the headwind) is $r-50$, and the average speed from New Orleans to Houston (with the tailwind) is $r+50$. The distance between Houston and

New Orleans is 300 miles. The time in the air from Houston to New Orleans is $\frac{300}{r-50}$, and the time in the air from New Orleans to Houston is $\frac{300}{r+50}$. The time in the air from Houston to New Orleans plus the time in the air from New Orleans to Houston is $1\frac{3}{4}=\frac{7}{4}$ hours. The equation to solve is $\frac{300}{r-50}+\frac{300}{r+50}=\frac{7}{4}$. The LCD is $4(r-50)(r+50)$.

$$4(r-50)(r+50)\frac{300}{r-50}+$$

$$4(r-50)(r+50)\frac{300}{r+50}=4(r-50)(r+50)\frac{7}{4}$$

$$1200(r+50)+1200(r-50)=7[(r-50)(r+50)]$$

$$1200r+60{,}000+1200r-60{,}000=7(r^2-2500)$$

$$2400r=7r^2-17{,}500$$

$$0=7r^2-2400r-17{,}500$$

$$r=\frac{-(-2400)\pm\sqrt{(-2400)^2-4(7)(-17{,}500)}}{2(7)}$$

$$=\frac{2400\pm\sqrt{5{,}760{,}000+490{,}000}}{14}$$

$$=\frac{2400\pm\sqrt{6{,}250{,}000}}{14}=\frac{2400\pm 2500}{14}$$

$$=350\left(\frac{2400-2500}{14}\text{ is negative}\right)$$

The average speed of the plane without the wind was 350 mph. We want the time in the air from Houston to New Orleans: $\frac{300}{r-50}=\frac{300}{350-50}=\frac{300}{300}=1$ hour. The plane was in flight from Houston to New Orleans for one hour.

3. Let r represent the boat's speed in still water. The average speed downstream is $r+4$ and the average speed upstream is $r-4$. The boat was in the water a total of 2 hours. The distance traveled in each direction is 15 miles. The time the boat traveled downstream is $\frac{15}{r+4}$ hours, and it traveled upstream $\frac{15}{r-4}$ hours. The time the

boat traveled upstream plus the time it traveled downstream equals 2 hours. The equation to solve is $\frac{15}{r+4}+\frac{15}{r-4}=2$. The LCD is $(r+4)(r-4)$.

$$(r+4)(r-4)\frac{15}{r+4}+(r+4)(r-4)\frac{15}{r-4}=2(r+4)(r-4)$$
$$15(r-4)+15(r+4)=2[(r+4)(r-4)]$$
$$15r-60+15r+60=2(r^2-16)$$
$$30r=2r^2-32$$
$$0=2r^2-30r-32$$
$$\frac{1}{2}(0)=\frac{1}{2}(2r^2-30r-32)$$
$$0=r^2-15r-16$$
$$0=(r-16)(r+1)$$

$r-16=0$ $\quad$ $r+1=0$ (This does not lead to a solution.)

$r=16$

The boat's average speed in still water is 16 mph.

4. Let r represent the plane's average speed without the wind. The plane's average speed from Denver to Indianapolis is $r+20$, and the plane's average speed from Indianapolis to Denver is $r-20$. The total time in flight is $5\frac{1}{2}$ hours and the distance between Denver and Indianapolis is 1000 miles. The time in the air from Denver to Indianapolis is $\frac{1000}{r+20}$ hours and the time in the air from Indianapolis to Denver is $\frac{1000}{r-20}$ hours. The time in the air from Denver to Indianapolis plus the time in the air from Indianapolis to Denver is 5.5 hours. The equation to solve is $\frac{1000}{r+20}+\frac{1000}{r-20}=5.5$. The LCD is $(r+20)(r-20)$.

$$(r+20)(r-20)\frac{1000}{r+20}+$$
$$(r+20)(r-20)\frac{1000}{r-20}=(r+20)(r-20)(5.5)$$
$$1000(r-20)+1000(r+20)=5.5[(r-20)(r+20)]$$

$$1000r - 20{,}000 + 1000r + 20{,}000 = 5.5(r^2 - 400)$$
$$2000r = 5.5r^2 - 2200$$
$$0 = 5.5r^2 - 2000r - 2200$$

$$r = \frac{-(-2000) \pm \sqrt{(-2000)^2 - 4(5.5)(-2200)}}{2(5.5)}$$
$$= \frac{2000 \pm \sqrt{4{,}000{,}000 + 48{,}400}}{11}$$
$$= \frac{2000 \pm \sqrt{4{,}048{,}400}}{11} \approx \frac{2000 \pm 2012.063617}{11} \approx 365,$$
$$\left(\frac{2000 - 2012.063617}{11} \text{ is negative}\right)$$

The plane would have averaged about 365 mph without the wind.

5. Let r represent the plane's average speed without the wind. The plane's average speed from Minneapolis to Atlanta (against the headwind) is $r - 30$. The plane's average speed from Atlanta to Minneapolis (with the tailwind) is $r + 30$. The total time in the air is $5\frac{1}{2}$ hours and the distance between Atlanta to Minneapolis is 900 miles. The time in the air from Minneapolis to Atlanta is $\frac{900}{r-30}$ hours, and the time in the air from Atlanta to Minneapolis is $\frac{900}{r+30}$ hours. The time in the air from Minneapolis to Atlanta plus the time in the air from Minneapolis to Atlanta is $5\frac{1}{2} = 5.5$ hours. The equation to solve is $\frac{900}{r-30} + \frac{900}{r+30} = 5.5$. The LCD is $(r-30)(r+30)$.

$$(r-30)(r+30)\frac{900}{r-30} +$$
$$(r-30)(r+30)\frac{900}{r+30} = (r-30)(r+30)(5.5)$$
$$900(r+30) + 900(r-30) = 5.5[(r-30)(r+30)]$$
$$900r + 27{,}000 + 900r - 27{,}000 = 5.5(r^2 - 900)$$
$$1800r = 5.5r^2 - 4950$$
$$0 = 5.5r^2 - 1800r - 4950$$

$$r = \frac{-(-1800) \pm \sqrt{(-1800)^2 - 4(5.5)(-4950)}}{2(5.5)}$$

$$= \frac{1800 \pm \sqrt{3{,}240{,}000 + 108{,}900}}{11} = \frac{1800 \pm \sqrt{3{,}348{,}900}}{11}$$

$$= \frac{1800 \pm 1830}{11} = 330 \left(\frac{1800 - 1830}{11} \text{ is negative}\right)$$

The plane's average speed without the wind was 330 mph.

If you need to find the distance between two bodies traveling at right angles away from each other, you must use the Pythagorean Theorem: $a^2 + b^2 = c^2$ in addition to $D = RT$.

Examples

A car passes under a railway trestle at the same time a train is crossing the trestle. The car is headed south at an average speed of 40 mph. The train is traveling east at an average speed of 30 mph. After how long will the car and train be 10 miles apart?

Let t represent the number of hours after the train and car pass each other. (Because the rate is given in miles per hour, time must be given in hours.) The distance traveled by the car after t hours is $40t$ and that of the train is $30t$.

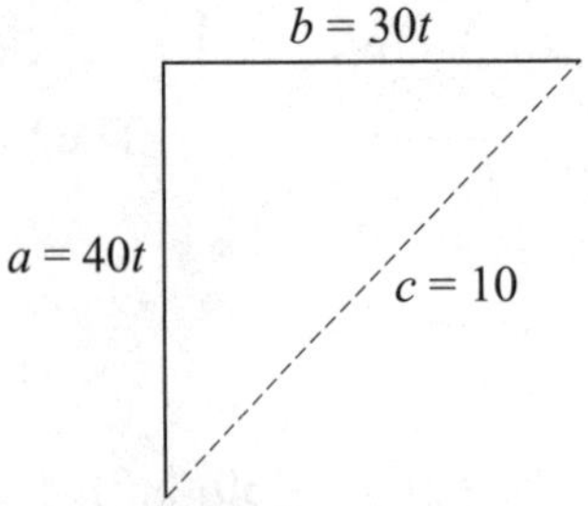

$b = 30t$

$a = 40t$

$c = 10$

$$a^2 + b^2 = c^2$$
$$(40t)^2 + (30t)^2 = 10^2$$

$$(40t)^2 + (30t)^2 = 10^2$$

$$1600t^2 + 900t^2 = 100$$

$$2500t^2 = 100$$

$$t^2 = \frac{100}{2500}$$

$$t = \sqrt{\frac{100}{2500}}$$

$$t = \frac{10}{50} = \frac{1}{5}$$

After $\frac{1}{5}$ of an hour (or 12 minutes) the car and train will be 10 miles apart.

Practice

1. A car and plane leave an airport at the same time. The car travels eastward at an average speed of 45 mph. The plane travels southward at an average speed of 200 mph. After how long will they be 164 miles apart?

2. Two joggers begin jogging from the same point. One jogs south at the rate of 8 mph and the other jogs east at a rate of 6 mph. When will they be five miles apart?

3. A cross-country cyclist crosses a railroad track just after a train passed. The train is traveling southward at an average speed of 60 mph. The cyclist is traveling westward at an average speed of 11 mph. When will they be 244 miles apart?

4. A motor scooter and a car left a parking lot at the same time. The motor scooter traveled north at 24 mph. The car traveled west at 45 mph. How long did it take for the scooter and car to be 34 miles apart?

5. Two cars pass each other at 4:00 at an overpass. One car is headed north at an average speed of 60 mph and the other is headed east at an average speed of 50 mph. At what time will the cars be 104 miles apart? Give your solution to the nearest minute.

Solutions

1. Let t represent the number of hours each has traveled. The plane's distance after t hours is $200t$ and the car's distance is $45t$.

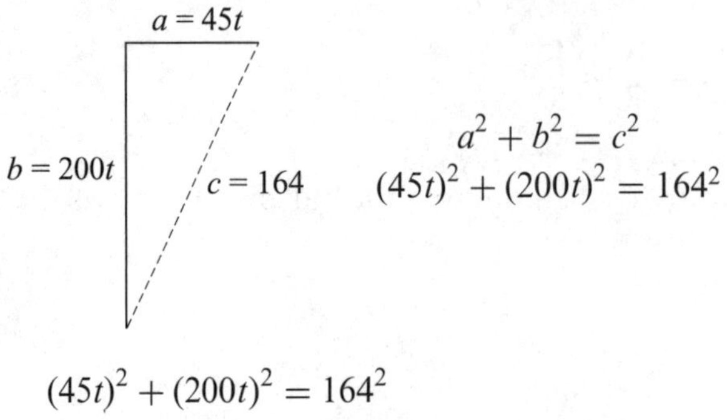

$$a^2 + b^2 = c^2$$
$$(45t)^2 + (200t)^2 = 164^2$$

$$(45t)^2 + (200t)^2 = 164^2$$
$$2025t^2 + 40{,}000t^2 = 26{,}896$$
$$42{,}025t^2 = 26{,}896$$
$$t^2 = \frac{26{,}896}{42{,}025}$$
$$t = \sqrt{\frac{26{,}896}{42{,}025}}$$
$$t = 0.80$$

The car and plane will be 164 miles apart after 0.80 hours or 48 minutes.

2. Let t represent the number of hours after the joggers began jogging. The distance covered by the southbound jogger after t hours is $8t$, and the distance covered by the eastbound jogger is $6t$.

$a = 6t$, $b = 8t$, $c = 5$

$$a^2 + b^2 = c^2$$
$$(6t)^2 + (8t)^2 = 5^2$$

$$(6t)^2 + (8t)^2 = 5^2$$
$$36t^2 + 64t^2 = 25$$
$$100t^2 = 25$$

$$t^2 = \frac{25}{100}$$
$$t^2 = \frac{1}{4}$$
$$t = \sqrt{\frac{1}{4}}$$
$$t = \frac{1}{2}$$

The joggers will be five miles apart after $\frac{1}{2}$ hour or 30 minutes.

3. Let t represent the number of hours after the cyclist crosses the track. The distance traveled by the bicycle after t hours is $11t$ and the distance traveled by the train is $60t$.

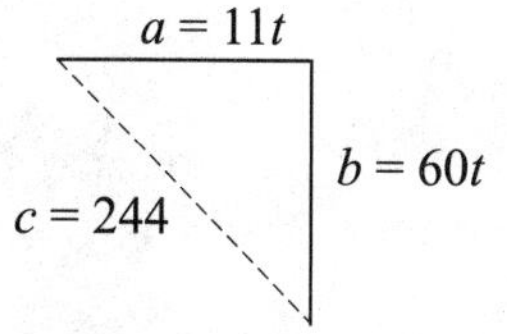

$a = 11t$

$b = 60t$

$c = 244$

$$a^2 + b^2 = c^2$$
$$(11t)^2 + (60t)^2 = 244^2$$

$$(11t)^2 + (60t)^2 = 244^2$$
$$121t^2 + 3600t^2 = 59{,}536$$
$$3721t^2 = 59{,}536$$
$$t^2 = \frac{59{,}536}{3721}$$
$$t^2 = 16$$
$$t = \sqrt{16}$$
$$t = 4$$

After four hours the cyclist and train will be 244 miles apart.

4. Let t represent the number of hours after the scooter and car left the parking lot. The car's distance after t hours is $45t$. The scooter's distance is $24t$.

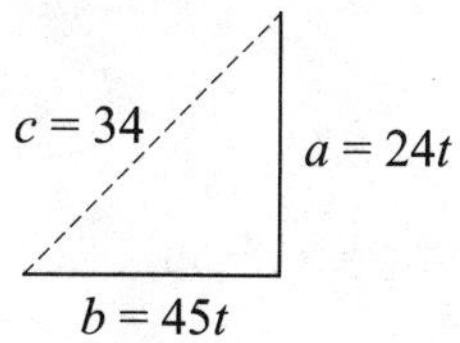

$c = 34$

$a = 24t$

$b = 45t$

$$a^2 + b^2 = c^2$$
$$(24t)^2 + (45t)^2 = 34^2$$

$$
\begin{aligned}
(24t)^2 + (45t)^2 &= 34^2 \\
576t^2 + 2025t^2 &= 1156 \\
2601t^2 &= 1156 \\
t^2 &= \frac{1156}{2601} \\
t^2 &= \frac{4}{9} \\
t &= \sqrt{\frac{4}{9}} \\
t &= \frac{2}{3}
\end{aligned}
$$

The car and scooter will be 34 miles apart after $\frac{2}{3}$ of an hour or 40 minutes.

5. Let t represent the number of hours after the cars passed the overpass. The northbound car's distance after t hours is $60t$ and the eastbound car's distance is $50t$.

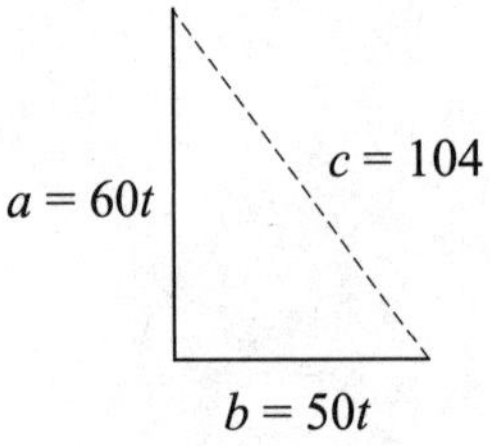

$a = 60t$

$c = 104$

$b = 50t$

$$
\begin{aligned}
a^2 + b^2 &= c^2 \\
(60t)^2 + (50t)^2 &= 104^2
\end{aligned}
$$

$$
\begin{aligned}
(60t)^2 + (50t)^2 &= 104^2 \\
3600t^2 + 2500t^2 &= 10{,}816 \\
6100t^2 &= 10{,}816 \\
t^2 &= \frac{10{,}816}{6100} \\
t^2 &= \frac{2704}{1525} \\
t &= \sqrt{\frac{2704}{1525}} \\
t &\approx 1.33
\end{aligned}
$$

The cars will be 104 miles apart after about 1.33 hours or 1 hour 20 minutes. The time will be about 5:20.

In the following problems people are making a round trip. The average speed in each direction will be different and the total trip time will be given. The equation to solve is

Time to destination + Time on return trip = Total trip time.

To get the time to and from the destination, use $D = RT$ and solve for T. The equation to solve becomes

$$\frac{\text{Distance}}{\text{Rate to destination}} + \frac{\text{Distance}}{\text{Rate on return trip}} = \text{Total trip time.}$$

Example

A jogger jogged seven miles to a park then jogged home. He jogged 1 mph faster to the park than he jogged on the way home. The round trip took 2 hours 34 minutes. How fast did he jog to the park?

Let r represent the jogger's average speed on the way home. He jogged 1 mph faster to the park, so $r + 1$ represents his average speed to the park. The distance to the park is 7 miles, so $D = 7$.

Time to the park + Time home = 2 hours 34 minutes

The time to the park is represented by $T = \frac{7}{r+1}$. The time home is represented by $T = \frac{7}{r}$. The round trip is 2 hours 34 minutes $= 2\frac{34}{60} = 2\frac{17}{30} = \frac{77}{30}$ hours. The equation to solve becomes

$$\frac{7}{r+1} + \frac{7}{r} = \frac{77}{30}.$$

The LCD is $30r(r + 1)$.

$$30r(r+1)\frac{7}{r+1} + 30r(r+1)\frac{7}{r} = 30r(r+1)\frac{77}{30}$$

$$210r + 210(r+1) = 77r(r+1)$$

$$210r + 210r + 210 = 77r^2 + 77r$$

$$420r + 210 = 77r^2 + 77r$$

$$0 = 77r^2 - 343r - 210$$

$$\frac{1}{7}(0) = \frac{1}{7}(77r^2 - 343r - 210)$$

$$0 = 11r^2 - 49r - 30$$

$$0 = (r - 5)(11r + 6)$$

$$r - 5 = 0 \qquad 11r + 6 = 0 \text{ (This does not lead to a solution.)}$$
$$r = 5$$

The jogger's average speed to the park was $5 + 1 = 6$ mph.

Practice

1. A man rode his bike six miles to work. The wind reduced his average speed on the way home by 2 mph. The round trip took 1 hour 21 minutes. How fast was he riding on the way to work?

2. On a road trip a saleswoman traveled 120 miles to visit a customer. She averaged 15 mph faster to the customer than on the return trip. She spent a total of 4 hours 40 minutes driving. What was her average speed on the return trip?

3. A couple walked on the beach from their house to a public beach four miles away. They walked 0.2 mph faster on the way home than on the way to the public beach. They walked for a total of 2 hours 35 minutes. How fast did they walk home?

4. A family drove from Detroit to Buffalo, a distance of 215 miles, for the weekend. They averaged 10 mph faster on the return trip. They spent a total of seven hours on the road. What was their average speed on the trip from Detroit to Buffalo? (Give your solution accurate to one decimal place.)

5. Boston and New York are 190 miles apart. A professor drove from his home in Boston to a conference in New York. On the return trip, he faced heavy traffic and averaged 17 mph slower than on his way to New York. He spent a total of 8 hours 5 minutes on the road. How long did his trip from Boston to New York last?

Solutions

1. Let r represent the man's average speed on the way to work. Then $r - 2$ represents the man's average speed on his way home. The distance each way is 6 miles, so the time he rode to work is $\frac{6}{r}$, and the time he rode home is $\frac{6}{r-2}$. The total time is 1 hour 21

minutes $= 1\frac{21}{60} = 1\frac{7}{20} = \frac{27}{20}$ hours. The equation to solve is $\frac{6}{r} + \frac{6}{r-2} = \frac{27}{20}$. The LCD is $20r(r-2)$.

$$20r(r-2)\frac{6}{r} + 20r(r-2)\frac{6}{r-2} = 20r(r-2)\frac{27}{20}$$
$$120(r-2) + 120r = 27r(r-2)$$
$$120r - 240 + 120r = 27r^2 - 54r$$
$$240r - 240 = 27r^2 - 54r$$
$$0 = 27r^2 - 294r + 240$$
$$\frac{1}{3}(0) = \frac{1}{3}(27r^2 - 294r + 240)$$
$$0 = 9r^2 - 98r + 80$$
$$0 = (r-10)(9r-8)$$

$r - 10 = 0 \quad 9r - 8 = 0$ (This does not lead to a valid solution.)
$r = 10$

The man's average speed on his way to work was 10 mph.

2. Let r represent the saleswoman's average speed on her return trip. Her average speed on the way to the customer is $r + 15$. The distance each way is 120 miles. She spent a total of 4 hours 40 minutes $= 4\frac{40}{60} = 4\frac{2}{3} = \frac{14}{3}$ hours driving. The time spent driving to the customer is $\frac{120}{r+15}$. The time spent driving on the return trip is $\frac{120}{r}$. The equation to solve is $\frac{120}{r+15} + \frac{120}{r} = \frac{14}{3}$. The LCD is $3r(r+15)$.

$$3r(r+15)\cdot\frac{120}{r+15} + 3r(r+15)\cdot\frac{120}{r} = 3r(r+15)\cdot\frac{14}{3}$$
$$360r + 360(r+15) = 14r(r+15)$$
$$360r + 360r + 5400 = 14r^2 + 210r$$
$$720r + 5400 = 14r^2 + 210r$$
$$0 = 14r^2 - 510r - 5400$$
$$\frac{1}{2}(0) = \frac{1}{2}(14r^2 - 510r - 5400)$$
$$0 = 7r^2 - 255r - 2700$$

$$r = \frac{-(-255) \pm \sqrt{(-255)^2 - 4(7)(-2700)}}{2(7)}$$

$$= \frac{255 \pm \sqrt{65{,}025 + 75{,}600}}{14} \frac{255 \pm \sqrt{140{,}625}}{14} = \frac{255 \pm 375}{14}$$

$$r = 45 \qquad \left(r = \frac{255 - 375}{14} \text{ is not a solution.}\right)$$

The saleswoman averaged 45 mph on her return trip.

3. Let r represent the couple's average rate on their way home, then $r - 0.2$ represents the couple's average speed to the public beach. The distance to the public beach is 4 miles. They walked for a total of 2 hours 35 minutes $= 2\frac{35}{60} = 2\frac{7}{12} = \frac{31}{12}$ hours.The time spent walking to the public beach is $\frac{4}{r - 0.2}$. The time spent walking home is $\frac{4}{r}$. The equation to solve is $\frac{4}{r - 0.2} + \frac{4}{r} = \frac{31}{12}$. The LCD is $12r(r - 0.2)$.

$$12r(r - 0.2)\frac{4}{r - 0.2} + 12r(r - 0.2)\frac{4}{r} = 12r(r - 0.2)\frac{31}{12}$$

$$48r + 48(r - 0.2) = 31r(r - 0.2)$$

$$48r + 48r - 9.6 = 31r^2 - 6.2r$$

$$96r - 9.6 = 31r^2 - 6.2r$$

$$0 = 31r^2 - 102.2r + 9.6$$

(Multiplying by 10 to clear the decimals would result in fairly large numbers for the quadratic formula.)

$$r = \frac{-(-102.2) \pm \sqrt{(-102.2)^2 - 4(31)(9.6)}}{2(31)}$$

$$= \frac{102.2 \pm \sqrt{10{,}444.84 - 1190.4}}{62} = \frac{102.2 \pm \sqrt{9254.44}}{62}$$

$$= \frac{102.2 \pm 96.2}{62} = \frac{16}{5}, \frac{3}{31} \qquad \left(\frac{3}{31} \text{ is not a solution.}\right)$$

The couple walked home at the rate of $\frac{16}{5} = 3.2$ mph.

4. Let r represent the average speed from Detroit to Buffalo. The average speed from Buffalo to Detroit is $r + 10$. The distance from Detroit to Buffalo is 215 miles and the total time the family spent driving is 7 hours. The time spent driving from Detroit to Buffalo is $\frac{215}{r}$. The time spent driving from Buffalo to Detroit is $\frac{215}{r+10}$. The equation to solve is $\frac{215}{r} + \frac{215}{r+10} = 7$. The LCD is $r(r + 10)$.

$$r(r+10)\frac{215}{r} + r(r+10)\frac{215}{r+10} = r(r+10)7$$

$$215(r+10) + 215r = 7r(r+10)$$

$$215r + 2150 + 215r = 7r^2 + 70r$$

$$430r + 2150 = 7r^2 + 70r$$

$$0 = 7r^2 - 360r - 2150$$

$$r = \frac{-(-360) \pm \sqrt{(-360)^2 - 4(7)(-2150)}}{2(7)}$$

$$= \frac{360 \pm \sqrt{129{,}600 + 60{,}200}}{14} = \frac{360 \pm \sqrt{189{,}800}}{14}$$

$$\approx \frac{360 \pm 435.66}{14} \approx 56.8 \text{ mph} \quad \left(\frac{360 - 435.66}{14} \text{ is not a solution.}\right)$$

The family averaged 56.8 mph from Detroit to Buffalo.

5. Let r represent the average speed on his trip from Boston to New York. Because his average speed was 17 mph slower on his return trip, $r - 17$ represents his average speed on his trip from New York to Boston. The distance between Boston and New York is 190 miles. The time on the road from Boston to New York is $\frac{190}{r}$ and the time on the road from New York to Boston is $\frac{190}{r-17}$. The time on the road from Boston to New York plus the time on the road from New York to Boston is 8 hours 5 minutes $= 8\frac{5}{60} = 8\frac{1}{12} = \frac{97}{12}$ hours. The equation to solve is $\frac{190}{r} + \frac{190}{r-17} = \frac{97}{12}$. The LCD is $12r(r - 17)$.

$$12r(r-17)\frac{190}{r} + 12r(r-17)\frac{190}{r-17} = 12r(r-17)\frac{97}{12}$$

$$12(190)(r-17) + 12(190)r = 97r(r-17)$$

$$2280(r-17) + 2280r = 97r^2 - 1649r$$

$$2280r - 38{,}760 + 2280r = 97r^2 - 1649r$$

$$4560r - 38{,}760 = 97r^2 - 1649r$$

$$0 = 97r^2 - 6209r + 38{,}760$$

$$r = \frac{-(-6209) \pm \sqrt{(-6209)^2 - 4(97)(38{,}760)}}{2(97)}$$

$$= \frac{6209 \pm \sqrt{38{,}551{,}681 - 15{,}038{,}880}}{194} = \frac{6209 \pm \sqrt{23{,}512{,}801}}{194}$$

$$= \frac{6209 \pm 4849}{194} = \frac{6209 + 4849}{194}, \frac{6209 - 4849}{194} = 7\tfrac{1}{97}, 57$$

The rate cannot be $7\frac{1}{97}$ because the total round trip is only 8 hours 5 minutes. The professor's average speed from Boston to New York is 57 mph. We want his time on the road from Boston to New York. His time on the road from Boston to New York is $\frac{190}{r} = \frac{190}{57} = 3\frac{1}{3}$ hours or 3 hours 20 minutes.

Chapter Review

1. Two hoses working together can fill a pool in four hours. One hose working alone can fill the pool in 6 hours less than the other hose. How long would it take the slower hose, working alone, to fill the pool?

   (a) 10 hours (b) 12 hours (c) 14 hours (d) 16 hours

2. Two walkers left an intersection at the same time. One walked northward at an average speed of 3 mph. The other walked westward at an average speed of 4 mph. After how long will they be one mile apart?

(a) About 23 minutes (b) About 38 minutes
(c) 20 minutes (d) 12 minutes

3. The area of a circle is 48 cm^2. The radius (rounded to one decimal place) is

(a) 15.3 cm (b) 3.9 cm (c) 6.9 cm (d) 7.6 cm

4. The sum of two positive integers is 35 and their product is 304. What is the smaller number?

(a) 16 (b) 18 (c) 19 (d) 17

5. An object is dropped from a height of 116 feet. It will hit the ground after about

(a) 10.8 seconds (b) 7.25 seconds (c) 2.7 seconds
(d) 9 cannot be determined

6. The volume of a rectangular box is 192 cubic inches and the height is four inches. The box's width is three-fourths the box's length. What is the length of the box?

(a) 8 inches (b) 6 inches (c) 4.5 inches (d) 9 inches

7. A salesman drove 200 miles to visit a client then returned home. He averaged 10 mph faster on his trip to the client than on his way home. The total trip took 7 hours 20 minutes. How long did he spend driving on his way home?

(a) 3 hours 20 minutes (b) 4 hours (c) 5 hours
(d) cannot be determined

8. A cotton candy vendor sells an average of 100 cones per day when the price of each cone is $1.50. The vendor believes that for each 10-cent drop in the price, she will sell 10 more cones. What should the price of the cones be if she wants a revenue of $156?

(a) $1.30 (b) Leave at $1.50 (c) $1.70 (d) $1.40

9. The diagonal of a rectangle is 29 inches. The length is one inch longer than the width. What is the width of the rectangle?

(a) 18 inches (b) 19 inches (c) 20 inches (d) 21 inches

10. An object is tossed up (from the ground) at the rate of 24 feet per second. After how long will the object be eight feet high?

(a) 0.75 seconds (b) 0.5 seconds (c) 1 second
(d) 0.5 seconds and 1 second

11. A small motorboat traveled downstream for 10 miles then turned around and traveled back upstream for 10 miles. The total trip took 3.5 hours. If the stream's rate is 3 mph, how fast would the motorboat have traveled in still water?

(a) 4 mph (b) 10 mph (c) 7 mph (d) 6 mph

Solutions

1. (b)	**2.** (d)	**3.** (b)	**4.** (a)
5. (c)	**6.** (a)	**7.** (b)	**8.** (a)
9. (c)	**10.** (d)	**11.** (c)	

Factoring is a skill that is developed with practice. The only surefire way to factor numbers into their prime factors is by trial and error. There are some number facts that will make your job easier. Some of these facts should be familiar.

- If a number is even, the number is divisible by 2.
- If a number ends in 0 or 5, the number is divisible by 5.
- If a number ends in 0, the number is divisible by 10.
- If the sum of the digits of a number is divisible by 3, then the number is divisible by 3.
- If a number ends in 5 or 0 and the sum of its digits is divisible by 3, then the number is divisible by 15.
- If a number is even and the sum of its digits is divisible by 3, then the number is divisible by 6.
- If the sum of the digits of a number is divisible by 9, then the number is divisible by 9.
- If the sum of the digits of a number is divisible by 9 and the number is even, then the number is divisible by 18.

Examples

126 is even and the sum of its digits is divisible by 9: $1 + 2 + 6 = 9$, so 126 is divisible by 18.
4545 is divisible by 5 and by 9 ($4 + 5 + 4 + 5 = 18$ and 18 is divisible by 9).

To factor a number into its prime factors (those which have no divisors other than themselves and 1), start with a list of prime numbers (a short list can be found on the last page of this appendix). Begin with the smallest prime number and keep dividing the prime numbers into the number to be factored. It might be that a prime number divides a number more than once. Stop dividing when the square of the prime number is larger than the number. The previous list of number facts can help you ignore 2 when the number is not even; 5 when does not end in 5; and 3 when the sum of its digits is not divisible by 3.

Examples

120: The prime numbers to check are 2, 3, 5, 7. The list stops at 7 because 120 is smaller than $11^2 = 121$.

249: The prime numbers to check are 3, 7, 11, 13. The list does not include 2 and 5 because 249 is not even and does not end in 5. The list stops at 13 because 249 is smaller than $17^2 = 289$.

608: The prime numbers to check are 2, 7, 11, 13, 17, 19, 23. The list does not contain 3 because $6 + 0 + 8 = 14$ is not divisible by 3 and does not contain 5 because 608 does not end in 5 or 0. The list stops at 23 because 608 is smaller than $29^2 = 841$.

342: The prime numbers to check are 2, 3, 7, 11, 13, 17. The list does not contain 5 because 342 does not end in 5 or 0. The list stops at 17 because 342 is smaller than $19^2 = 361$.

Practice

List the prime numbers to check.

1. 166

2. 401

3. 84

4. 136

5. 465

Solutions

1. 166: The list of prime numbers to check are 2, 7, 11.
2. 401: The list of prime numbers to check are 7, 11, 13, 17, 19.
3. 84: The list of prime numbers to check are 2, 3, 7.
4. 136: The list of prime numbers to check are 2, 7, 11.
5. 465: The list of prime numbers to check are 3, 5, 7, 11, 13, 17, 19.

To factor a number into its prime factors, keep dividing the number by the prime numbers in the list. A prime number might divide a number more than once. For instance,

$12 = 2 \times 2 \times 3.$

Examples

Factor 1224.

The prime factors to check are 2, 3, 7, 11, 13, 17, 19, 23, 29, 31.
$1224 \div \mathbf{2} = 612$
$612 \div \mathbf{2} = 306$
$306 \div \mathbf{2} = 153$
$153 \div \mathbf{3} = 51$
$51 \div \mathbf{3} = \mathbf{17}$
$1224 = 2 \times 2 \times 2 \times 3 \times 3 \times 17$

Factor 300.

The prime factors to check are 2, 3, 5, 7, 11, 13, 17
$300 \div \mathbf{2} = 150$
$150 \div \mathbf{2} = 75$
$75 \div \mathbf{3} = 25$
$25 \div \mathbf{5} = \mathbf{5}$
$300 = 2 \times 2 \times 3 \times 5 \times 5$

Factor 1309.

The prime factors to check are 7, 11, 13, 17, 19, 23, 29, 31
$1309 \div \mathbf{7} = 187$

$187 \div \mathbf{11} = \mathbf{17}$
$1309 = 7 \times 11 \times 17$

Factor 482.

The prime factors to check are 2, 3, 7, 11, 13, 17, 19.
$482 \div \mathbf{2} = \mathbf{241}$
$482 = 2 \times 241$

Practice

Factor each number into its prime factors.

1. 308
2. 136
3. 390
4. 196
5. 667
6. 609
7. 2679
8. 1595
9. 1287
10. 540

Solutions

1. $308 = 2 \times 2 \times 7 \times 11$
2. $136 = 2 \times 2 \times 2 \times 17$
3. $390 = 2 \times 3 \times 5 \times 13$

4. $196 = 2 \times 2 \times 7 \times 7$

5. $667 = 23 \times 29$

6. $609 = 3 \times 7 \times 29$

7. $2679 = 3 \times 19 \times 47$

8. $1595 = 5 \times 11 \times 29$

9. $1287 = 3 \times 3 \times 11 \times 13$

10. $540 = 2 \times 2 \times 3 \times 3 \times 3 \times 5$

What happens if you need to factor something like 3185? Do you really need *all* the primes up to 59? Maybe not. Try the smaller primes first. More than likely, one of them will divide the large number. Because 3185 ends in 5, it is divisible by 5: $3185 \div 5 = 637$. Now all that remains is to find the prime factors of 637, so the list of prime numbers to check stops at 23. The reason this trick works is that a number will not divide 637 unless it also divides 3185. In other words, every divisor of 637 is a divisor of 3185. Once you divide the large number, the list of prime numbers to check will be smaller.

The First Sixteen Prime Numbers

Prime Number	Square of the Prime Number
2	4
3	9
5	25
7	49
11	121
13	169
17	289
19	361
23	529
29	841
31	961
37	1369
41	1681
43	1849
47	2209
53	2809

FINAL REVIEW

1. The grade in a psychology class is determined by three tests and a final exam. The final exam counts twice as much as a test. A student's three test grades are 78, 82, and 100. What does he need to score on his final exam to bring his average up to 90?

   (a) 83 (b) 95 (c) 50 (d) 190

2. If $2(x-3)-4(x+5)=7x+1$, then

   (a) $x=\frac{13}{9}$ (b) $x=\frac{1}{9}$ (c) $x=-\frac{2}{3}$ (d) $x=-3$

3. $\frac{x-3}{2x^2-x-15}=$

   (a) $\frac{-3}{2x-16}$ (b) $\frac{1}{2x-1}$ (c) $\frac{1}{2x+5}$

   (d) Cannot be reduced

4. $(2x^3)^2=$

   (a) $2x^5$ (b) $2x^6$ (c) $4x^5$ (d) $4x^6$

5. A rectangle's length is three times its width. The area is 108 square inches. How wide is the rectangle?

   (a) 6 inches (b) 5 inches (c) 4 inches (d) 3 inches

6. If $\frac{2}{3}x - \frac{1}{9} = \frac{1}{2}x$, then

(a) $x = -\frac{1}{7}$ (b) $x = \frac{3}{2}$ (c) $x = \frac{2}{3}$ (d) $x = \frac{1}{3}$

7. A coin bank contains \$1.65 in nickels, dimes, and quarters. There are twice as many nickels as dimes and one more quarter than nickels. How many quarters are there?

(a) 3 (b) 4 (c) 5 (d) 6

8. $\sqrt{x^3} =$

(a) $x^{1/3}$ (b) $x^{2/3}$ (c) $x^{3/2}$ (d) $x^{-1/3}$

9. A small boat traveled five miles upstream and later traveled back downstream. The stream's current was 4 mph, and the boat spent a total of 1 hour 40 minutes traveling. What was the boat's speed in still water?

(a) 5 mph (b) 6 mph (c) 9 mph (d) 8 mph

10. If $2x^2 - x - 2 = 0$, then

(a) $x = \frac{-1 \pm \sqrt{17}}{4}$ (b) $1 \pm \frac{\sqrt{17}}{4}$ (c) $x = \frac{1 \pm \sqrt{17}}{4}$

(d) There are no solutions.

11. Melinda walked a six-mile path in 1 hour 45 minutes. At first she walked at the rate of 4 mph then she finished her walk at the rate of 3 mph. How far did she walk at 4 mph?

(a) 1 mile (b) $1\frac{1}{2}$ miles (c) 2 miles (d) 3 miles

12. What is the solution to $4 - x > 1$?

(a) $(-\infty, 3)$ (b) $(-\infty, 3]$ (c) $(3, \infty)$ (d) $[3, \infty)$

13. $x^2 - \frac{4}{9} =$

(a) $\left(x - \frac{2}{3}\right)\left(x + \frac{2}{3}\right)$ (b) $\left(x - \frac{2}{3}\right)^2$ (c) $(x - 4)\left(x + \frac{1}{9}\right)$

(d) Cannot be factored

14. A businessman has a choice of two rental cars. One costs \$40 per day with unlimited mileage. The other costs \$25 per day plus 30 cents per mile. For what mileage is the \$40 plan no more expensive than the other plan?

(a) At least 50 miles per day
(b) No more than 50 miles per day
(c) At least 133 miles per day
(d) No more than 133 miles per day

15. $3x^8 =$

(a) $3(x^4)^2$ (b) $3(x^4)^4$ (c) $(3x^4)^2$ (d) $(3x^4)^4$

16. If $\frac{1}{2}(x+1) = \frac{2}{3}$, then

(a) $x = \frac{2}{3}$ (b) $x = -\frac{2}{3}$ (c) $x = \frac{1}{3}$ (d) $x = -\frac{1}{3}$

17. A small group has \$100 to spend for lunch. They plan to tip 20% (before tax). The sales tax is $7\frac{1}{2}$%. What is the most they can spend on their order?

(a) \$72.50 (b) \$78.43 (c) \$79.52 (d) \$77.52

18. If $x^2 - 3x + 2 = 0$, then

(a) $x = -1, -2$ (b) $x = 1, 2$ (c) $x = 3\frac{1}{2}, 2\frac{1}{2}$

(d) No solution

19. The sum of two consecutive positive integers is 61. What is their product?

(a) 930 (b) 870 (c) 992 (d) 960

20. $(x+4)(2x-3) =$

(a) $2x^2 + 5x - 12$ (b) $2x^2 - 5x - 12$ (c) $7x - 12$

(d) $7x + 12$

21. $-x^3 - 2x\sqrt{x} + 5x^2 =$

(a) $-x(x^2 - 2\sqrt{x} + 5x)$ (b) $-x(x^2 + 2\sqrt{x} - 5x)$

(c) $-x(x^2 + 2\sqrt{x} + 5x)$ (d) $-x(-x^2 - 2\sqrt{x} + 5x)$

22. One pipe can fill a tank in three hours. A larger pipe can fill the tank in two hours. How long would it take for both pipes, working together, to fill the tank?

(a) 5 hours (b) 50 minutes (c) 1 hour 20 minutes

(d) 1 hour 12 minutes

23. $\dfrac{x}{x^2 - 2x - 8} + \dfrac{3}{x^2 + 3x + 2} =$

(a) $\dfrac{x+6}{(x+1)(x-4)}$ (b) $\dfrac{x+3}{(x+2)(x+1)(x-4)}$

(c) $\dfrac{x^2+4x-12}{(x+2)(x+1)(x-4)}$ (d) $\dfrac{x+3}{2x^2+x-6}$

24. The division problem $0.415\overline{)3.72}$ can be rewritten as

(a) $4150\overline{)372}$ (b) $415\overline{)372}$ (c) $4150\overline{)3720}$

(d) $415\overline{)3720}$

25. What is the complete factorization of $2x^3 + 3x^2 - 18x - 27$?

(a) $(x^2 - 9)(2x + 3)$ (b) $(x^2 + 9)(2x - 3)$

(c) $(2x + 3)(x - 3)(x + 3)$ (d) cannot be factored

26. The difference of two positive numbers is 12. Their product is 405. What is the smaller number?

(a) 13 (b) 16 (c) 15 (d) 17

27. $14x^2y - 21xy^2 + 7x =$

(a) $7x(2xy - 3y^2)$ (b) $7x(2xy - 3y^2 + 1)$

(c) $-7x(-2xy - 3y^2 + 1)$ (d) $-7x(-2xy + 3y^2 + 1)$

28. A small college received a gift of \$150,000. The financial officer will deposit some of the money into a CD, which pays 4% annual interest. The rest will go to purchase a bond that pays $6\frac{1}{2}$% annual interest. If \$7,890 annual income is required, how much should be used to purchase the bond?

(a) \$90,000 (b) \$74,400 (c) \$75,600 (d) \$86,000

29. $\dfrac{1}{3x^4} =$

(a) $\dfrac{1}{3}x^{-4}$ (b) $\dfrac{1}{3}x^4$ (c) $(3x)^{-4}$ (d) $\dfrac{1}{3x^{-4}}$

30. $-6x^2 - 9x + 12 =$

(a) $-3(2x^2 - 3x + 4)$ (b) $-3(2x^2 + 3x + 4)$

(c) $-3(2x^2 - 3x - 4)$ (d) $-3(2x^2 + 3x - 4)$

31. A snack machine has \$8.75 in nickels, dimes, and quarters. There are five more dimes than nickels and four more quarters than dimes. How many dimes are there?

(a) 15 (b) 24 (c) 19 (d) 20

32. $\frac{1}{2} + \frac{5}{x-1} =$

(a) $\frac{6}{x+1}$ (b) $\frac{x+4}{2x-2}$ (c) $\frac{3}{x-1}$ (d) $\frac{x+9}{2x-2}$

33. If $1.18x - 0.2 = 1.2x - 0.3$, then

(a) $x = -\frac{1}{106}$ (b) $x = 5$ (c) $x = \frac{17}{106}$ (d) $x = \frac{1}{2}$

34. A room is five feet longer than it is wide. Its area is 300 square feet. What is the width of the room?

(a) 15 feet (b) 12 feet (c) 20 feet (d) 25 feet

35. $\dfrac{\frac{(x-1)^2}{4}}{\frac{x^2-1}{6}} =$

(a) $\frac{3x-3}{2x+2}$ (b) $\frac{2}{3}$ (c) $\frac{(x-1)^3}{24}$ (d) $\frac{3}{2}$

36. To use the quadratic formula on $2x^2 - x = 4$, let

(a) $a = 2, b = 0, c = 4$ (b) $a = 2, b = 1, c = 4$

(c) $a = 2, b = -1, c = 4$ (d) $a = 2, b = -1, c = -4$

37. An experienced worker can unload a truck in one hour forty minutes. When he works together with a trainee, they can unload the truck in one hour. How long would the trainee need to unload the truck if he works alone?

(a) 1 hour 50 minutes (b) 40 minutes (c) 36 minutes

(d) 2 hours 30 minutes

38. $3(7x-4)^5 + 8x(7x-4)^4 =$

(a) $56x^2 - 11x - 12$ (b) $(7x-4)^4(29x-12)$

(c) $(3+8x)(7x-4)^4$ (d) $(3+8x)(7x-4)^5$

39. If the radius of a circle is increased by 4 meters then the circumference is increased to 18π meters. (Recall: $C = 2\pi r$.) What is the original radius?

(a) 6 meters (b) 5 meters (c) 4 meters (d) 3 meters

40. What is the solution for $2x + 5 \geq 9$?

(a) $(2, \infty)$ (b) $(-\infty, 2]$ (c) $[2, \infty)$ (d) $(-\infty, 2)$

41. $2x^{-3} =$

(a) $\dfrac{1}{2x^3}$ (b) $\dfrac{2}{x^3}$ (c) $\dfrac{2}{x^{-3}}$ (d) $\dfrac{1}{8x^3}$

42. A rectangular box is 10 inches tall. Its width is three-fourths as long as its length. The box's volume is 1080 cubic inches. What is the box's width?

(a) 9 inches (b) 10 inches (c) 11 inches (d) 12 inches

43. $\dfrac{1}{2(x-5)} =$

(a) $\dfrac{1}{2}(x-5)$ (b) $\dfrac{1}{2} \cdot \dfrac{1}{x-5}$ (c) $2\left(\dfrac{1}{x-5}\right)$ (d) $2(x-5)$

44. If $\dfrac{2x+1}{x-5} = \dfrac{x+8}{-7x}$, then

(a) $x = \dfrac{-2 \pm \sqrt{619}}{15}$ (b) $x = \dfrac{-7 \pm \sqrt{2449}}{30}$

(c) $x = -2$ only (d) $x = -2, \dfrac{4}{3}$

45. A mixture containing 16% of a drug is to be combined with another mixture containing 28% of a drug to obtain 15 ml of a 24% mixture. How much 16% mixture is required?

(a) 5 ml (b) 7 ml (c) 10 ml (d) 12 ml

46. Reduce $\dfrac{4}{2+x}$.

(a) $\dfrac{2}{1+x}$ (b) $\dfrac{2}{x}$ (c) $2 + \dfrac{4}{x}$ (d) Cannot be reduced

47. The sum of two numbers is 14, and their product is 24. What is the smaller number?

(a) 1 (b) 2 (c) 3 (d) 4

48. If $2x^2 + 4x - 1 = 0$, then

(a) $x = \pm\sqrt{2}$ (b) $x = \frac{-2 \pm \sqrt{2}}{2}$ (c) $x = \frac{-2 \pm \sqrt{6}}{2}$

(d) $x = -1 \pm \sqrt{6}$

49. Daniel is twice as old as Jimmy. Terry is one year younger than Daniel. The sum of their ages is 44. How old is Daniel?

(a) 16 years (b) 18 years (c) 20 years (d) 22 years

50. $\frac{4}{x^2 + 1} =$

(a) $\frac{1}{4}(x^2 + 1)$ (b) $4\left(\frac{1}{x^2 + 1}\right)$ (c) $4^{-1}(x^2 + 1)^{-1}$

(d) $4^{-1}(x^2 + 1)$

51. A jogger left a park at 6:00. He jogged westward at the rate of 5 mph. At the same time a cyclist left the park traveling southward at the rate of 12 mph. When were they $6\frac{1}{2}$ miles apart?

(a) 6:20 (b) 6:30 (c) 6:45 (d) 7:00

52. The manager of an office building can rent all 40 of its offices when the monthly rent is $1600. For each $100 increase in the monthly rent, one tenant is lost and is not likely to be replaced. The manager wants $68,400 in monthly revenue. What rent should he charge?

(a) $1800 (b) $1900 (c) $1700 (d) $2000

53. If $x^2 - 3x = -2$, then

(a) $x = -2, -1$ (b) $x = 2, 1$ (c) $x = 3 \pm \frac{\sqrt{17}}{2}$

(d) $x = \frac{3 \pm \sqrt{17}}{2}$

54. The perimeter of a right triangle is 24 inches. One leg is two inches longer than the other leg. The hypotenuse is two inches longer than the longer leg. What is the length of the hypotenuse?

(a) 12 inches (b) 10 inches (c) 8 inches (d) 6 inches

55. $\frac{36x^2y^4z^2}{5} \cdot \frac{15xz}{4xy^3} =$

(a) $27x^2z^3$ (b) $27x^3yz^3$ (c) $27yz^3$ (d) $27x^2yz^3$

56. When the radius of a circle is increased by two inches, its area is increased by 24π inches^2. What is the radius of the larger circle?

(a) 4 inches (b) 5 inches (c) 6 inches (d) 7 inches

57. If $4x(x-3) - 5(3x-6) = (2x-3)^2$, then

(a) $x = -\frac{13}{5}$ (b) $x = -\frac{7}{9}$ (c) $x = \frac{7}{5}$ (d) No solution

58. If $-2(x-5) = 12$, then

(a) $x = -\frac{17}{2}$ (b) $x = -11$ (c) $x = -1$ (d) $x = 11$

59. A math student has a 100 homework average and test grades of 99, 100, and 97. The homework average counts 15%, each test counts 20%, and the final exam counts 25%. What is the lowest grade the student can get on the final exam and still get an A (an average of 90 or better) in the class?

(a) 62 (b) 64 (c) 54 (d) 75

60. If $x^2 + 8x + 1 = 0$, then

(a) $x = -8 \pm \frac{\sqrt{60}}{2}$ (b) $x = -8 \pm \frac{\sqrt{68}}{2}$

(c) $x = -4 \pm \sqrt{15}$ (d) $x = \frac{-8 \pm \sqrt{68}}{2}$

61. Linda has \$16,000 to invest. She plans to invest part of the money in a bond that pays 5% and the rest in a bond that pays $6\frac{1}{4}$%. She wants \$937.50 in annual interest payments. How much should she invest in the $6\frac{1}{4}$% bond?

(a) \$5,000 (b) \$7,000 (c) \$9,000 (d) \$11,000

62. $\dfrac{1}{\sqrt{x^2+4}} =$

(a) $(x^2+4)^{1/2}$ (b) $\dfrac{\sqrt{x^2-4}}{x^2+4}$ (c) $(x^2+4)^{-1/2}$

(d) $\dfrac{1}{x+2}$

63. A woman paid \$21.56 (including sales tax) for a book that was marked 20% off. The sales tax was 8%. What was the cover price of the book?

(a) \$24.50 (b) \$26.95 (c) \$24.95 (d) \$24.15

64. $\dfrac{5}{1-x} + \dfrac{1}{x-1} =$

(a) $\dfrac{4}{1-x}$ (b) $\dfrac{6}{0}$ (c) $\dfrac{6}{x-1}$ (d) $\dfrac{4}{x-1}$

65. The height of a rectangular box is 8 inches. The length is one-and-one-half times the width. The volume is 192 cubic inches. What is the box's width?

(a) 4 inches (b) 6 inches (c) 8 inches (d) 10 inches

66. $\sqrt[3]{16x^{11}y^2} =$

(a) $4x^2y^3\sqrt[3]{x}$ (b) $2x^2\sqrt[3]{2x^2y^2}$ (c) $2x^2\sqrt[3]{2xy^2}$

(d) $2x^3\sqrt[3]{2x^2y^2}$

67. 72 increased by 25% is

(a) 97 (b) 90 (c) 18 (d) 72.25

68. $\dfrac{x^2+x-6}{x^2-4x+4} =$

(a) $\dfrac{x-3}{-2x+4}$ (b) $\dfrac{-3}{-x+2}$ (c) $\dfrac{x+3}{x-2}$

(d) Cannot be reduced

69. A highway and train track run parallel to each other. At 5:00 a train crosses a river. Fifteen minutes later a car, traveling in the same direction, crosses the river. If the train's average speed is 52 mph and the car's average speed is 64 mph, when will the car pass the train?

(a) 6:00 (b) 6:20 (c) 6:30 (d) 6:40

70. Completely factor $81 - x^4$.

(a) $(3-x)^2(3+x)^2$ (b) $(9+x^2)(3-x)(3+x)$

(c) $(9+x^2)(9-x^2)$ (d) Cannot be factored

71. Peanuts and a nut mixture containing 40% peanuts will be mixed together to produce eight pounds of a 50% peanut mixture. What quantity of peanuts should be used?

(a) 1 lb. (b) $1\frac{1}{3}$ lbs (c) 2 lbs (d) $2\frac{1}{2}$ lbs

72. $(2x-1)(3x+4) =$

(a) $6x^2 - 4$ (b) $6x^2 + 11x - 4$ (c) $11x - 4$

(d) $6x^2 + 5x - 4$

73. $(3x^3y^2)^2 =$

(a) $3x^6y^4$ (b) $9x^6y^4$ (c) $3x^5y^4$ (d) $9x^5y^4$

74. A department store sells 60 personal CD players per week when the price is \$40. For each \$2 increase in the price, three fewer players per week will be sold. What should the price of the players be if the store manager needs \$2346 per week in revenue?

(a) \$50 (b) \$46 (c) \$48 (d) \$51

75. Completely factor $x^3 + 3x^2 - 4x - 12$.

(a) $(x+3)(x-3)$ (b) $(x^2-4)(x+3)$

(c) $(x-2)(x+2)(x+3)$ (d) Cannot be factored

76. A man is dividing \$15,000 between two investments. One will pay 8% annual interest, and the other will pay $6\frac{1}{2}$% annual interest. If he requires at least \$1000 in annual interest payments, how much money can he invest at $6\frac{1}{2}$%?

(a) At least \$1666.67 (b) At most \$1666.67

(c) At least \$13,333.33 (d) At most \$13,333.33

77. Factor $x^4 + 5x^2 - 36$.

(a) $(x^2+9)(x-2)(x+2)$ (b) $(x+4)(x-9)$

(c) $(x-4)(x+9)$ (d) $(x-2)(x+2)(x-3)(x+3)$

78. $\left(\frac{3x^2}{5y}\right)^{-2} =$

(a) $\frac{25y^2}{9x^4}$ (b) $\frac{3x^{-4}}{5y^{-2}}$ (c) $\frac{9x^4}{25y^2}$ (d) $\frac{9x^{-4}}{25y^{-2}}$

79. If $x^2 - x - 2 = 0$, then

(a) $x = 2, -1$ (b) $x = -2, 1$ (c) $x = -\frac{1}{2}, \frac{5}{2}$

(d) $x = \frac{1 \pm \sqrt{7}}{2}$

80. A pair of boots is sale priced at \$78.40, which is 30% off the original price. What is the original price?

(a) \$112 (b) \$101.92 (c) \$104.53 (d) \$98

81. $\frac{2x^2 - x - 1}{x^2 + 2x - 3} =$

(a) 1 (b) $\frac{2x+1}{x+3}$ (c) $\frac{2x-1}{x-3}$ (d) -1

82. What is the interval notation for $x \geq 4$?

(a) $(4, \infty)$ (b) $[4, \infty)$ (c) $(-\infty, 4)$ (d) $(-\infty, 4]$

83. A salesman earns \$12,000 annual base salary plus 8% commission on sales. If he wants an annual salary of at least \$45,000, what should his annual sales be?

(a) At least \$712,500 (b) At most \$712,500

(c) At least \$412,500 (d) At most \$412,500

84. $\frac{9}{\sqrt{2}} =$

(a) $\frac{9\sqrt{2}}{2}$ (b) 3 (c) $\frac{3}{2}$ (d) $\frac{3\sqrt{2}}{2}$

85. A real estate agent drove to a remote property. She averaged 50 mph to the property and 48 mph on the return trip. Her total driving time was 4 hours 54 minutes. How far was the property?

(a) 100 miles (b) 110 miles (c) 115 miles (d) 120 miles

86. $(5x - 2)^2 =$

(a) $25x^2 - 20x + 4$ (b) $5x^2 + 4$ (c) $25x^2 + 4$

(d) $25x^2 - 4$

87. A company that manufactures calculators wants \$18,000 monthly profit. Each calculator costs \$6 to produce. The selling price is \$11. Monthly overhead runs to \$150,000. How many calculators should be produced and sold each month?

(a) 13,636 (b) 3,600 (c) 26,400 (d) 33,600

88. $2x^3y^{-4}(x^{-2} - x^{-1}y^3 + y^4) =$

(a) $2x^3y^{-4} - 2x^{-3}y^{-12} + 2x^3y^{-16}$ (b) $2xy^{-4} - 2x^2y^{-1} + 2x^3$

(c) $2x^2y^{-1} + 2x^3$ (d) $-2x^2y^{-1} + 2x^3$

89. How much skim milk (0% milk fat) should be added to 4 gallons of 2% milk to obtain $\frac{1}{2}$% milk?

(a) 12 gallons (b) 8 gallons (c) 10 gallons

(d) 6 gallons

90. $\dfrac{5x^3y^{-1}}{15x^2y^{-4}} =$

(a) $\dfrac{xy^4}{3}$ (b) $\dfrac{y^3}{3x}$ (c) $\dfrac{1}{3}xy^4$ (d) $\dfrac{xy^3}{3}$

91. $\dfrac{4}{x^2(x^2 - 2x - 3)} + \dfrac{9}{x(x^2 + 3x + 2)} =$

(a) $\dfrac{9x^2 - 23x + 8}{x^2(x-3)(x+1)(x+2)}$ (b) $\dfrac{13}{x(x^3 - x^2 + 2)}$

(c) $\dfrac{13}{x^2(x^2 - 2x - 3)(x^2 + 3x + 2)}$ (d) $\dfrac{13x - 9}{x^2(x-3)(x+1)(x+2)}$

92. The diameter of a rectangular room is 20 feet. The room is four feet longer than it is wide. How wide is the room?

(a) 15 feet (b) 16 feet (c) 14 feet (d) 12 feet

93. $\sqrt[3]{\sqrt[4]{x}} =$

(a) $\sqrt[12]{x}$ (b) $\sqrt[7]{x}$ (c) x (d) x^{-1}

94. Working together, Matt and Juan can restock a store's shelves in 2 hours 24 minutes. Alone, Juan needs two hours longer than Matt needs. How long does Matt need to restock the shelves when working alone?

(a) 48 minutes (b) 1 hour (c) 4 hours (d) 6 hours

95. $\dfrac{1}{2x} + \dfrac{1}{6x^2} + \dfrac{1}{9} =$

(a) $2x^2 + 9x + 3$ (b) $\dfrac{3}{6x^2 + 2x + 9}$ (c) $\dfrac{2x^2 + 9x + 3}{18x^2}$

(d) $\dfrac{9x + 5}{18}$

96. A car and small airplane leave an airport at the same time. The car is traveling northward at an average speed of 64 mph. The plane is flying eastward at an average speed of 120 mph. When will the car and plane be 102 miles apart?

(a) 45 minutes (b) 30 minutes (c) 1 hour
(d) 1 hour 15 minutes

97. The Holt family pays a monthly base charge of \$12 for electricity plus 5 cents per kilowatt-hour. If they want to keep monthly electric costs between \$80 and \$100, how many kilowatt-hours can they use each month to stay within their budget?

(a) Between 1600 and 2000 kilowatt hours
(b) Between 1333 and 1667 kilowatt hours
(c) Between 1840 and 2240 kilowatt hours
(d) Between 1360 and 1760 kilowatt hours

98. If $\sqrt{3x - 5} = 5$, then

(a) $x = \dfrac{10}{3}$ (b) $x = 0$ (c) $x = 10$ (d) $x = 5$

99. 150 is what percent of 40?

(a) $26\frac{2}{3}\%$ (b) 375% (c) 110% (d) $126\frac{2}{3}\%$

100. A small object is dropped from the top of a 40-foot building. How long will it take the object to hit the ground?

(a) About 1.58 seconds (b) About 0.4 seconds
(c) 2.5 seconds (d) 1.6 seconds

101. What is the solution for $-1 \le 4 - x \le 10$?

(a) $[5, -6]$ (b) $[6, -5]$ (c) $[-6, 5]$ (d) $[-5, 6]$

Solutions

1. (b)	**2.** (d)	**3.** (c)	**4.** (d)
5. (a)	**6.** (c)	**7.** (c)	**8.** (c)
9. (d)	**10.** (c)	**11.** (d)	**12.** (a)
13. (a)	**14.** (a)	**15.** (a)	**16.** (c)
17. (b)	**18.** (b)	**19.** (a)	**20.** (a)
21. (b)	**22.** (d)	**23.** (c)	**24.** (d)
25. (c)	**26.** (c)	**27.** (b)	**28.** (c)
29. (a)	**30.** (d)	**31.** (d)	**32.** (d)
33. (b)	**34.** (a)	**35.** (a)	**36.** (d)
37. (d)	**38.** (b)	**39.** (b)	**40.** (c)
41. (b)	**42.** (a)	**43.** (b)	**44.** (d)
45. (a)	**46.** (d)	**47.** (b)	**48.** (c)
49. (b)	**50.** (b)	**51.** (b)	**52.** (a)
53. (b)	**54.** (b)	**55.** (d)	**56.** (d)
57. (c)	**58.** (c)	**59.** (b)	**60.** (c)
61. (d)	**62.** (c)	**63.** (c)	**64.** (a)
65. (a)	**66.** (d)	**67.** (b)	**68.** (c)
69. (b)	**70.** (b)	**71.** (b)	**72.** (d)
73. (b)	**74.** (b)	**75.** (c)	**76.** (d)
77. (a)	**78.** (a)	**79.** (a)	**80.** (a)
81. (b)	**82.** (b)	**83.** (c)	**84.** (a)
85. (d)	**86.** (a)	**87.** (d)	**88.** (b)
89. (a)	**90.** (d)	**91.** (a)	**92.** (d)
93. (a)	**94.** (c)	**95.** (c)	**96.** (a)
97. (d)	**98.** (c)	**99.** (b)	**100.** (a)
101. (c)			

INDEX

age problems, 222–224
area
  circle, 280–282, 394
  rectangle, 277–280
  triangle, 390–391, 393
  see also geometric figures

box volume (*see* geometric figures)

canceling in fractions (*see* reducing fractions)
circle (*see* geometric figures)
clearing decimals
  in division problems, 61–63
  in equations, 177–181, 324–330
  in fractions, 60–61
clearing fractions in equations, 172–177, 187–188, 190, 324–329
coefficient, 117
coin problems, 229–233
combining like terms, 117–119
compound fractions
  with mixed numbers, 44–45
  with variables, 44–45
  without variables, 22–23, 29–30
consecutive number problems, 214–217, 353–354, 355–356
cost problems, 207–208, 211, 295–297, 312, 313

decimal numbers, 55–56
  adding and subtracting, 56–58
  dividing, 61–63
  in fractions, 59–61
  multiplying, 58–59
  terminating and nonterminating, 56
  see also clearing decimals
denominator, 1
  with decimals, 55, 59–61
  in improper fractions, 23
diagonal (*see* geometric figures)
distance problems
  moving at right angles, 404–408
  moving in opposite directions (at different times), 268–271
  moving in opposite directions (at the same time), 263–267
  moving in the same direction, 261–263
  round trip (non-stream), 271–275, 409–414
  round trip (stream), 398–404
distributive property, 113–114
  distributing minus signs, 115–116
  distributing negative quantities, 116–117
  the FOIL method, 133–136
double inequalities (*see* inequalities)
double negative, 69

equations
  linear equations, leading to, 187–189, 190–193
  quadratic equations, leading to, 189–190, 193–194, 342–350
  solving linear equations, 165–181
  solving quadratic equations
    by factoring, 319–330
    by taking square roots, 330–332
    by using the quadratic formula, 333–341
  with square roots, 189, 194–195
  see also linear equations; quadratic equations; rational equations
exponents
  fractions as, 105–108

exponents (*contd.*)
negative numbers as, 84–88, 88–91
properties 79, 80, 84, 88, 89
roots expressed as, 105–108
zero as, 80

factoring
algebraic expressions, 122–127
quadratic expressions, 136–143, 145–147
quadratic type expressions, 143–145, 147–150
by grouping, 128–129
into prime factors, 14–15, 417–422
negative quantities, 123–124, 131–132
to find the LCD
with algebraic expressions, 152–155
with variables, 45–48, 82–84
without variables, 13–18
to reduce a fraction, 5–10, 129–132, 150–152
to solve a quadratic equation, 319–330
falling object, height of
object dropped, 375–381
object thrown/fired upward, 381–384
FOIL method, 133–136
formulas
geometric figures, 385–386
solving for a variable in, 181–186
fractions
addition
three or more, 17–18
with algebraic expressions, 119–121, 131–132, 155–159, 342–350
with variables, 45–48, 82–84
without variables, 10–17, 19–20, 71–72
compound fractions
with variables, 44–45
without variables, 22–23, 29–30
division
with decimals, 61–63
with variables, 44
without variables, 4–5
as exponents, 105–107
multiplication
with variables, 41–42
without variables, 1–4
reducing

fractions (*contd.*)
algebraic expressions in fractions, 129–132, 150–152
using exponent properites, 80–82
with variables, 38–40
without variables, 5–10
subtraction
with algebraic expressions, 129–132
with decimals, 56–58
with exponents, 79–81, 82–84, 91–94
with negative numbers, 71–72
with variables, 45–48
without variables, 10–16
whole numbers and fractions, 20–21
see also LCD; reducing fractions; simplifying fractions

GCD (greatest common divisor), 7–10
geometric figures, 276–282, 385–397
box, volume of, 209–210, 211, 392
can and cup, volume of, 391–392, 392–393
circle, area of, 280–282, 394
formulas, 385–386
hypotenuse, 391, 393
rectangle (and square), area of, 277–280
rectangle (and square), diagonal of, 386–390
rectangle, perimeter of, 210, 276–277
right circular cylinder (*see* can and cup, volume of)
sphere, surface area of, 392
triangle, area of, 390–391, 393
grade problems, 224–229

hypotenuse (*see* geometric figures)

improper fractions, 23–30
inequalities
double inequalities, 301–311
linear inequalities, 287–290
interest problems, 233–235, 294–297
interval, 285–287, 291, 292, 302–303
interval notation
finite, 302–303
infinite, 290–292

LCD (least common denominator)
with algebraic expressions, 152–159

LCD (least common denominator) (*contd.*)
to clear fractions, 172–177, 187–188, 190
with exponents, 82–84
with variables, 46–48, 82–84
without variables, 13–18
linear equations, 165–181
linear inequalities (*see* inequalities)

mixed numbers, 23–30
mixture problems, 235–244

negating variables, 69–70, 74–76
negative numbers, 65
addition and subtraction, 65–69
double negative, 69
in exponents, 84–91
multiplication and division, 72–76
rewriting subtraction as addition, 69–70
number line, intervals on, 285–287, 289–290, 292, 302–303
number sense problems, 217–222, 354–357
see also consecutive number problems
numerator, 1, 55
with decimals, 59–61
in improper fractions, 23
simplifying, 119–121

operations, order of, 163–165

percent, 197–207
see also interest problems; mixture problems
perimeter (*see* geometric figures)
prime factorization, 419–421
profit problems, 208–209, 211, 294, 297
Pythagorean theorem
distance problems, used in, 404–408
formula, 385
rectangles (and squares), used in, 386–390
triangles, used in, 391, 393

quadratic equations, 319
solved by factoring, 319–330
solved by taking square roots, 330–332

quadratic equations (*contd.*)
solved using the quadratic formula, 333–341
quadratic formula, 332
see also quadratic equations

rational equations, 187–188, 190, 342–350
rectangle (*see* geometric figures)
reducing fractions
with algebraic expressions, 129–132, 150–152
exponent properties, using, 80–82
with variables, 38–40
without variables, 5–10
revenue problems, 357–367
roots
fraction exponents, as, 105–107
multiple roots, 107–108
properties, 96–97
simplifying, 97–101
simplifying quadratic equation solutions, 335–337
simplifying roots in denominators, 101–104

salary problems, 211, 298, 312, 313
sale price problems, 200–201
simplifying fractions
compound fractions, 22–23
with exponent properties, 91–94
with roots in the denominator, 98–104
simplifying the numerator, 119–121
simplifying quadratic equation solutions, 335–337
see also reducing fractions
sphere (*see* geometric figures)
square (*see* geometric figures)
square roots (*see* roots)

temperature problems, 182, 210, 211, 313
triangle (*see* geometric figures)

volume (*see* geometric figures)

work problems, 244–260, 367–374

Rhonda Huettenmueller has taught mathematics at the college level for over ten years. Popular with students for her ability to make higher math understandable and even enjoyable, she incorporates many of her teaching techniques in this book. She received her Ph.D. in mathematics from the University of North Texas.